Integrated Assessment of Water Resources and Global Change

Integrated Assessment of Water Resources and Global Change

A North-South Analysis

Edited by

ERIC CRASWELL
Global Water System Project, Bonn, Germany

MIKE BONNELL
UNESCO Division of Water Sciences, Paris, France

DEBORAH BOSSIO
International Water Management Institute, Colombo, Sri Lanka

SIEGFRIED DEMUTH
Federal Institute of Hydrology, Koblenz, Germany and
University of Freiburg, Germany

NICK VAN DE GIESEN
Delft University of Technology, The Netherlands

Reprinted from *Water Resources Management*, Volume 21(1), 2007

Springer

A C.I.P. catalogue record for this book is available from the Library of Congress

ISBN 978-1-4020-5590-4 (HB)
ISBN 978-1-4020-5591-1 (e-book)

Published by Springer,
P.O. Box 17, 3300 AA Dordrecht, The Netherlands.

www.springer.com

Cover illustration. *Upper part*: Power plant (photograph from morgueFile.com, taken by V. Nyberg); *Lower part*: Tamale, Ghana (photograph taken by Center for Development Research, University of Bonn, Germany)

Printed on acid-free paper

Table of Contents

Water Resour Manage (2007) 21:1–2
DOI 10.1007/s11269-006-9058-7

ORIGINAL ARTICLE

Preface

Paul L. G. Vlek · Eric T. Craswell

Received: 20 April 2006

This Volume contains selected papers from a conference held in Bonn, Germany in February 2005. The topic was *Integrated Assessment of Water Resources and Global Change: A North-South Analysis* and the 130 participants came from 29 countries.

The conference was organised by the Global Water System Project (GWSP) in co-operation with the Project on Global Change in the Hydrological Cycle (GLOWA) at the Centre for Development Research (ZEF), University of Bonn, Germany; the German Federal Ministry of Education and Research (BMBF); the German National Committee of the International Hydrological Programme (IHP) and the Hydrology and Water Resources Programme (HWRP); the initiative Hydrology for the Environment, Life and Policy (HELP) of the United Nations Educational, Scientific and Cultural Organization (UNESCO); the Challenge Program on Water and Food (CGIAR); and the International Association of Hydrological Sciences (IAHS).

We acknowledge the financial support from UNESCO, the German Federal Ministry of Education and Research (BMBF), the government of North Rhine Westphalia, the City of Bonn, and INWENT (Capacity Building International, Ministry for Development Cooperation (BMZ) Germany).

The main themes of the papers published here include:

- Water science and policy interactions
- Stakeholder perspectives
- Water resource data
- Scaling
- Integration

P. L. G. Vlek
Centre for Development Research, Chair, Conference Organising Committee

E. T. Craswell
Global Water System Project, Chair, Guest Editors

We hope that readers, especially water scientists and water resource managers from both industrialised and developing countries, will benefit from the new insights that abound in the papers published here. Hard work by the Guest Editors – Mike Bonell, Deborah Bossio, Nick van de Giesen, and Siegfried Demuth – and the many anonymous reviewers are also gratefully acknowledged.

Water Resour Manage (2007) 21:3–18
DOI 10.1007/s11269-006-9037-z

ORIGINAL ARTICLE

Shift in thinking to address the 21st century hunger gap

Moving focus from blue to green water management

Malin Falkenmark

Received: 20 April 2006

Facts are facts but perceptions guide approaches

Abstract The present water policy debate is dominated by the 30 yr old mission to secure water supply and sanitation to all people. The water needed to produce a nutritionally acceptable diet for one person is however 70 times as large as the amount needed for domestic water supply. The food security dilemma is largest in arid climate regions, a situation constituting a formidable challenge. It is suggested that an additional 5 600 km^3/yr of consumptive water use will be needed to produce an adequate amount of food by 2050 – i.e almost a doubling of today's consumptive use of 6800 km^3/yr. Past misinterpretations and conceptual deficiencies show the importance of a shift in thinking. Combining the scale of the challenge and the time scale of the efforts to feed humanity and eradicate hunger leads to an impression of great urgency. This urgency strengthens the call for international research both for supporting agricultural upgrading, and for much better handling of issues of environmental sustainability. What stands out is the need of a new generation of water professionals, able to handle complexity and able to incorporate water implications of land use and of ecosystem health in integrated water resources management. It will for those reasons be essential and urgent to upgrade the educational system to producing this new generation.

Keywords Water perceptions · Consumptive water use · Global food security · Water losses · Rainwater partitioning · Blue water · Green water · Hydroclimatic differences · Environmental sustainability · Water management

M. Falkenmark
Stockholm International Water Institute (SIWI), Drottninggatan 33, SE-111 51 Stockholm, Sweden
e-mail: malin.falkenmark@siwi.org

Introduction[1]

The international water debate has tended to discuss one issue at a time. In the present water policy debate, 90 percent of the interest goes to 10 percent of the problematique (citation from A. Berntell, SIWI). What is being referred to is the dominating role in the debate of the 30 yr old mission to secure household water provision to all people. This task originates from the UN Water Conference in Mar del Plata in 1977 and its implementation has been ongoing since the start of the International Drinking Water Supply and Sanitation Decade 1980, but still remains unfulfilled.

The hunger alleviation dilemma

By concentrating so much of the policy debate on this issue, an even more daunting water-related effort tends to remain in the shadow: the water required for feeding an expanding humanity and to eradicate hunger in line with the Millennium Development Goals for 2015 and beyond. The water required to produce a nutritionally acceptable diet for one person amounts – with present level of water productivity – to 70 times as much on a per capita basis as the amount seen as needed for domestic water supply on the 50 l/p d level, often referred to as a human right (Gleick, 1996).

In spite of the lack of debate, the global food security issue is sharpening in view of a set of conflicting tendencies:

- on the one hand, food needs are increasing in order both to raise the diets to nutritionally acceptable levels and to feed the additional world population. At the same time, food consumption is moving towards more water-consuming items (meat)
- on the other hand, possibilities to expand irrigation are shrinking due to groundwater decline, streamflow depletion, and urban expansion and water source appropriation. At the same time also agricultural land is shrinking due to erosion/salinisation and to urban expansion.

This dilemma is largest in arid climate regions, where potential evaporation is larger than precipitation. This situation constitutes a formidable challenge and largely an issue of learning how to better live with water scarcity. This particular dilemma is characteristic of many of the countries with lowest human development index, referred to as top/high priority countries in the Millennium project.

The fact that this dilemma is not more discussed is quite remarkable, especially in view of the huge amounts of water involved in producing food for the growing populations in the arid climate region. Not even the water professionals themselves seem very concerned.

In the water debate, when discussing food security, countries beyond self sufficiency potential are simplistically referred to socalled virtual water trade, i.e. to import from better endowed regions. There is however to my knowledge no serious efforts to assess the potential of and implications for the export regions. In other words, where is the food going to come from, and what will be the consequences for the envisaged source regions?

[1] This paper is an expansion of the author's contribution to the working document produced for CSD 13. "Let it reign: the new water paradigm for global food security" (SIWI, IWMI, IFPRI & IUCN 2005)

Irrigated or rainfed?

The discussions of the linkages between food production and water have until recently been limited to discussions of plausible irrigation possibilities, given an assumed market development. The problem with this approach is the gap between plausible future food production and future food requirements to achieve food security for the world population. What is left is a "hunger gap" (Conway, 1997), mainly in non-irrigated dry climate regions in Subsaharan Africa and S Asia.

Since however most of the crop production in the world takes place in rainfed agriculture, that discussion is far too limited. The crops don't mind what water is available to the roots: whether infiltrated rain or applied irrigation water. The situation therefore indicates that there has to be a shift in thinking when discussing the formidable task of feeding a growing humanity. Expanded irrigation can only solve part of the problem. Already today, there is a large scale overappropriation of river flow over 15 percent of the land area (Smakthin *et al.*, 2004). In addition there is a huge overuse of groundwater beyond the renewal rate, leading to declining water tables, more and more difficult to reach for the individual farmer. The present irrigation is in other words not sustainable. According to the Millennium Ecosystem Assessment as much as 1000 km^3/yr, out of the overall some 4000 km^3/yr withdrawn for societal water needs, is being non-sustainable.

The limited expansion potential for irrigation calls for a need to turn attention to the potential of upgrading rainfed agriculture. Earlier studies suggest a considerable potential, provided that the crops can be protected against dryspell damage (Rockström and Rouw, 1997; Rockström, 2003).

This paper will have its focus on the shift in thinking and the conceptual framework required to clarify the issue of feeding humanity on an acceptable nutritional level in line with FAO's projections. FAO foresees an average calory level in the developing world by 2030 of 2980 kcal/person day, i.e. almost 3000 kcal/person day (FAO, 2002). How much more water will have to be reserved for crop production to produce that amount of food? What has to be analysed is the net requirements in terms of consumptive water use; the possible savings by maximising "crop per drop"-productivity, i.e. reducing "true losses" and finally the potential water sources that remain by which the remaining water requirements can be met. In addition it will be essential to analyse also the *environmental sustainability aspects*: what environmental problems can be foreseen and which ones can be avoided and how? And finally – respecting the existing human right for food in an International Covenant on Economic, Social and Cultural Rights, what side effects will be unavoidable and will therefore have to be balanced against human needs by trade-off approaches?

Getting concepts right

A first condition to be able to address these issues is to have words for the different phenomena. One has to be clear about the basic truth that concepts are much more important than theories, since theories are formulated in concepts. The same holds for problem definition which means that concepts guide also the way we try to solve the problems as they have been identified.

In the scientific water community, there is for some reason an astonishingly slow tendency to update concepts and the conceptual framework that interlinks humanity and the life support system that provides human livelihoods. What will first be highlighted here are some misleading perceptions. Moreover, a set of fundamental regional differences will be discussed that are seldom being focused clearly enough, probably due to a general endeavour to keep the debate generic and not "embarrass" any particular region. By that attitude, regional particularities tend to remain in the dark and international recommendations not always be all that reliable. The reliance on the Kuznet curve is an excellent example (Arrow *et al.*, 1995; Falkenmark, 2005).

Irrigated versus rainfed agriculture

As already indicated, the issue of water and food production has until recently been concentrated on irrigated agriculture, based on addition of liquid, socalled *blue water*. Most countries in the world however depend for more that 60 percent of their cereal supply on naturally infiltrated rain, socalled *green water* (Rockström, 2001). There is also a decreasing relevance of the dichotomy between irrigated and rainfed agriculture which will have to be addressed. The concepts are increasingly difficult to separate (Rockström and Barron, 2004). In fact, one has to admit that irrigated agriculture is partly dependent on infiltrated rain. And the opposite is equally true: in the present upgrading of rainfed agriculture, small-scale farming is being increasingly supported by supplementary irrigation for the purpose of dryspell mitigation. Therefore, future development solutions tend to to be found in-between the two extremes of purely irrigated and purely rainfed agriculture. The main solutions of future agriculture will be different forms of in-between varieties.

Water losses

Since so much attention currently goes into finding out the implications of reducing water losses in low efficiency irrigation systems, through efforts to increase the amount of crop produced per drop of water, the concept water *losses* has to be properly clarified. In its present use, it is diffuse and partly misleading. On the one hand, it may refer to blue water losses from canals and irrigation fields that return to the basin and can be reused. On the other hand, it may refer to green water losses in terms of pure non-productive evaporation losses from canals and from irrigation fields. What we need is therefore to get a clear picture of what are '*true*' as opposed to '*imaginary*' losses in agriculture. On a catchment scale, it is only the green water losses which are true losses, while the return flows are only imaginary losses.

Also the concept *water use* is diffuse. It often refers to water withdrawals, irrespective of whether part of that water is going back to the water system after use as *return flow*, or it is turned into *consumptive water use* and vanishes from the area. For instance, (Shiklomanov, 2000) has assessed water withdrawals for municipal, industrial and agricultural uses to 3900 km^3/yr, out of which only 1800 are being referred to as consumptive water use. In order to avoid double-counting, focus should be given to the consumptive water use of irrigation water, rather than the amount withdrawn from the river, since it implies *a blue-to-green redirection* of the water flow, that will basically involve a corresponding depletion of the streamflow.

Rockström *et al.* (1999) has estimated the consumptive/depletive water use involved in current food production at 6 800 km^3/yr, out of which some 1800 originates from irrigation (blue-to-green redirection), whereas 5000 originates from naturally infiltrated rain. Applying today's crop water productivity, consumptive water use in agriculture includes *avoidable losses* in the sense of water use beyond the biologically controled transpiration needs, or "unnecessary" evaporation, amounting to maybe one third of the 6 800 km^3/yr or about 2300 km^3/yr, a sizeable amount.

Some regional particularities

Differences in hydroclimate are reflected in large differences in terms of both human livelihood and dominating vegetation patterns (Falkenmark and Chapman, 1989). A factor of dominating importance is the evaporative demand of the atmosphere and how it relates to precipitation. In fact, precipitation over populated agricultural regions in the temperate climate zone does not differ very much from the situation in corresponding areas in the tropics. What is different is the evaporative demand (Falkenmark and Lindh, 1976).

The implications are illustrated for three different hydroclimatic situations in Figure 1 (Falkenmark and Rockström, 2004)

- the temperate region is least complicated, as there is enough precipitation, moderate evaporative demand and therefore a precipitation surplus left to generate runoff
- in the semiarid tropics, the rainfall is similar but the evaporative demand returns almost all rainfall to the atmosphere, leaving only a minimal amount to generate runoff. This complicates irrigation in areas devoid of rivers entering from remote mountain regions
- in the humid tropics, both rainfall and evaporative demand are high but there still remains a large surplus generating runoff.

Although the semiarid tropics are characterised by highly vulnerable ecosystems, they combine at the same time rapid population growth, poverty and land use as a base for life support. They can therefore be seen as the global hot spot region in terms of hunger alleviation challenges. Although they are often rather misleadingly referred to as 'marginal drylands', the term *savanna* better reflects the fact that these drylands are not as dry as often perceived (Falkenmark and Rockström, 2004): there is basically rainfall enough to support crop production during the wet season. *Many of the top/high priority countries highlighted in the Millennium Project are in this region.*

Poorly adaptable ecological concepts

There is finally reason to address the issue of environmental sustainability and the concepts involved. Basically, ecological concepts are based on biological phenomena while hydrologists – in order to enter attention to vital ecosystems in their water management efforts – need to put focus on the water determinants of the ecosystems (Falkenmark *et al.*, 2003; GWP, 2003). For hydrologists, *terrestrial* ecosystems where the soil moisture is a key determinant have to be properly distinguished from *aquatic* ecosystems for which the water in the river is the determinant. *Wetlands* – although hydrologically quite different – all combine a biological meaning in the sense that the soil is wet and oxygen free (Pielou, 1998). They may however differ considerably in terms of the type of water that keeps the wetland wet: condensation, rainfall, soil moisture, groundwater discharge, inundating surface water, streamflow, etc.

Water requirements to feed tomorrow's humanity

Forecasting versus backcasting

Past studies on future relations between water and agriculture have tended to have their focus on irrigation. They have started from projections of plausible food consumption needs, paying attention to an assumed income growth and market development, and analysed how those needs could be met by an increased food production (FAO, 2003; Rosegrant *et al.*, 2002).

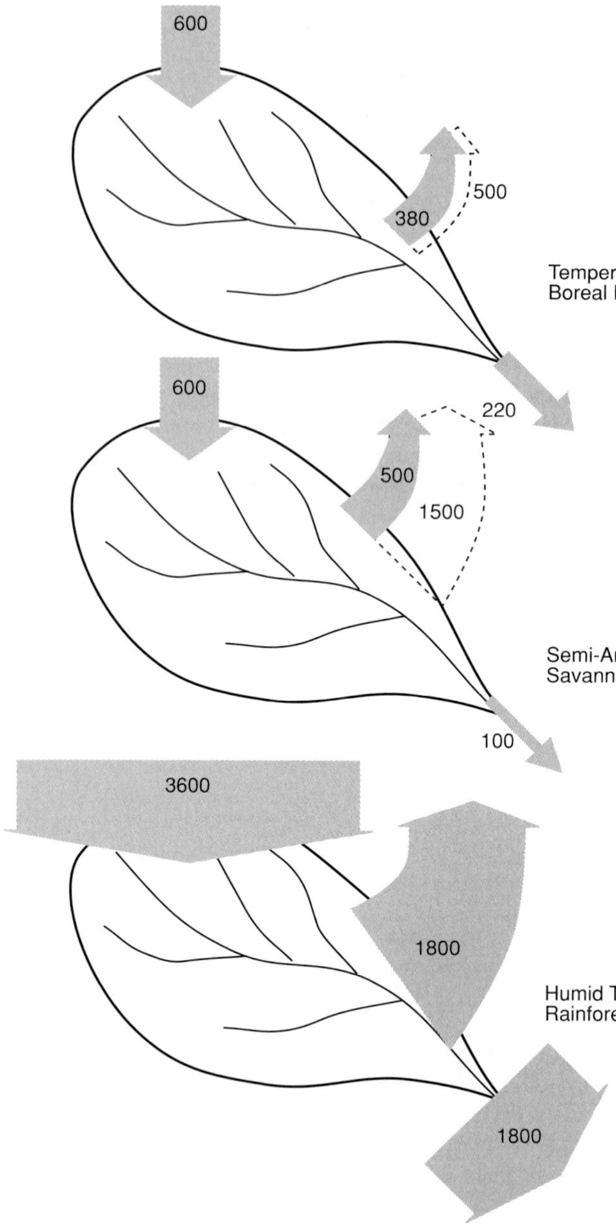

Fig. 1 Differences in water-related livelihood determinants (precipitation, potential and actual evaporation, and runoff generated). Comparison for three ecological regions based on data from Lvovich

It has however been revealed that even if foreseen food consumption needs will be met, the result will not be food security, due to a remaining "hunger gap" in poor regions (Conway, 1997).

Assuming that the international ambitions to alleviate hunger and undernutrition as reflected in the goals of the World Food Summit in 1996 and in the MDG's are serious, it is possible to take the opposite approach: to estimate what consumptive use will be needed to feed humanity as a whole on an acceptable nutritional level and to estimate what would be the freshwater implications. In other words taking a *backcasting approach.*

Additional consumptive use requirements

Both Gleick (2000) and Rockström (2003) have analysed the consumptive water use involved in producing today's diets. According to Rockström, they amount to 690 m^3/p yr in Subsaharan Africa and 820 m3/p yr in Asia (except the former Soviet Union). He has also shown that to produce a diet of 3000 kcal/p d, which is the average nutrition level in developing countries foreseen by FAO by 2030, will correspond to a consumptive use of 1300 m^3/p yr (including 20 percent animal protein).

Combining these data with population increase as foreseen til 2050 AD – when world population is expected to have more or less stabilised – suggests that to produce the food needed on the one hand for raising the regional diets to this level, and on the other to feed the additional population, *we can foresee the need for an additional consumptive water use for food production of altogether 5 600* km^3/yr (Rockström, 2003) – assuming no change in water productivity. This is almost a doubling from the current 6 800 km^3/yr. Comparing the regional freshwater needs with current consumptive water use in the two hot spot regions indicates that for food selfsufficiency, Subsaharan Africa would need to increase the consumptive use by a factor 3.1 (from 465 km^3/yr to 1450) and S and E Asia by a factor 2.2 (from 2830 km^3/yr to 6210).

There are of course numerous options to find these additional amounts of water that will have to be appropriated from other current uses by humans or by ecosystems (Figure 2). The first option is of course *to increase water productivity* by reducing true losses, i.e. transfer non-productive evaporation into productive transpiration (maximise crop per transpired drop by vapour shift).

Rockström (2003) has assessed the different options as follows:

- *loss reduction*, in irrigated agriculture maybe 200 km^3/yr, in rainfed maybe 1500 at the most
- *additional irrigation*, to be limited due to the streamflow depletion that might follow – scarcely more than 600 km^3/yr

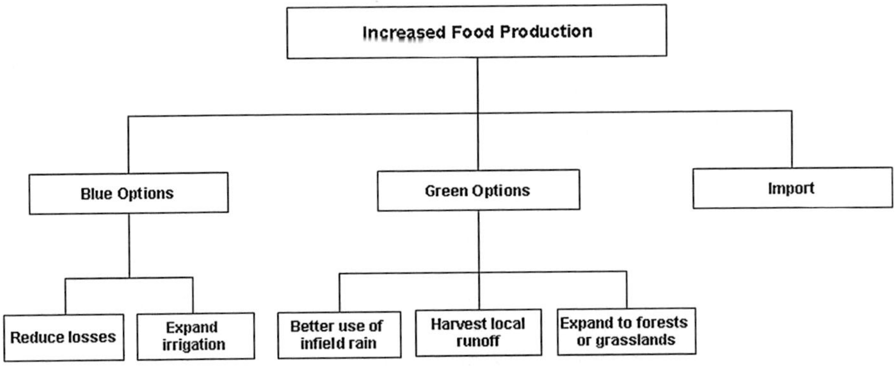

Fig. 2 Three basic ways to capture the additional water needed to meet consumptive water use of increased food production: blue water options, green water options, and import/virtual water option

- the rest or 3 300 km^3/yr will have to be covered by *expanding rainfed agriculture* into forests and grasslands, by *import from cooler areas* (saving maybe some 400 km^3/yr, Oki and Kanae, 2004), or by altering the water requirements of the crops through *biotechnology* influencing the time to ripening etc.
- reducing the consumptive demands by *diet changes*, basically covering the highly water consuming meat component by less water demanding protein sources (soya beans, etc.).

Implications of current environmental degradation

Living with change

The task to mobilise the water needed to meet the water requirements discussed above, to feed the growing humanity and alleviate undernutrition evidently involves large changes in terms of both land use and water partitioning into green vapour and blue liquid water flows. These changes will all have ecological consequences which will be the price to be paid for food security. A crucial challenge for humanity will therefore be to find out how to live with change while protecting environmental sustainability. And what should be meant by environmental sustainability?

First of all, it will be essential to try to get out of the current environmental degradation, caused by today's non-sustainable agricultural practices. There is in fact a large scale undermining of the biophysical resource base going on in the world. It has moreover to be realised that some of the undermining phenomena are *unavoidable* (streamflow depletion due to reduced runoff generation or consumptive use of irrigation water; clearing of new croplands), whereas others are at least theoretically *avoidable* (groundwater overexploitation, nutrient leaching, and land productivity decline by erosion and salinisation).

Especially critical for future food production is the current overdraft beyond acceptable blue water withdrawal quantities in rivers and aquifers. Such overdraft involves not only the *loss of water sources in current use for irrigation purposes,* but limits also the possibility to expand irrigation withdrawals.

Reduction of river flow

The consequences of large consumptive use in irrigation-dependent areas is a widespread streamflow depletion through the blue-to-green redirection of the water flows involved. These consequences are particularly troublesome in socalled closed river basins, where the remaining unappropriated flow has to be protected for the aquatic ecosystems (Lannerstad, 2002; Falkenmark and Lannerstad, 2005).

Ecologists have estimated the minimum water flow required for aquatic ecosystems at some 30 percent of the annual flow, for regions where there is a dry season to which biota are already used (Smakthin *et al.*, 2004). In a recent world overview these authors estimate, as already indicated, that over altogether 15 percent of the land area, hosting some 1.4 billion people, is streamflow already overappropriated. River depletion implies a reduced river flow relative to the long term average, in some rivers even a change from perennial to intermittent flow. Reduced flow involves also reduced wastewater dilution capacity, and of course sharpens the conflict of interest between upstream and downstream water uses.

The most evident example of river depletion. is the tributaries to the Aral Sea, where the result gets particularly evident since the Aral Sea is a closed lake without outlet (Falkenmark and Lannerstad, 2005). The only way the lake can respond to a reduced inflow is by evaporation in order to diminish the lake surface, until there is balance between the evaporation from the shrunken surface and the new inflow. The Yellow river is another example where the mouth first dried up in 1972, continuously increasing so that by 1997 it was dry for altogether 7 months with the dry-up reaching 700 km upstream.

Groundwater overexploitation

Also the groundwater overdraft is cause for concern (Foster and Chilton, 2003) since it indicates that a non-sustainable source of water is being consumed for irrigation, a source that will no longer be available in the long-term. Rosegrant *et al.* (2002a) estimate groundwater overdraft beyond groundwater recharge to 200 km^3/yr, with India, China and USA as the major groundwater depleting nations. Seckler assessed a quarter of India's harvest to rely on such overdraft (Shah *et al.*, 2003).

Unless reduced to sustainable levels, present groundwater use in NE China will effectively destroy the groundwater dependent agriculture base, cause massive subidence and sea water intrusion, and involve the loss of 'insurance' water for the future generations (Chinese Ministry of Water Resources in Moench *et al.*, 2003).

Nutrient leaching and eutrophication of surface water bodies

Leakage of the nutrients *N* and *P* from agricultural systems is causing major environmental problems at present. It is estimated that only about half of these fertilisers are captured in harvested crops (Tilman *et al.*, 2001). Should current trends continue in terms of global *N*-fertilisation, this would add 60 percent more fertilisers by 2025 and 170 percent more by 2050.

One consequence of the large scale leaching of N from areas where groundwater is being recharged is rising nitrate levels in groundwater, a phenomenon widely spread under agricultural regions in for instance Europe (Vogel and Grath, 1998). In semiarid climates with low groundwater recharge, even small losses of N are enough to rise to high concentrations of nitrate in groundwater (professor G. Jacks, personal comm.). Not only commercial fertilisers but also use of large amounts of manure involves high risk for large leaching losses.

Consequences are widespread also in surface water bodies where they are easily detectable in ecosystem switches with time, through the generated eutrophication problems. In the long term the bottom waters develop oxygen free conditions, like in the Baltic Sea, the Black Sea and the Mexican Gulf.

Reduction of land productivity

Over 80 percent of arable land worldwide is affected by soil degradation, reducing its productivity (Millstone and Lang, 2003). Contributing factor in the tropics has been population growth and the related collapse of the fallow-based production system of shifting cultivation. While an *erosion* of 10 tons/ha yr is considered an absolute limit for sustained agriculture, erosion in semiarid tropics with intense seasonal rains may be three times as high (Jacks, 2004).

An important determinant of land productivity is the *nutrient* level in the soil. While in many industrialised countries, due to high level fertiliser application, nutrients are accumulated in soils, building up pools of "chemical bombs" (Hekstra, 1995), there is the opposite situation in Africa where a kind of 'soil mining' is ongoing. This is due i.a. to the reduction of fallow periods forced by population growth, not compensated by addition of the nutrients needed, following the example of regions in the industrialised world when going into continuous agriculture (Falkenmark and Rockström, 2004).

Moreover, semiarid tropical countries are highly vulnerable to altered land use due to the changes generated in terms of the consumptive water use. When runoff generation is low, even moderate changes in green vapour flow may be reflected in alteration of runoff that may be large in a relative sense. Such changes have been generated in Australia which now suffers from serious *salinisation* problems caused by the land use change during the immigration period, when the eucalyptus woodlands were cleared for crops and grazing (Gordon *et al.*, 2003). These transitions generated water logging and salinisation of both soil and groundwater, and now adversely affects agricultural and pastoral yields. The damage now encompasses some 3.3 Mha. Some 5.7 Mha are considered at risk, predicted to increase further to 17 Mha in 50 years time.

Salinisation is a widespread problem also on irrigated land when there is absence of appropriate drainage, especially under excessive irrigation. Such absence is usually due to the large expenses linked to digging of drainage canals. According to WWAP (2005), poor drainage and irrigation practices have led to waterlogging and salinisation of approximately 10 percent of the world's irrigated lands.

The environmental protection challenge

In order to minimise the ongoing undermining of the resource base it will have to be clarified what the criteria are for "environmental protection" and "protection of ecosystems". Protection from what? To achieve what? First of all, the word ecosystem carries no scale, which leads to the question what scale of ecosystems that we are referring to: a particular component of the landscape regarded as an iconic site, a biodiversity reserve etc., or the catchment as a whole seen as an ecosystem (GWP, 2003).

A key function to secure for future generations is the capacity of the life support system to deliver food and biomass, ecological services of various kinds while enduring disturbances and variability without shifting to a non-desirable state of the system (Gordon, 2003), for example unproductive soils, savannisation of the rainforest, collapsing farming system, eutrophication of a lake etc. Ecological systems in the landscape are linked by flows of water in an upstream/downstream pattern. Freshwater flows, crop production and other terrestrial ecosystem services are interconnected and interdependent. Aquatic ecosystems downstream respond to the integrated result of all upstream acvtivities.

One way of seeing the linkage between integrated water resources management and ecological services is to manage catchments as an asset that delivers a bundle of water and ecological goods and services (GWP, 2003). Some of these services work in synergy, others in conflict. Criteria have to be developed for the protection of the capacity for sustainable production of life support, i.e. identification of what key functions are essential for the production of terrestrial and aquatic ecosystem's goods of social and economic importance, and terrestrial and aquatic ecosystem's services of ecological importance from different aspects. Humanity, through its activities tends to alter disturbance regimes with which organisms have evolved over time. There is therefore a need to secure enough "elasticity" (resilience) of ecosystems to change in the surrounding conditions like storms, fire, drought, pollution events, or creeping pollution. What has to be protected is the capacity to absorb continuous change without loss of the dynamic capacity of vital ecosystems to uphold the supply of ecological goods and services.

As pointed out by Gordon (2003), freshwater redistribution may reduce resilience of ecosystems in two main ways: both through a change of the role of freshwater as an internal *structuring variable* (water quantity and quality), and as a *disturbance regime* (temporal variability and timing). Land use and land cover change that alters the fluxes of water to and from the soil can change the soil moisture at a local scale. Land cover change that modifies the fluxes of water to the atmosphere can cause effects on scales, ranging from local to global. Ecosystems may have multiple equilibrium and internally changing variables within the system, like changes in quality or quantity of freshwater, reducing its resilience and causing a transition from a desirable to a non-desirable state.

Towards food production sustainability

Main challenges

It is evident from the above that global food security is an enormous challenge not only in terms of today's weakness – food distribution – but in terms of the water implications if FAO's projections of food consumption by 2030 would materialise. It has already been shown that producing the food needed to reach a world nutrition average of 3000 kcal/p d – because of the consumptive water use involved – would have major water implications. One has to consider the problematic fact that on the one hand river flow is already overused in many of the irrigation-dependent regions (15 percent of land surface), and on the other there is a large scale use of non-renewable groundwater over essential food producing regions in India and China.

Groundwater overexploitation has to stop before it leads to foreseeable societal collapses when the water source for irrigated agriculture gets out of reach. Such calamities will be unavoidable in heavily groundwater-dependent regions in India and NE China. Where groundwater use cannot be controled in time due to millions of farmers like in India (Shah *et al.*, 2003), an agricultural restructuring will be unavoidable towards upgraded rainfed agriculture supported by protective irrigation, especially during dryspells, and based on water harvesting, or close to cities on reuse of recirculated urban wastewater.

This basically means that future food production will have to benefit maximally from rainfall rather than from irrigation. As already indicated, climatic data show that there is, also in the semiarid regions, generally rainwater enough during the rainy season to meet consumptive water requirements from one crop, provided that the roots get access to that water and plants can be protected from dryspell damages (Rockström and Falkenmark, 2000).

Since the approach taken here is based on today's water productivity, considerable reduction of water requirements would be possible by reducing water losses. Primarily, it will be essential to maximise productivity per drop of water transpired, i.e. minimise evaporation losses. Productivity increase is in other words equivalent to loss reduction or "regaining" the water involved in evaporative losses to cover part of the additional consumptive water use (Falkenmark and Lannerstad, 2005). Minimising runoff losses is a more complicated issue, since these losses are no true losses but form blue water flows available elsewhere.

At the same time, attention has to be paid to the fact that today's agriculture is undermining its own resource base. A sustainable future agriculture therefore implies mitigation of avoidable problems linked both to land productivity (nutrient leaching, erosion, salinisation) and to water productivity (groundwater overexploitation), Figure 3.

But it also involves acceptance of streamflow depletion as a principal consequence of on the one hand reduced runoff generation as a result of land use alteration, and of on the other hand blue-to-green redirection of water withdrawn for irrigated agriculture. Also loss of terrestrial ecosystems will have to be accepted where necessary to alleviate hunger and no other alternatives remain.

Key measures needed to make agriculture more environmentally sustainable involves in other words two categories of management activities: on the one hand *mitigation* of avoidable problems or at least ceasing to aggravate them, and on the other hand balancing conflicting land and water interests by *trade offs* in regard to the risk of further streamflow depletion or lack of alternatives. When situations have already gone too far in terms of river depletion, the first step may be to buy back allocated irrigation water from irrigated farms in the way that is now practised in Australia (Scanlon, 2004).

Water management implications

On the most general level, one can say (Falkenmark and Rockström, 2004) that a successful management will have to incorporate efforts to

– *secure* water and related services, in this case water-dependent food production
– *avoid* degradation of water and land resources and of ecosystem integrity
– *foresee* changes (climate, population, diet preferences etc).

To achieve this, an integrated approach will have to be taken to blue and green water, seeing precipitation as the basic resource, and to water quantity and quality (GWP, 2003). The best way will be to benefit from present focus on integrated water resources management (IWRM) on a catchment/river basin basis (blue water approach) but to incorporate green water through its interaction with the blue water, Figure 4. This would lead to an integrated land and water resources management (ILWRM). Efforts towards such management approach is being practised in several transnational rivers supported by the Global Environment Facility (Duda, 2003).

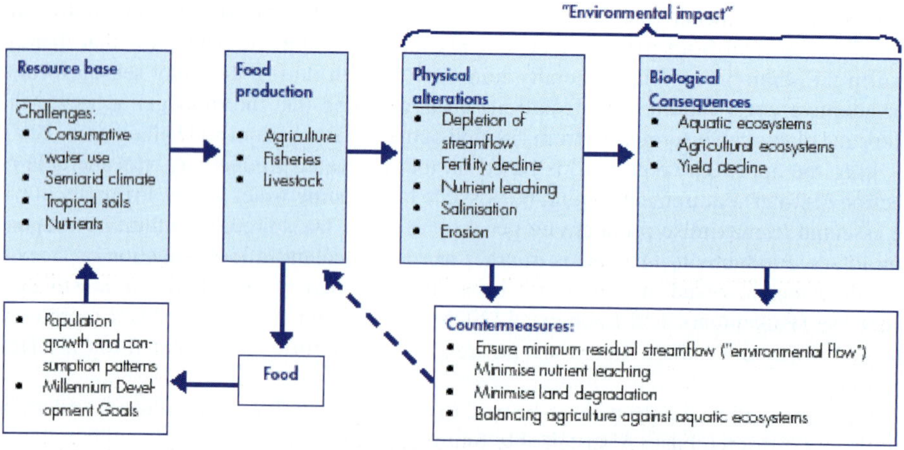

Fig. 3 Links between food production to meet global food needs, environmental impacts generated by agricultural practices, and possible countermeasures towards ecologically sustainable food production

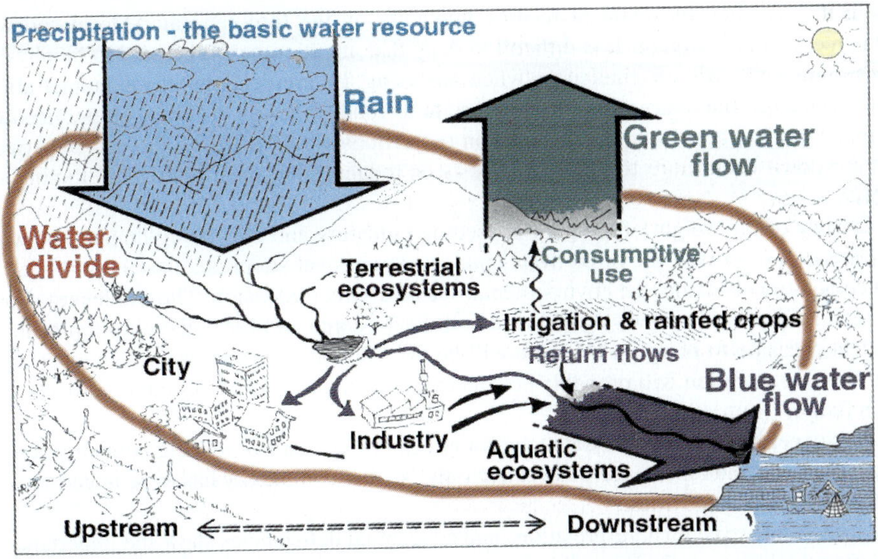

Fig. 4 The catchment allows an integrated approach to all water-related phenomena within the water divide. All the rain falling within the water divide is being partitioned between humans and ecosystems, terrestrial as well as aquatic; between land use and water; and between upstream and downstream

Minimising ecosystem degradation will involve both water pollution abatement to protect the habitats of aquatic ecosystems, and securing minimum streamflow ("environmental flow"). Minimising nutrient leaching and land fertility decline will be a major part of management efforts. What it all boils down into is an *integrated land/water/ecosystem approach.*

Catchment routing

Taking a truly integrated land/water/ecosystem approach will demand a stepwise routing procedure with water accounting (GWP, 2003) – stretch by stretch – down the catchment, or conversely from the river mouth up to the water divide (FAO, 2000). Attention has in each stretch to be paid to runoff added to different river stretches by runoff generation from surrounding land; to water demand sites and withdrawal needs; to pollution sites and amounts added; assessing consumptive water use involved, return flows, available amounts of dilution water and water quality implications.

The process has to pay adequate attention to upstream/downstream relations, resilience criteria both for iconic ecological sites and for downstream aquatic ecosystems, and to related bottom lines in terms of downstream streamflow (GWP, 2003). In the analysis, value routing may be useful to guide in the necessary trade off striking (Falkenmark and Rockström, 2004). Throughout the process, social acceptability will have to be secured through a legally achnowledged stakeholder participation in the trade off process.

Conclusions

FAO foresees a rapid improvement of diets with calory levels increasing to an average for the developing world of 3000 kcal/p day in 2030. What will be needed to complement these studies is an analysis of the water implications; this paper has given a first idea about the unbelievable scale of feeding humanity on the foreseen calory level and with 20 percent animal protein. It is difficult to deny that attention has to be paid to biophysical constraints and trade off challenges when analysing the implications of the hunger alleviation, linked to the conception of human right to food. In terms of the most extreme consequences, it suggests that FAO's nutrition projections would bring on the choice between cutting down even more tropical rainforests or reducing the meat content of tomorrow's diets.

In view of the weight of the goal of feeding humanity and alleviating hunger, the water implications discussed above motivates major international attention, especially in view of the choices involved and the environmental sustainability dimension. The challenges should be seen as issues of evident significance both for research and for policy development. An evident occasion to bring up the issue will be the Millennium Summit in September 2005, but there the attention will probably be concentrated on the 2015 MDG-targets, i.e. the near term future.

This paper has also demonstrated the complexity of the issue of feeding humanity already when seen in a natural science perspective only. To that complexity has to be added societal and economic complexities.

Furthermore, past misinterpretations and conceptual deficiencies show the importance of a *shift in thinking* to make it possible to address the complexity. The water community has to get out of the "eddying discourse" that now characterises the international policy debate. This debate tends to circulate around a limited number of issues – however important: water supply and sanitation, privatisation, dams etc. What is neglected is the water-related implications of the very basic issue of human right to food.

The scale of the challenge and the time scale of the efforts to feed humanity and eradicate hunger, gives an impression of *urgency*. It is quite remarkable that such urgency does not characterise the general debate. This debate continues more or less along the lines of the 1980's and 1990's. This urgency strengthens the call for international research both for supporting agricultural upgrading, and for much better handling of issues of environmental sustainability.

What also stands out is the need for a new generation of water professionals and hydroecologists, able to handle complexity and able to incorporate water implications of land use and of ecosystem health. It will for those reasons be essential and urgent to upgrade the educational system for producing this new generation.

References

Arrow K et al. (1995) Economic growth, carrying capacity, and the environment. Science 268:520–521

Conway G (1997) The doubly green revolution. Food for all in the twenty-first century. Penguin Books, USA, p 334

Duda AM (2003) Integrated management of land and water resources based on collective approach to fragmented international conventions. Philosophical Transactions Royal Society London B 358:2051–2062

Falkenmark M (2003) Freshwater as shared between society and ecosystems: from divided approaches to integrated challenges. Philosophical Transactions Royal Society London B 358:2037–2049

Falkenmark M (2005) Water usability degradation – Economist wisdom or societal madness? Water International 30:136–146

Falkenmark M, Lindh G (1976) Water for a starving world. WestView Press, Boulder, Colorado, p. 201

Falkenmark M, Chapman T (eds) (1989) Comparative Hydrology – An ecological approach to land and water resources. Unesco, Paris, France, p 475

Falkenmark M, Rockström J (2004) Balancing water for humans and nature. The new approach in ecohydrology. Earthscan, London

Falkenmark M, Lannerstad (2005) Consumptive water use to feed humanity – Curing a blindspot. Hydrology and Earth Sciences, pp 15–28
FAO (2000) New dimensions in water security. In: Lundqvist J (ed) AGL/MISC/25/2000. FAO Rome
FAO (2003) World Agriculture: towards 2015/2030. In: Bruisma J (ed) Earthscan, London
Foster SSD, Chilton PJ (2003) Groundwater: the processes and global significance of aquifer degradation. Philosophical Transactions, Royal Society London B 358:1957–1972
Gleick P (1996) Basic water requirements for human activities. Meeting basic needs. Water International 21:83–92
Gleick P (2000) The World's Water 2000–2001. The biennial report on freshwater resources. Island Press, USA
Gordon L, Dunlop M, Foran B (2003) Land cover change and water vapour flows: learning from Australia. Philosophical Transactions, Royal Society London B 358:1973–1984
GWP (2003) Water management and ecosystems. Living with change. TEC Background Report No.9. Global Water Partnership, Stockholm
Hekstra GP (1995) Delayed effects of pollutants in soils and sediments: understanding and handling of chemical time bombs in Europe. Ecoscript 56. Stichtung Mondiaal Alternatiet, Amsterdam
Jacks G (2004) Personal communication. Professor Gunnar Jacks, Royal Institute of Technology, Stockholm
Lannerstad M (2001) Consumptive water use feeds the world and makes rivers run dry. MSc thesis, TRITA-LWR 02-13, Royal Institute of Technology (KTH), Stockholm
Lvovich MI (1979) World water resources and their future, (Translation by the American Geophysical Union). LithoCrafters Inc., Chelsea, Michigan
Millstone E, Lang T (2003) The atlas of food. Who eats what, where and why. Earthscan, London
Moench M, Burke J, Moench Y (2003) Rethinking the approach to groundwater and food security. FAO Water Report 24, FAO, Rome
Oki T, Kanae S (2004) Virtual water trade and world water resources. In: Drainage basin security – Balancing production, trade and water use. Stockholm Water Symposium 2003. Water Science and Technology 49:203–209
Pielou EC (1998) Fresh water. The University of Chicago Press, Chicago
Rockström J (2001) Green water security for the food makers of tomorrow: windows of oppportunity in drought-prone savannahs. In: Water security for the 21st century – Innovative approaches. Stockholm Water Symposium 2000. Water Science and Technology 43:71–78
Rockström J, Rouw de A (1997) Water, nutrients, and slope position in on-farm pearl millet cultivation in the Sahel. Plant and Soil 195:311–327
Rockström J (2003) Water for food and nature in drought-pronme tropics: vapour shift in rainfed agriculture. Philosophical Transactions, Royal Society London B 358:1997–2010
Rockström J, Gordon L, Falkenmark M, Folke C, Engwall M (1999) Humanity's dependence on water vapour flows for terestrial ecosystem services and food production. Conservation Ecology 3:5. On line. URL: http://www.consecol.org/vol3/iss2/art5
Rockström J, Barron J (2004) Water productivity in rainfed systems: Overview of challenges and analysis of opportunities in water scarcity prone savannas. Special issue. Irrigation Science (forthcoming)
Rockström J, Falkenmark M (2000) Semi-arid crop production from a hydrological perspective – Gap between potential and actual yields. Critical Rev Plant Sc 19:319–346
Rosegrant MW, Cai SA, Cline Nakagawa N (2002)The Role of Rainfed Agriculture in the Future of Global Food Production. Environment and Production. Technology Division Discussion Paper No. 90. International Food Policy Research Institute, Washington, DC
Rosegrant MV, Cai X, Cline SA (2002) World water and food by 2025. IFPRI, Washington, DC
Scanlon J (2004) Negotiating tradeoffs between competing water claims – the critical role of 'environmental flows' in basin management. In: Drainage basin mangement – Regional approaches for food and urban security. Stockholm Water Symposium 16–20 August, 2004, Abstract volume. Stockholm International Water Institute, Stockholm, pp 235–236
Shah T, Roy AD, Qurershi AS, Wang J (2003) Sustaining Asia's groundwater boom: an overview of issues and evidence. Natural Resources Forum 27:130–141
Shiklomanov IA (2000) Appraisal and assessment of world water resources. Water International 25:11–32
SIWI, IWMI, IFPRI, IUCN (2005) Let it reign: the new water paradigm for global food security. Final Report to CSD 13. Stockholm International Water Institute, Stockholm
Smakthin V, Revenga C, Döll P (2004) Taking into account environmental water requirements in global-scale water resources assessments. Comprehensive Assessment Research. Report 2. IWMI, Colombo, Sri Lanka

Tilman D, Fargione J, Wolff B, D'Antonmio C, Dobson A, Howarth R, Schindler D, Schlesinger WH, Simberloff D, Swackhamer D (2001) Forecasting agriculturally driven global environmental change. Science 292:281–284

Vogel WR, Grath J (1998) Groundwater in Europe – State and trends in quality and quantity. In: Hilding-Rydevik T, Johansson I (eds) How to cope with degrading groundwater quality in Europe. Swedish Council for Planning and Coordination of Research. Stockholm, FRN-Report 98:4

WWAP (2005) Currents – news from the World Water Assessment Programme. Issue no 13, February 2005

Water Resour Manage (2007) 21:19–33
DOI 10.1007/s11269-006-9038-y

ORIGINAL ARTICLE

A grid-based assessment of global water scarcity including virtual water trading

Md. Sirajul Islam · Taikan Oki · Shinjiro Kanae · Naota Hanasaki · Yasushi Agata · Kei Yoshimura

Received: 20 April 2006

Abstract A 0.5-degree grid-based assessment of the scarcity of global water resources including virtual water trading has been made. The three components of water availability considered for each grid were local runoff, routed flow from upstream and virtual water trading. Several assumptions were postulated to convert country-base estimations of virtual water trading to grid values. The results show that unequal spatial distribution of global water resources had been considerably neutralized by virtual water trading. A large proportion of people in the Middle-East, North-Africa and Sub-Sahara region are able to relieve their water stress through virtual water import. The paper also reports two hypothetical scenarios with extremes of natural flow availability based on the presence and absence of routed upstream flow.

Keywords Global water scarcity · Virtual water trading · Grid-based analysis · Water stress index · Spatial distribution of stress

Introduction

Global water resources in the 21st century are an increasingly important concern for the sustenance of human life, ecosystems and economic progress. The issue is a matter of international interest since the origin and movement of water is interlinked globally amongst different parts of the world; it therefore needs to be addressed in a holistic way. In this context, the availability of global water resources and its movement can be defined by the following three components; (1) Local runoff; (2) Exogenous runoff (routed runoff from upstream); and (3) Virtual Water flow.

The first two components can be considered as part of the natural movement of water. Along with its natural movement or availability, the third component of global water resources i.e. virtual water flow between countries through trading of food products is also playing an important role in global water balance. Theoretically, the total water resource available

Md. S. Islam (✉) · T. Oki · S. Kanae · N. Hanasaki · Y. Agata · K. Yoshimura
Institute of Industrial Science (IIS), The University of Tokyo, 4-6-1 Komaba, Meguro-ku, Tokyo 153-0041, Japan
e-mail: s_islam@iis.u-tokyo.ac.jp

in the world is still enough to support the existing population, if it could be distributed homogeneously. However, in reality water resources are unevenly distributed and at current population levels many countries are already suffering from water scarcity. Direct import of water is generally out of the question, so the concept of virtual water trading has emerged. The term *Virtual Water* is defined as the amount of water needed to produce a commodity or service of any kind, although it is most commonly used in relation to the water required to produce agricultural commodities. The idea was first introduced by J.A. Allan in 1993, and later received attention by many water resources experts.

Established approaches to global water resources assessment aimed at defining annual water scarcity focus on estimating either the natural availability of water resources compared with demand, or the size of population per unit of water flow (Kulshreshtha, 1993; Raskin *et al.*, 1997; Vörösmarty *et al.*, 2000; Oki *et al.*, 2001, 2003; Alcamo *et al.*, 2003; Döll *et al.*, 2003 and Arnell, 1999b, 2004). These narrow definitions apply easily to a country or a region that has no global communications, or an isolated economy, and must achieve its development goals through its own resources. However, in the current era of globalization and abundant communication facilities, most of water-scarce countries have ample scope to compensate their shortcomings by importing food or other water-intensive products. Exchange of virtual water between different parts of the world is already playing a significant role in redressing the unequal spatial distribution of global water resources, especially in the case of countries in arid regions (Oki *et al.*, 2003).

In the present study, global estimates of virtual water trading have been superimposed on a spatially distributed grid-based analysis of global water availability. To the best of our knowledge, no other study has yet attempted to consider these two factors together, i.e. natural water availability in relation to virtual water trade at a global scale. A number of recent studies have estimated virtual water trading between countries, or between regions, including Hakimain (2003), Turton (2000), Wickelns (2001, 2004), Fadel and Maroun (2003), Yegnes-Botzer (2001), Parveen and Faisal (2004) and Yang and Zehnder (2001, 2002). Hoekstra and Hung (2002, 2004) reports the only study that estimated the global balance of virtual water trading of food and other products among different countries. These studies compared country-level virtual water trading with present water resources availability and demand by introducing indices such as *water self sufficiency* and *water dependency*, i.e. the percentage of water demand a country meets by its own resources and the percentage that it covers by importing virtual water. However, these studies did not combine virtual water trading with that of natural availability of flow explicitly in a spatially distributed manner. Basic country-based data on natural water availability was not estimated directly but collected from secondary sources as FAO (2004).

In our study, the global estimate of natural water availability has been made directly from 11 Land Surface Models (LSM), under Global Soil Wetness Project-2 (GSWP-2). The superiority of direct grid-based estimates over country-based studies has been recognized, since it can better describe the spatial variability in water resources availability. Some of the contemporary studies on global water resources availability thus concentrated on smaller scales such as basin- or sub-basin scale, or on grid-based studies. The state-of-the art approaches for these studies use a Macro-Scale Hydrological Model (MSHM) to estimate water balance components, as in the cases of Macro-PDM by Arnell (1999a,b, 2004), Variable Infiltration Capacity (VIC) model by Nijssen *et al.* (2001), Water Balance Model (WBM) by Vorosmarty (2000), and WaterGAP-2 by Alcamo (2003) and Döll *et al.* (2003). It is true that MSHMs are simpler than LSMs, and a calibrated MSHM can reproduce hydrographs better than an un-calibrated LSM. However, compared to MSHM, an LSM not only considers hydrological water balance, but details energy and water balances including hydrological, radiative and even plant physiological processes (Sellers *et al.*, 1986, 1996; Dickinson *et al.*, 1986, 1998). Currently available LSMs can simulate monthly river runoff quite well, provided that the precipitation and other forcing input data for the LSMs are accurate enough (Oki *et al.*, 1999). In fact, all the global circulation models (GCMs) that provide future climate projections use some kind of LSMs. So it is likely that LSMs will be also used directly for water resources projections in future GCMs to simulate the hydrological cycle (Oki *et al.*, 2001).

The method for estimating virtual water in the current study also differs from the methodology of recently published global scale studies. In estimating virtual water trading, Hoekstra (2004) used the virtual water content of the products of the exporting countries for both export and import process. However, in this study the virtual water content of a product was estimated separately for exporting and for importing countries. Usually, the crop yield in the exporting country is higher than the importing country, so the virtual water content of a product in the exporting country is less than that of the importing country. Differences in water contents of the product between the exporting and importing countries should be of interest to estimate how much water the importing country is saving by importing that product. So, when estimating the grid-based water balance between natural flow availability and virtual water trading, it is more realistic to consider the virtual water content of products separately for importing and exporting countries.

Virtual water trading in a broad sense includes both food products as well as many other industrial products that require water in their production process. It is thus a limitation of the study that it considered only the food products, i.e. some major crops and livestock. However, as an attempt to couple virtual water trading with natural availability of flow in a spatially distributed manner, the study still provides a useful estimate of water scarcity around the world.

Global water availability estimation

Local runoff

One of the main objectives of the second phase of the Global Soil Wetness Project (GSWP-2) was to produce the best estimate of global water cycle components for the years from 1986 through 1995. Global distribution of runoff is one of the datasets GSWP2 is producing, which is used in this study to evaluate global water scarcity. Offline simulation of the energy and water balance at the land surface was calculated by 11 LSMs for the purposes. Detailed descriptions of the project are available at Dirmeyer *et al.* (2006).

The runoff dataset produced by the first Global Soil Wetness Project, GSWP1 was evaluated by Oki *et al.* (1999). Their study pointed out that it tends to underestimate stream flow, especially in northern mid- to high-latitude, probably due to gauge under-catch in strong wind conditions. Overcoming the problem is one of the motivations of GSWP2. An empirical technique to correct gauge under-catch was proposed and adopted in the process of producing GSWP2 precipitation data. To examine the reliability of the GSWP2 Baseline (B0) results, the average LSM output of annual runoff was compared with other estimates as shown in Table 1.

It has been evident from the results that GSWP2 B0 runoff data produced by 11 GSWP2 participating LSMs are much higher than earlier studies, especially in the northern mid- to high-latitude. This is a completely opposite result from GSWP1. It might be due to the over-correction of the gauge under-catch of GSWP2 B0 precipitation dataset. An alternative approach was adopted in this case to improve the average output by excluding those extreme

Table 1 Continental runoff (km^3/year)

Region	Raskin *et al.* (1997)	Vörösmarty *et al.* (2000)	GSWP1	GSWP2-B0	GSWP2-B0-CT
Africa	4050	4520	3616	4473	4533
Asia	13510	13700	9385	15902	10797
Europe	2900	2770	2191	9827	5093
Oceania	2404	714	1680	1943	1879
North America	7890	5890	3824	10713	6456
South America	12030	11700	8789	9799	10183
Total	42784	39394	29485	52657	38941

values for each grid out of 12 models, i.e. the maximum and minimum values of two models. This version of GSWP2 output was named as GSWP2-B0-CT. Even though such a measure has no strong scientific rationalale, still it improved the runoff estimates significantly as shown in Table 1. The only problem that remains is in data for Europe, which are still very high. It is assumed that possibly the original precipitation data in the corresponding European region had already corrected for wind under catch, and the GSWP-2 forcing data in the region was over-corrected. Further improvement of the issue is under consideration. In this study we used the output from the GSWP2-B0-CT version for local runoff estimate.

Exogenous runoff from upstream

Present output from the GSWP2-B0-CT version can be further routed to global river networks to produce estimates of the river discharge. Total Runoff Integrated Pathways (TRIP), developed by Oki and Sud (1998), was used for this purpose at a resolution of 0.5 × 0.5 degree grids. A detailed description of the TRIP model is available at Oki *et al.* (1999) or Okada (2000). Actually, the grid runoff estimated from those LSMs is the *local runoff* as shown in Figure 1a. Use of a river routing model TRIP added the *exogenous runoff* available at each grid from upstream to that of local runoff. The difference between TRIP-routed runoff with that of LSM runoff is actually the net contribution to a cell of exogenous runoff from the upstream as shown in Figure 1b.

Theoretically, all of this exogenous runoff should be available at the downstream grid points, as assumed in previous studies by Vörösmarty *et al.* (2000) and Alcamo *et al.* (2003). However, in reality because of upstream withdrawal, a significant portion of this exogenous runoff is not available at the downstream grids. Especially in the case of long trans-boundary rivers, in arid or semiarid regions, the problem is more serious. This exogenous runoff is actually an uncertain amount of water for a particular grid. This study thus critically assessed the routed amount of flow and accounted for the effect by introducing the following formula:

$$Q = R + \alpha \sum D_{\mathrm{up}}$$

where, Q is the water available at a particular grid, R is the grid runoff from the LSM output, D is the discharge from other upstream grids due to routing, α is the ratio of water from outside the grid to that of the water resources inside the grid. Here $\alpha = 1.0$, means that all of the exogenous runoff generated upstream from the routing scheme can be used at the downstream grid, i.e. the TRIP-routed discharge. When $\alpha = 0.0$, it assumes that no upstream contribution

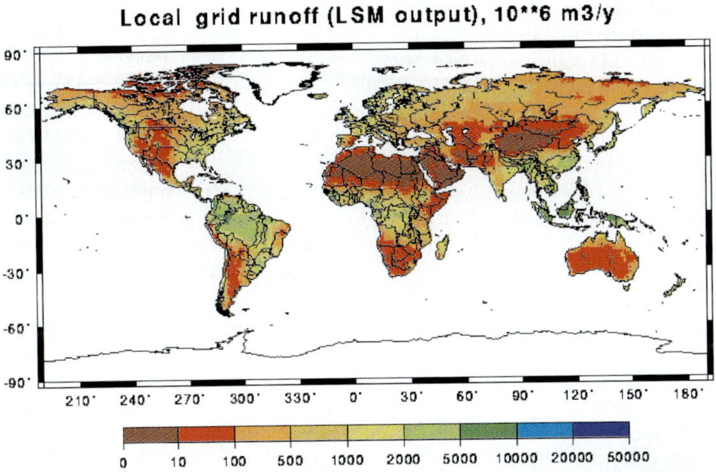

(a)

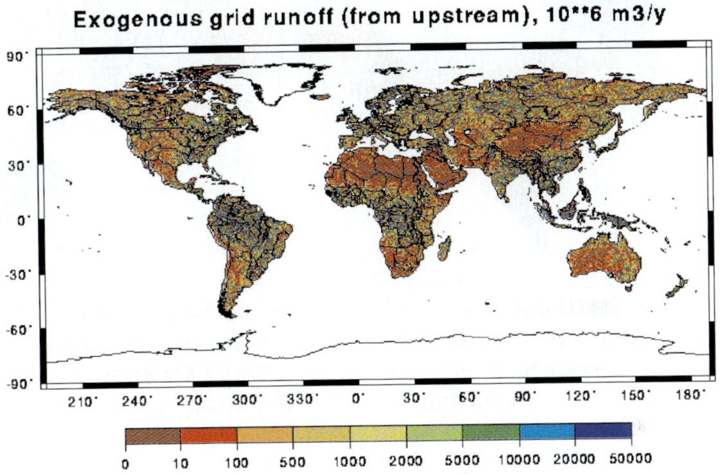

(b)

Fig. 1 Global distribution of runoff (a) local runoff or vertical component of flow (direct output of runoff from LSMs without routing) (b) Exogenous or horizontal component of flow (net upstream contribution of flow due to routing effect)

is available, so the water resources availability at a grid is just what it is available within the grid, i.e. only the local runoff. Depending on a number of factors as climatic, socio-economic, land use, or topographical factors that affect upstream water withdrawal, the α value varies among different regions in the world.

Estimating virtual water exchanges

Unit requirement of water resources to produce each commodity (hereafter called *UW*) is the starting point for the quantification of virtual water trading. Some estimates of *UW* are available from Wichelns (2001), Hoekstra and Hung (2002), and Oki *et al.* (2003). However, there are still a lot of uncertainties in determining *UW*, probably because alternative rational

Table 2 Estimated values of UW for different products by Oki *et al.* (2003)

Products	Unit Water Content, UW (m^3/ton)
Rice	3200
Wheat	1600
Maize	900
Soybean	2500
Barley	1200
Chicken	4500
Pork	5900
Beef	20700
Egg	3200
Milk	560

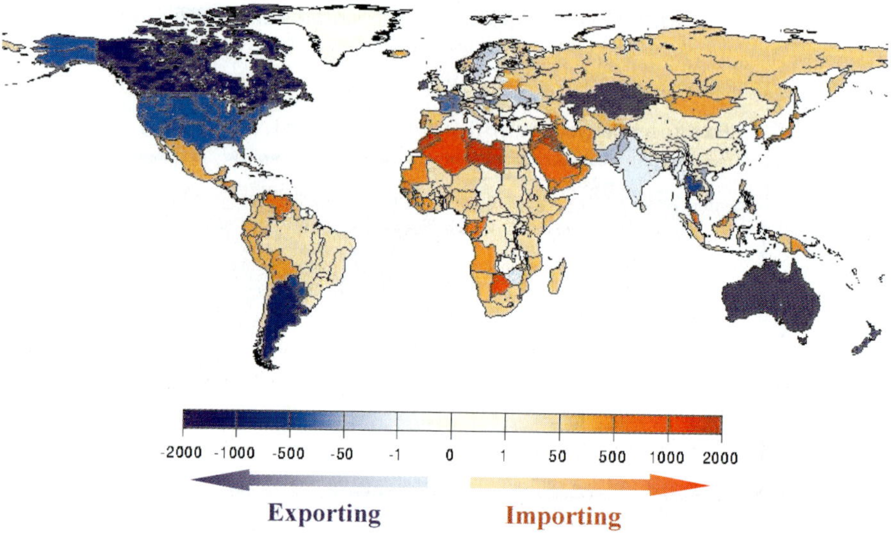

Fig. 2 Net trading of virtual water among the countries (m^3/c/y)

definitions of virtual water have been made. Previous studies have estimated *UW* for grains, and livestock individually as described detail in the reference Oki *et al.* (2004). *UW* estimates for individual products were made based on the available information or experience in Japan. A list of estimated UW for different products is shown in Table 2.

As mentioned above, Oki *et al.* (2003) differentiated *virtual water content* of a product for both the exporting and importing countries. Usually it varies due to the difference in crop yields between the two respective countries. FAO data sets of crop yields for different countries in the world (FAO, 2000) were used to estimate the UW requirements of products for each country, and compared to Standard UW estimates and crop yield. The modified UW requirement is then termed the virtual water content of a product for that country. FAO estimates of global trading of crops and livestock between different countries for the year 2000 (FAO, 2004) were utilized to estimate the country-level annual virtual water flows between countries. To make the figure comparable among different countries, the total amount of export and import was divided by year 2000 population and the country-based per capita export or import is shown in Figure 2.

Table 3 Comparison of food consumption rate among rural and urban people (modified from Regmi and Dyck, 2001)

	Urban				Rural			
	Cereals	Roots & tubers	Fruit & vegetables	Meats & offals	Cereals	Roots & tubers	Fruit & vegetables	Meats & offals
				Kg/capita/year				
China[a]								
1988	199	NA	195	29	208	NA	135	12
1998	140	NA	166	30	200	NA	128	16
				Grams/capita/week				
Indonesia[b]								
1978	2.165	275	1.005	64	2.560	810	975	36
1987	2.182	279	1.275	108	2.579	612	1.364	16
				Kg/capita/month				
Pakistan[b]								
1979	10.59	0.73	NA	0.85	13.66	0.72	NA	0.46
1987/88	9.75	0.68	NA	0.76	12.69	0.68	NA	0.51

[a]Economic Research Service, USDA
[b]FAO, 1993

Because, FAO estimates for crop trading were available at country level only, virtual water trading was estimated at country scale. This is a usual procedure followed by other global scale studies as well (Hoekstra and Hung, 2002, 2004; Oki *et al.*, 2003). However, this study assumes that virtual water export and import should have specific spatial variability within the country based on the nature of land use and population density. The first scientific question was how to match these country level estimates of virtual water trading to that of the grid runoff values from GSWP2.

In this connection, the following assumptions were made:

- virtual water export from a country is spatially distributed among the grids in proportion to the density of agricultural areas.
- virtual water import by a country is distributed to its grids in proportion to population density.

In the case of virtual water export, it can be reasonably assumed that crops or livestock are collected from those agricultural areas utilizing local water resources. Therefore the virtual water export for a particular grid can be estimated as:

$$= (\text{Agricultural area of the grid}/\text{Total Agricultural area of the country})$$
$$* \text{Total amount of virtual water exported from the country}$$

In the case of the country-based virtual water import value, however, allocations among grids are not so simple. For a particular country, there are at least two questions that need to be answered as follows – whether there is any significant difference in food consumption patterns amongst different parts of the country, and which parts of the country depend heavily on imported foods. As shown in Table 3, the per capita food consumption among urban and rural areas actually does not differ much (Regmi and Dyck, 2001). The only difference is the combination of foods as cereal, meat or vegetables. Converting the total food consumption into equivalent virtual water content, the difference is again negligible. Another study by Ozcan (2003) on food consumption patterns in Turkey reported the same results.

Regarding the other issue – i.e. the dependence in different parts of a country (e.g. urban and rural) on imported food products – the most important problem is the extent of internal trading, data on which are rarely available, and vary significantly among countries. It is rational to assume that urban people are mostly dependent on imported food products, while the rural people mostly produce their own food. However, other factors such as storage facilities, economic conditions and food habits, etc. may significantly affect dependence on imported food. For a 0.5-degree grid-based analysis, in the absence of adequate information, such factors are difficult to quantify accurately. Under the same program, studies are in progress to consider such effects in as much detail as possible. At this stage of the present study, we assumed simplistically that all the imported food products for a particular country are equally distributed along different grids based on population density. Based on the above assumptions, the grid based distribution of virtual water exports and imports is shown in Figure 3a and b.

Water stress level

There are a number of indices used to define water resources stress. Two of the widely used indices are the *use- to- resources ratio* and *per capita water availability*. The ratio of water use, or withdrawal, to runoff was used as an indicator in many studies such as in the UN Comprehensive Assessment of the Freshwater Resources by Raskin *et al.* (1997), and in Alcamo *et al.* (2003); Vorosmarty *et al.* (2000), Oki *et al.* (2001), and Arnell (1999b). However, the difficulty with this indicator is correctly estimating water use. Another index developed by Falkenmark *et al.* (1989) is simpler, and defines water stress by estimating the number of people per flow unit (i.e. 10^6 m^3) annually. It differentiates four stress levels based on per capita water availability as follows:

Per capita water availability (m^3/c/y)	Stress level
>1700	No stress
1000–1700	Moderate stress
500–1000	High stress
<500	Extreme stress

Several studies designed for global water resources assessment (Arnell, 2004; Revenga *et al.*, 2000; Kulshreshtha, 1993) also adopted this index.

In the present study, the Falkenmark index was adopted because it is simple and comprehensive to couple with virtual water export and import values. The per capita virtual water exports from different grids in the exporting country were deduced from the per capita runoff availability in those grids. Likewise, per capita virtual water import was added to the per capita annual runoff availability in the importing country grids. The above four stress levels were adopted and the number of people under each stress level estimated. To differentiate the effect of virtual water trading on relieving stress, two estimates were made: one for water stress using the GSWP2 results without virtual water trading; and the other for GSWP-2 results combined with the net virtual water trade.

One other important concern is the exogenous runoff availability. As mentioned in the previous section, this flow is actually uncertain so that a coefficient α was introduced. Here, two extreme scenarios of α value have been considered as $\alpha = 1.0$, i.e. full availability

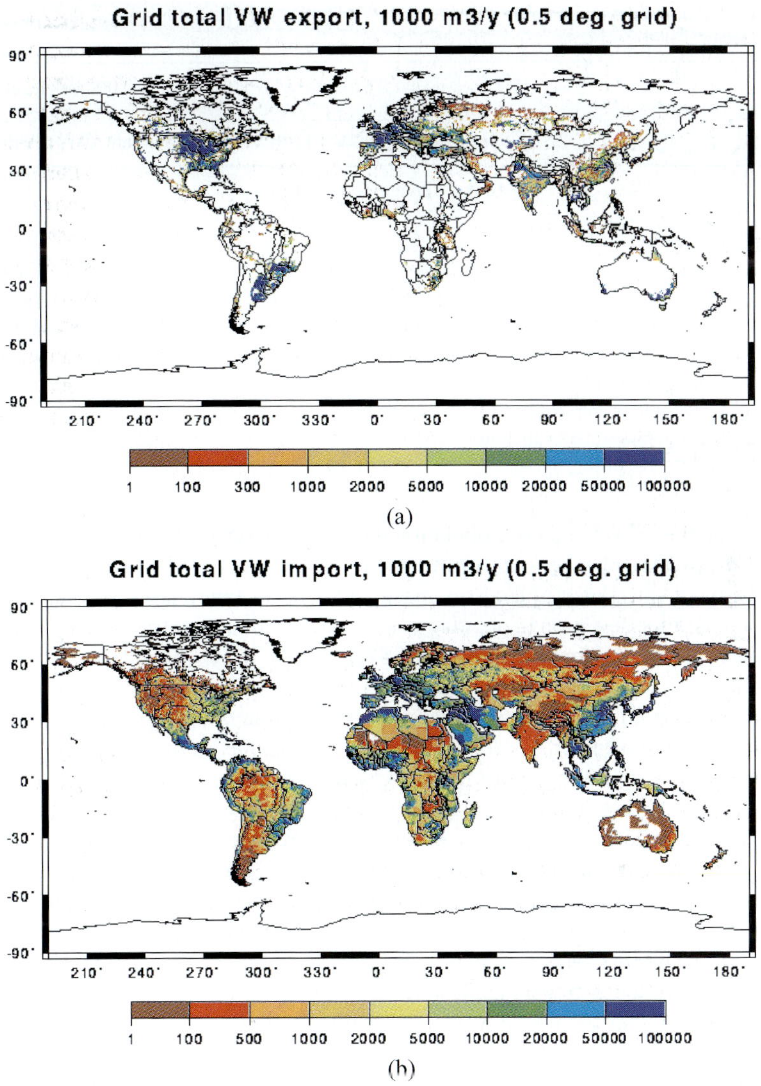

Fig. 3 Distribution of country-based virtual water trading among grids; (**a**) Export of virtual water (based on real water requirements of the country) (**b**) Import of virtual water

of routed flow from upstream grids; and $\alpha = 0.0$, i.e. no upstream flow available at those downstream grids but only the LSM grid runoff. Availability of this exogenous flow in the real world would be between these two α values.

Global figures

Table 4, shows detailed results of the number of people at different stress levels under different scenarios. Population data used for this purpose are derived from the CIESIN 2.5-minute grid data aggregated to 0.5 degree. Because of differences in the land-sea mask between CIESIN and GSWP2 grids, special adjustments were made to relocate CIESIN

Table 4 Global population under different stress levels and scenarios, including virtual/real water flows and exogenous runoff

	$\alpha = 1.0$ Stress level (m^3/c/y)		$\alpha = 0.0$ Stress level (m^3/c/y)	
	Without VW	With VW	Without VW	With VW
Global population in millions				
No stress: >1700	3768	3931	2179	2347
Moderate: 1000–1700	524	564	756	879
High: 500–1000	632	705	996	1103
Extreme: <500	1142	851	2111	1672
Population below 1000 m^3/c/y level	1774	1556	3107	2775
Population below 1700 m^3/c/y level	2298	2120	3863	3654
Virtual water trading derived from				
–Increase in per capita water availability	4160		4160	
–Decrease in per capita water availability	1525		1525	

grids to match the GSWP2 grids. Population data for boundary grids between two countries also needed adjustment when aggregated from 2.5-minute to 0.5-degree grids. However, both the cases of difference in land-sea mask and country boundaries, still some problems remain so that minor deviation in population count for smaller countries might occur. Further improvement of the problem by adjusting GSWP2 and CIESIN boundaries is underway in future studies.

It can be seen from Table 4 that virtual water trading plays an important role in relieving pressure for a large number of global population. The maximum number of people that reduced their stress are in the group with the extreme stress level of 500 m^3/c/y. Arid countries with very low water availability belong to this group as they have no option but to survive by importing water-intensive products. Gradually, for reduced stress levels, such imports of virtual water declined. However, for the 1000 m^3/c/y or 1700 m^3/c/y categories, the total number of people relieved from stress through virtual water trading is still significant. As shown at the bottom of Table 4, the number of people benefiting from an increase in per capita water availability due to virtual water trading is higher than the number of people suffered from a reduction in per capita water availability.

Table 4 shows the difference in water availability under two extreme α values. The exact value of α is difficult to estimate and varies between different regions. Here $\alpha = 1.0$ is the best scenario when all the upstream flow is available for downstream users. However, in the changing world with increasing population and industrial activities, it is already evident that the upstream water withdrawal rates are increasing gradually, thus decreasing the value of α over time. For trans-boundary rivers, a decreasing trend in α might portend increasing conflict over water sharing between countries. So, the $\alpha = 0.0$ figure can be seen as the worst scenario at some hypothetical future time.

Regional figures

The spatial distribution of water stress levels affected by virtual water trading on a regional basis can be seen in Figures 4 and 5. These Figures indicate that countries in the Middle-East and North-African are relieving their water stress significantly by importing virtual water. To explore this phenomenon further, Table 5 shows a quantitative analysis based on six different regions. On the top of the list, the North-Africa and Middle-East countries benefit the most

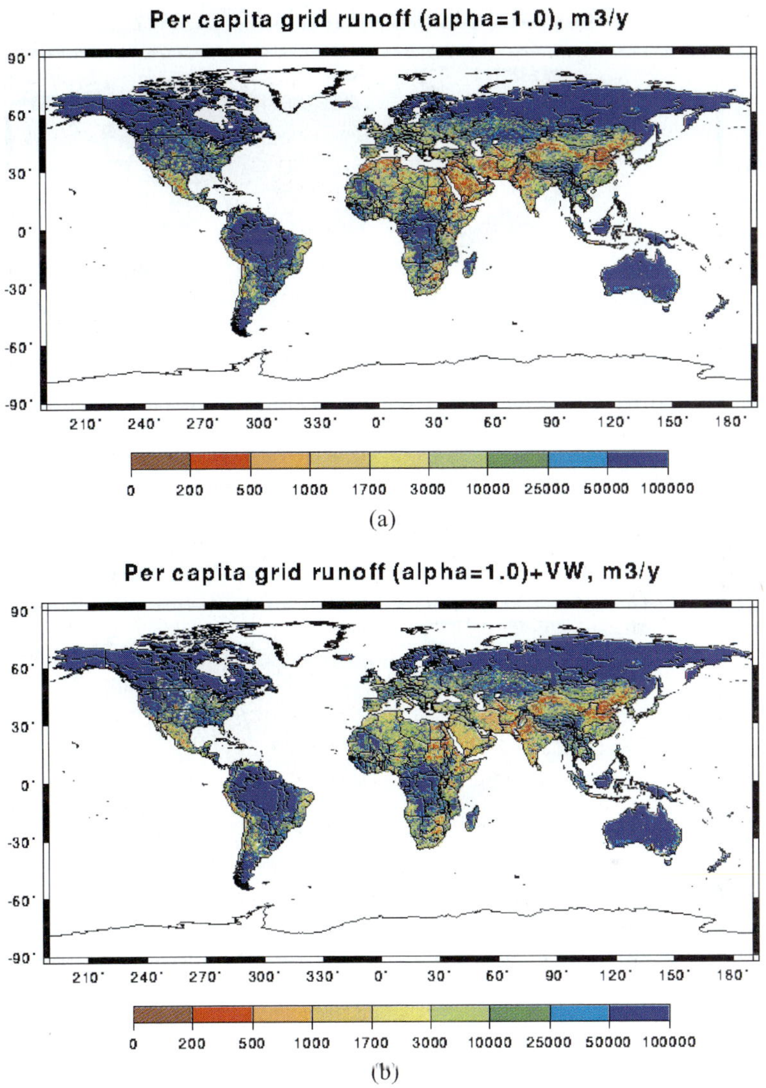

Fig. 4 Effect of virtual water trading on the present state of water stress ($m^3/c/y$) level for $\alpha = 1.0$ scenario (i.e. full use of exogenous flow)

from the virtual water trading process. Approximately 70% of the 166 million people of this region under the extreme water stress scenario could be upgraded to reduced stress conditions through virtual water trading. In total almost 90% of the population could increase their per capita water availability in the region through virtual water trading. Next to this region is Latin America, which also improved per capita water availability through virtual water trading. For Asia the net upgrading of per capita water availability is the lowest. This result is to some extent confusing considering the country-based data shown in Figure 2, which shows that most of Asian countries are importing virtual water. Because of the larger import of virtual water, the per capita water availability should increase for a larger number of people.

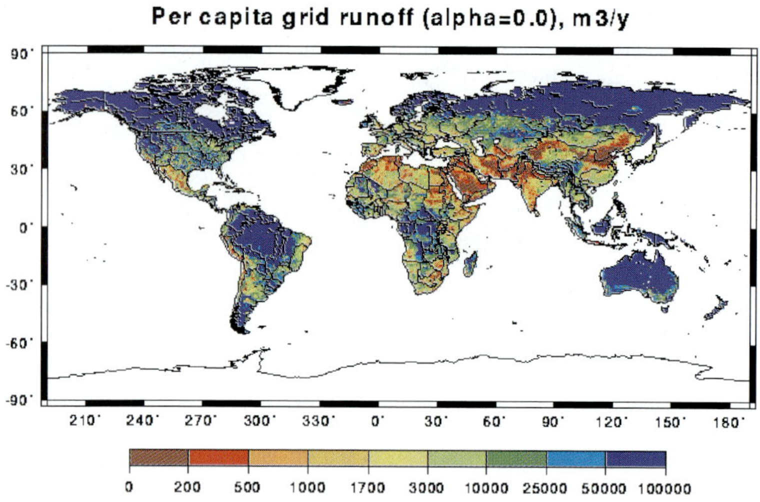

(a)

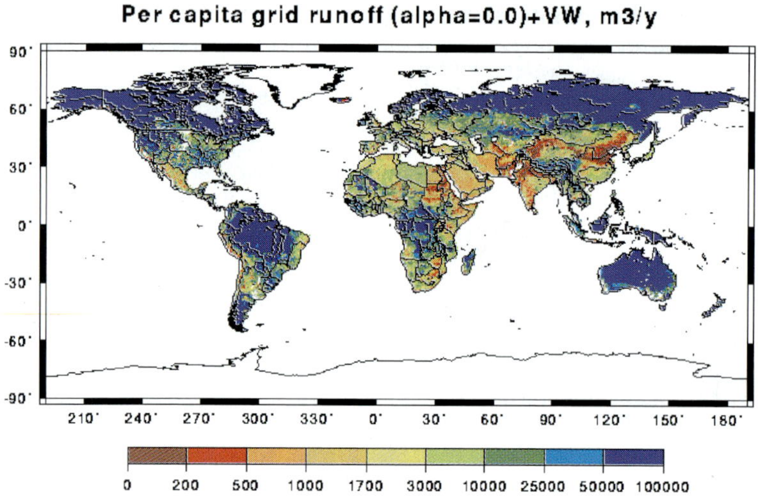

(b)

Fig. 5 Effect of virtual water trading on the present state of water stress ($m^3/c/y$) level for $\alpha = 0.0$ scenario (i.e. no exogenous flow)

However, this actually happened because of the fact that most Asian people live in rural areas or agricultural districts. Therefore the export of virtual water from those agricultural grids affected a large number of people in the form of reduced per capita water availability. On the other hand, even though the OECD countries are larger exporters, the population density in their agricultural grids are lower than Asian countries, so that the percentage of people suffering from reduced per capita water availability is comparatively lower than Asian countries.

Regarding the effect of exogenous runoff, the region FSU is affected most seriously for the case where $\alpha = 0$. This is because of the large number of long rivers in the FSU region where

Table 5 Regional disaggregation of water stress affected by virtual water (VW) trading and exogenous flow

	Population (million)					
Stress level (m3/c/y)	Asia	FSU	Latin America	N. Africa & Mid. East	OECD	SubSahara
		$\alpha = 0.0$, without VW				
<500	1346.79	72.7830	132.960	242.080	149.876	153.552
500–1000	596.735	40.4396	65.3456	46.6223	148.526	86.4631
1000–1700	404.240	33.5412	50.6976	28.2680	135.057	94.6533
>1700	833.344	140.632	246.496	51.9847	520.554	328.044
		$\alpha = 0.0$, with VW				
<500	1260.48	47.7393	84.7941	80.4676	96.3944	101.258
500–1000	589.564	55.8949	87.2469	77.9914	151.068	127.415
1000–1700	437.615	38.7019	58.4312	79.2682	151.965	96.2663
>1700	843.970	143.938	261.603	149.346	544.242	343.043
		$\alpha = 1.0$, without VW				
<500	681.028	15.0467	107.178	166.351	71.5279	88.3487
500–1000	394.449	25.0310	27.8109	41.0389	69.2100	66.2937
1000–1700	294.123	22.3815	37.4982	29.2238	81.0784	53.9800
>1700	1812.41	225.031	323.184	150.061	732.252	458.195
		$\alpha = 1.0$, with VW				
<500	637.054	10.07678	60.3195	49.7634	30.2300	62.7386
500–1000	383.744	20.0299	58.1494	52.1955	98.2730	80.1350
1000–1700	320.609	30.2506	34.7646	49.2471	60.8454	57.8338
>1700	1829.93	226.673	340.330	235.935	762.129	467.448
			Total			
Increase in per capita water availability	1868.29	206.110	408.677	345.076	665.238	580.066
Decrease in per capita water availability	1178.82	33.9605	44.0391	17.0300	238.578	8.47740

the routing effect therefore produces a large percentage of flow which is affected in the case where $\alpha = 0$. In practice the α value should vary within 0 to 1 so that the tabulated numbers here provide a complete range of future possibilities of the exogenous runoff availability in different parts of the world.

Conclusion

This study demonstrated a state-of-the art approach of preparing a grid-based estimation of virtual water trading in different parts of the world, coupled with the latest estimation of runoff availability from the GSWP2 project. The percentage of population that benefited from virtual water trading is the highest for the Middle East, North African, and Sub-Saharan African regions. Considering the Falkenmark index, the population below the worst stress scenario, i.e. 500 m^3/c/y water availability, are the greatest beneficiaries of this virtual water trading. Around 25% of the total global population suffering at this acute shortage level could upgrade themselves to the upper level.

The difference in the number of affected people for country-level and grid-level studies is another important finding to note here. From the country-based estimates it was seen that some OECD countries including USA, Canada, Australia, and New Zealand are the largest exporters of virtual water, and Asia is the greatest net importer. However, the regional estimate of affected population based on grid-base calculations shows that the percentage of people suffering from a reduction in per capita water availability due to export of virtual water is still higher for Asia compared to that of OECD countries. This is because of the difference in the spatial distribution of land use and population density among different regions.

Because of the upstream water withdrawal effect, availability of the routed runoff is actually quite uncertain, so two different scenarios of natural runoff, as with or without routed flow, were made. This analysis showed that the global population under the 500 m^3/c/y stress level would be doubled under the worst scenario of non-availability of upstream routed flow. In the changing world with increased population and water demand, conflicts on water sharing issues along the trans-boundary rivers are already evident in many parts of the world, and these might be aggravated in future.

The main limitation of the study is that the estimated grid values of virtual water trading are based on country-level estimations of virtual water trading from the FAO food trading data base, modified by some assumptions. True grid-based estimates of virtual water trading would require a direct estimation of virtual water flows among different grids, irrespective of national boundaries between countries. Unfortunately such detailed information is not available on a global scale; nevertheless, the output of this study can be considered useful as an improved and more realistic analysis of global water scarcity incorporating virtual water trading.

References

Alcamo J, Doll P, Heinrichs T, Kaspar F, Lehner B, R.osch T, Siebert S (2003) Global estimates of water withdrawals and availability under current and future business-as-usual conditions. Hydrol Sci J 48:339–348

Allan JA (1998) Moving water to satisfy uneven global needs: trading water as an alternative to engineering it. ICID J 47(2):1–8

Allan JA (2003) Virtual water – the water, food, and trade nexus: useful concept or misleading metaphor? Water Int 28(1):106–113

Allan JA (1998) Virtual water: a strategic resource. Global solutions to regional deficits. Ground Water 36(4):545–546

Allan JA (2002) The Middle East Water Question: Hydropolitics and the Global Economy. L.B. Tauris Publishers, London

Arnell NW (1999a) A simple water balance model for the simulation of streamflow over a large geographic domain. J Hydrol 217:314–335

Arnell NW (1999b) Climate change and global water resources. Global Environmental Change 9:31–49

Arnell NW (2004) Climate change and global water resources: SRES emissions and socio-economic scenarios. Global Environmental Change 14:31–52

Center for International Earth Science Information Network (CIESIN) (2000) Columbia University; International Food Policy Research Institute (IFPRI); and World Resources Institute (WRI) (2000) Gridded Population of the World (GPW) Version 2. CIESIN, Columbia University, Palisades, NY, available at http://sedac.ciesin.columbia.edu/plue/gpw

Dickinson RE, Henderson-Sellers A, Kennedy PJ, Wilson MF (1986) Biosphere – atmosphere transfer scheme (BATS) for the NCAR community climate model. Tech. Report no. TN275+STR, National Center for Atmospheric Research, Boulder, Colorado, USA

Dickinson RE, Shaikh M, Bryant R, Graumlich L (1998) Interactive canopies for a climate model. J Climate 11:2823–2836

Dirmeyer PA, Gao X, Zhao M, Guo Z, Oki T, Hanasaki N (2006) GSWP2 multimodel analysis and implications for our perception of the land surface. Bulletin of the American Meteorological Society 87(10):1381–1397

Döll P, Frank K, Bernhard L (2003) A global hydrological model for deriving water availability indicators: model tuning and validation. J Hydrol 270:105–134

El Fadel M, Maroun R (2003) The concept of 'virtual water' and its applicability in Lebanon. In: Hoekstra AY (ed) VirtualWater Trade: Proceedings of the International Expert Meeting on VirtualWater Trade, Research Report Series No 12. IHE Delft, The Netherlands

FAO (2004) Online AQUASTAT 2000 database
Falkenmark M, Lundquist J, Widstrand C (1989) Macro-scale water scarcity requires micro-scale approaches: aspects of vulnerability in semi-arid development. Natural Resources Forum 13:258–267
Hakimain H (2003) Water scarcity and food imports: an empirical investigation of the "virtual water" hypothesis in the MENA region. Review of Middle East Economics and Finance 1(1):71–85
Hoekstra AY, Hung PQ (2002) Virtual Water Trade: A Quantification of Virtual Water Flows between Nations in relation to International Crop Trade, Value of Water. Research Report Series 11, IHE Delft, The Netherlands
Hoekstra AY, Hung PQ (2004) Globalisation of water resources: international virtual water flows in relation to crop trade. Global Environmental Change Part A 15(1):45–56
Kulshreshtha SN (1993) World Water Resources and Regional Vulnerability: Impact of future Changes, IIASA Research Report RR-93-010, June 1993, pp 137
Okada Y (2000) Simulating discharge of major rivers in the world through development of a global river channel network at 0.5 resolution. Master's thesis, University of Tokyo, Tokyo, Japan
Oki T, Agata Y, Kanae S, Saruhashi T, Yang D, Musiake K (2001) Global assessment of current water resources using total runoff integrating pathways. Hydrol. Sci. J. 46:983–996
Oki T, Sud YC (1998) Design of Total Runoff Integrating Pathways (TRIP) – A global river channel network. Earth Interactions 2 (http://EarthInteractions.org)
Oki T, Nishimura T, Dirmeyer P (1999) Assessment of land surface models by runoff in major river basins of the globe using Total Runoff Integrating Pathways (TRIP). J Met Soc Jpn 77:235–255
Oki T, Sato M, Kawamura A, Miyake M, Kanae S, Musiake K (2003) Virtual water trade to Japan and in the world. In: Hoekstra AY (ed) Virtual Water Trade: Proceedings of the International Expert Meeting on Virtual Water Trade, Research Report Series No. 12. IHE Delft, The Netherlands
Ozcan KM (2003) Basic food consumption in Turkey: effects of income, price and family size in Urban areas, occasional paper available at www.bilkent.edu.tr/~kivilcim/foodpaper.pdf
Nijssen B, Greg M, O'Donnell AF, Dennis P (2001) Hydrological sensitivity of global rivers to climate change. Climatic Change 50:143–175
Parveen S, Faisal IM (2004) Trading virtual water between Bangladesh and India: a Politico – economic dilemma. Water Policy 6:1–10
Raskin P, Gleick P, Kirshen P, Pontius G, Strzepek K (1997) Comprehensive Assessment of the Freshwater Resources of the World. Stockholm Environment Institute, Stockholm
Regmi A, Dyck J (2001) Effects of urbanization on global food demand changing structure of global food consumption and trade. In: Regmi A (ed) Market and Trade Economics Division, Economic Research Service, U.S. Department of Agriculture, Agriculture and Trade Report, WRS-01-1.
Revenga C, Brunner J, Henninger N, Kassem K, Payne N (2000) Pilot analysis of global ecosystems: freshwater ecosystems. World Resources Institute and Worldwatch Institute, Washington, DC
Sellers PJ, Mintz Y, Sud YC, Dalcher A (1986) A simple biosphere model (SiB) for use within general circulation models. J Atmos Sci 43:505–531
Sellers PJ, Randall DA, Collatz GJ, Berry JA, Field CB, Dazlich DA, Zhang C, Collelo GD, Bounoua L (1996) A revised land surface parameterization (SiB2) for atmospheric GCMs. Part I: model formulation. J Climate 9:676–705
Turton AR (1999) Precipitation, people, pipelines and power: towards a 'virtual water' based political ecology discourse. MEWREW Occasional Paper No. 11, Available at http://www.soas.ac.uk/waterissues/occasionalpapers/OCC11.PDF
Vorosmartsy C, Green P, Salisbury J, Lammers R (2000) Global water resources: vulnerability from climate change and population growth. Science 289:284–288
Wichelns D (2001) The role of 'virtual water' in efforts to achieve food security and other national goals, with an example from Egypt. Agric Water Manage 49:131–151
Yang H, Zehnder AJB (2002) Water scarcity and food import: a case study for Southern Mediterranean Countries. World Development 30(8):1413–1430
Yang H, Zehnder A (2001) China's regional water scarcity and implications for grain supply and trade. Environ Plan A 33:79–95
Yegnes-Botzer A (2001) Virtual water export from Israel: quantities, driving forces and consequences. M.Sc. Thesis, DEW 166, IHE, Delft, The Netherlands

Water Resour Manage (2007) 21:35–48
DOI 10.1007/s11269-006-9039-x

ORIGINAL ARTICLE

Water footprints of nations: Water use by people as a function of their consumption pattern

A. Y. Hoekstra · A. K. Chapagain

Received: 18 January 2005 / Accepted: 12 October 2005

Abstract The water footprint shows the extent of water use in relation to consumption of people. The water footprint of a country is defined as the volume of water needed for the production of the goods and services consumed by the inhabitants of the country. The internal water footprint is the volume of water used from domestic water resources; the external water footprint is the volume of water used in other countries to produce goods and services imported and consumed by the inhabitants of the country. The study calculates the water footprint for each nation of the world for the period 1997–2001. The USA appears to have an average water footprint of 2480 m^3/cap/yr, while China has an average footprint of 700 m^3/cap/yr. The global average water footprint is 1240 m^3/cap/yr. The four major direct factors determining the water footprint of a country are: volume of consumption (related to the gross national income); consumption pattern (e.g. high versus low meat consumption); climate (growth conditions); and agricultural practice (water use efficiency).

Keywords Water footprint · Consumption · Virtual water · Indicators · Water use efficiency · External water dependency

Introduction

Databases on water use traditionally show three columns of water use: water withdrawals in the domestic, agricultural and industrial sector respectively (Gleick, 1993; Shiklomanov, 2000; FAO, 2003). A water expert being asked to assess the water demand in a particular country will generally add the water withdrawals for the different sectors of the economy. Although useful information, this does not tell much about the water actually needed by the people in the country in relation to their consumption pattern. The fact is that many goods

A. Y. Hoekstra (✉)
University of Twente, Enschede, The Netherlands
e-mail: a.y.hoekstra@utwente.nl

A. K. Chapagain
UNESCO-IHE, Delft, The Netherlands

consumed by the inhabitants of a country are produced in other countries, which means that it can happen that the real water demand of a population is much higher than the national water withdrawals do suggest. The reverse can be the case as well: national water withdrawals are substantial, but a large amount of the products are being exported for consumption elsewhere.

In 2002, the water footprint concept was introduced in order to have a consumption-based indicator of water use that could provide useful information in addition to the traditional production-sector-based indicators of water use (Hoekstra and Hung, 2002). The water footprint of a nation is defined as the total volume of freshwater that is used to produce the goods and services consumed by the people of the nation. Since not all goods consumed in one particular country are produced in that country, the water footprint consists of two parts: use of domestic water resources and use of water outside the borders of the country.

The water footprint has been developed in analogy to the ecological footprint concept as was introduced in the 1990s (Rees, 1992; Wackernagel and Rees, 1996; Wackernagel *et al.*, 1997). The 'ecological footprint' of a population represents the area of productive land and aquatic ecosystems required to produce the resources used, and to assimilate the wastes produced, by a certain population at a specified material standard of living, wherever on earth that land may be located. Whereas the 'ecological footprint' thus quantifies the *area* needed to sustain people's living, the 'water footprint' indicates the *water* required to sustain a population.

The water footprint concept is closely linked to the virtual water concept. Virtual water is defined as the volume of water required to produce a commodity or service. The concept was introduced by Allan in the early 1990s (Allan, 1993, 1994) when studying the option of importing virtual water (as opposed to real water) as a partial solution to problems of water scarcity in the Middle East. Allan elaborated on the idea of using virtual water import (coming along with food imports) as a tool to release the pressure on the scarcely available domestic water resources. Virtual water import thus becomes an alternative water source, next to endogenous water sources. Imported virtual water has therefore also been called 'exogenous water' (Haddadin, 2003).

When assessing the water footprint of a nation, it is essential to quantify the flows of virtual water leaving and entering the country. If one takes the use of domestic water resources as a starting point for the assessment of a nation's water footprint, one should subtract the virtual water flows that leave the country and add the virtual water flows that enter the country.

The objective of this study is to assess and analyse the water footprints of nations. The study builds on two earlier studies. Hoekstra and Hung (2002, 2005) have quantified the virtual water flows related to the international trade of crop products. Chapagain and Hoekstra (2003) have done a similar study for livestock and livestock products. The concerned time period in these two studies is 1995–1999. The present study takes the period of 1997–2001 and refines the earlier studies by making a number of improvements and extensions.

Method

A nation's water footprint has two components, the internal and the external water footprint. The internal water footprint (*IWFP*) is defined as the use of domestic water resources to produce goods and services consumed by inhabitants of the country. It is the sum of the total water volume used from the domestic water resources in the national economy *minus* the volume of virtual water export to other countries insofar related to export of domestically produced products:

$$IWFP = AWU + IWW + DWW - VWE_{\text{dom}} \tag{1}$$

Here, *AWU* is the agricultural water use, taken equal to the evaporative water demand of the crops; *IWW* and *DWW* are the water withdrawals in the industrial and domestic sectors respectively; and VWE_{dom} is the virtual water export to other countries insofar related to export of domestically produced products. The agricultural water use includes both effective rainfall (the portion of the total precipitation which is retained by the soil and used for crop production) and the part of irrigation water used effectively for crop production. Here we do not include irrigation losses in the term of agricultural water use assuming that they largely return to the resource base and thus can be reused.

The external water footprint of a country (*EWFP*) is defined as the annual volume of water resources used in other countries to produce goods and services consumed by the inhabitants of the country concerned. It is equal to the so-called virtual water import into the country *minus* the volume of virtual water exported to other countries as a result of re-export of imported products.

$$EWFP = VWI - VWE_{\text{re-export}} \tag{2}$$

Both the internal and the external water footprint include the use of *blue water* (ground and surface water) and the use of *green water* (moisture stored in soil strata).

The use of domestic water resources comprises water use in the agricultural, industrial and domestic sectors. For the latter two sectors we have used data from AQUASTAT (FAO, 2003). Though significant fractions of domestic and industrial water withdrawals do not evaporate but return to either the groundwater or surface water system, these return flows are generally polluted, so that they have been included in the water footprint calculations. The total volume of water use in the agricultural sector has been calculated in this study based on the total volume of crop produced and its corresponding virtual water content. For the calculation of the virtual water content of crop and livestock products we have used the methodology as described in Chapagain and Hoekstra (2004). In summary, the virtual water content (m^3/ton) of primary crops has been calculated based on crop water requirements and yields. Crop water requirement have been calculated per crop and per country using the methodology developed by FAO (Allen *et al.*, 1998). The virtual water content of crop products is calculated based on product fractions (ton of crop product obtained per ton of primary crop) and value fractions (the market value of one crop product divided by the aggregated market value of all crop products derived from one primary crop). The virtual water content (m^3/ton) of live animals has been calculated based on the virtual water content of their feed and the volumes of drinking and service water consumed during their lifetime. We have calculated the virtual water content for eight major animal categories: beef cattle, dairy cows, swine, sheep, goats, fowls/poultry (meat purpose), laying hens and horses. The calculation of the virtual water content of livestock products is again based on product fractions and value fractions.

Virtual water flows between nations have been calculated by multiplying commodity trade flows by their associated virtual water content:

$$VWF[n_e, n_i, c] = CT[n_e, n_i, c] \times VWC[n_e, c] \tag{3}$$

in which *VWF* denotes the virtual water flow ($m^3 yr^{-1}$) from exporting country n_e to importing country n_i as a result of trade in commodity c; *CT* the commodity trade (ton yr^{-1}) from the exporting to the importing country; and *VWC* the virtual water content (m^3 ton^{-1}) of the commodity, which is defined as the volume of water required to produce the commodity in the exporting country. We have taken into account the trade between 243 countries for which international trade data are available in the Personal Computer Trade Analysis System of the International Trade Centre, produced in collaboration with UNCTAD/WTO. It covers trade data from 146 reporting countries disaggregated by product and partner countries (ITC, 2004). We have carried out calculations for 285 crop products and 123 livestock products. The virtual water content of an industrial product can be calculated in a similar way as described earlier for agricultural products. There are however numerous categories of industrial products with a diverse range of production methods and detailed standardised national statistics related to the production and consumption of industrial products are hard to find. As the global volume of water used in the industrial sector is only 716 Gm^3/yr ($\approx$10% of total global water use), we have – per country – simply calculated an average virtual water content per dollar added value in the industrial sector (m^3/US$) as the ratio of the industrial water withdrawal (m^3/yr) in a country to the total added value of the industrial sector (US$ /yr), which is a component of the Gross Domestic Product.

Water needs by product

The total volume of water used globally for crop production is 6390 Gm^3/yr at field level. Rice has the largest share in the total volume water used for global crop production. It consumes about 1359 Gm^3/yr, which is about 21% of the total volume of water used for crop production at field level. The second largest water consumer is wheat (12%). The contribution of some major crops to the global water footprint insofar related to food consumption is presented in Figure 1. Although the total volume of the world rice production is about equal to the wheat production, rice consumes much more water per ton of production. The difference is due

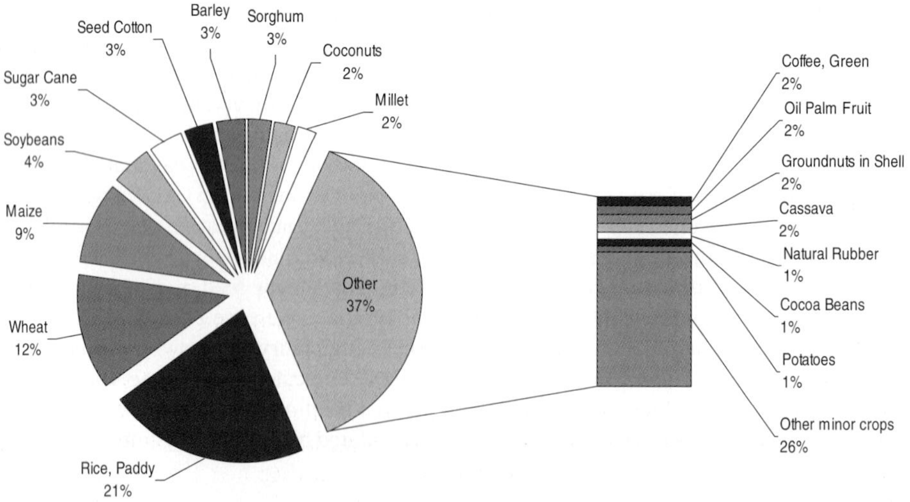

Fig. 1 Contribution of different crops to the global water footprint

to the higher evaporative demand for rice production. As a result, the global average virtual water content of rice (paddy) is 2291 m^3/ton and for wheat 1334 m^3/ton.

The virtual water content of rice (broken) that a consumer buys in the shop is about 3420 m^3/ton. This is larger than the virtual water content of paddy rice as harvested from the field because of the weight loss if paddy rice is processed into broken rice. The virtual water content of some selected crop and livestock products for a number of selected countries are presented in Table 1.

In general, livestock products have a higher virtual water content than crop products. This is because a live animal consumes a lot of feed crops, drinking water and service water in its lifetime before it produces some output. We consider here an example of beef produced in an industrial farming system. It takes in average 3 years before it is slaughtered to produce about 200 kg of boneless beef. It consumes nearly 1300 kg of grains (wheat, oats, barley, corn, dry peas, soybean meal and other small grains), 7200 kg of roughages (pasture, dry hay, silage and other roughages), 24 cubic meter of water for drinking and 7 cubic meter of water for servicing. This means that to produce one kilogram of boneless beef, we use about 6.5 kg of grain, 36 kg of roughages, and 155 l of water (only for drinking and servicing). Producing the volume of feed requires about 15340 l of water in average. With every step of food processing we loose part of the material as a result of selection and inefficiencies. The higher we go up in the product chain, the higher will be the virtual water content of the product. For example, the global average virtual water content of maize, wheat and rice (husked) is 900, 1300 and 3000 m^3/ton respectively, whereas the virtual water content of chicken meat, pork and beef is 3900, 4900 and 15500 m^3/ton respectively. However, the virtual water content of products strongly varies from place to place, depending upon the climate, technology adopted for farming and corresponding yields.

The units used so far to express the virtual water content of various products are in terms of cubic meters of water per ton of the product. A consumer might be more interested to know how much water it consumes per unit of consumption. One cup of coffee requires for instance 140 l of water in average, one hamburger 2400 l and one cotton T-shirt 2000 l (Table 2).

The global average virtual water content of industrial products is 80 l per US$. In the USA, industrial products take nearly 100 l per US$. In Germany and the Netherlands, average virtual water content of industrial products is about 50 l per US$. Industrial products from Japan, Australia and Canada take only 10–15 l per US$. In world's largest developing nations, China and India, the average virtual water content of industrial products is 20–25 l per US$.

Water footprints of nations

The global water footprint is 7450 Gm^3/yr, which is 1240 m^3/cap/yr in average. In absolute terms, India is the country with the largest footprint in the world, with a total footprint of 987 Gm^3/yr. However, while India contributes 17% to the global population, the people in India contribute only 13% to the global water footprint. On a relative basis, it is the people of the USA that have the largest water footprint, with 2480 m^3/yr per capita, followed by the people in south European countries such as Greece, Italy and Spain (2300–2400 m^3/yr per capita). High water footprints can also be found in Malaysia and Thailand. At the other side of the scale, the Chinese people have a relatively low water footprint with an average of 700 m^3/yr per capita. The average per capita water footprints of nations are shown in Figure 2. The data are shown in Table 3 for a few selected countries.

Table 1 Average virtual water content of some selected products for a number of selected countries (m^3/ton)

	USA	China	India	Russia	Indonesia	Australia	Brazil	Japan	Mexico	Italy	Netherlands	World average*
Rice (paddy)	1275	1321	2850	2401	2150	1022	3082	1221	2182	1679		2291
Rice (husked)	1656	1716	3702	3118	2793	1327	4003	1586	2834	2180		2975
Rice (broken)	1903	1972	4254	3584	3209	1525	4600	1822	3257	2506		3419
Wheat	849	690	1654	2375		1588	1616	734	1066	2421	619	1334
Maize	489	801	1937	1397	1285	744	1180	1493	1744	530	408	909
Soybeans	1869	2617	4124	3933	2030	2106	1076	2326	3177	1506		1789
Sugar cane	103	117	159		164	141	155	120	171			175
Cotton seed	2535	1419	8264		4453	1887	2777		2127			3644
Cotton lint	5733	3210	18694		10072	4268	6281		4812			8242
Barley	702	848	1966	2359		1425	1373	697	2120	1822	718	1388
Sorghum	782	863	4053	2382		1081	1609		1212	582		2853
Coconuts		749	2255		2071		1590		1954			2545
Millet	2143	1863	3269	2892		1951		3100	4534			4596
Coffee (green)	4864	6290	12180		17665		13972		28119			17373
Coffee (roasted)	5790	7488	14500		21030		16633		33475			20682
Tea (made)		11110	7002	3002	9474		6592	4940				9205
Beef	13193	12560	16482	21028	14818	17112	16961	11019	37762	21167	11681	15497
Pork	3946	2211	4397	6947	3938	5909	4818	4962	6559	6377	3790	4856
Goat meat	3082	3994	5187	5290	4543	3839	4175	2560	10252	4180	2791	4043
Sheep meat	5977	5202	6692	7621	5956	6947	6267	3571	16878	7572	5298	6143
Chicken meat	2389	3652	7736	5763	5549	2914	3913	2977	5013	2198	2222	3918
Eggs	1510	3550	7531	4919	5400	1844	3337	1884	4277	1389	1404	3340
Milk	695	1000	1369	1345	1143	915	1001	812	2382	861	641	990
Milk powder	3234	4648	6368	6253	5317	4255	4654	3774	11077	4005	2982	4602
Cheese	3457	4963	6793	6671	5675	4544	4969	4032	11805	4278	3190	4914
Leather (bovine)	14190	13513	17710	22575	15929	18384	18222	11864	40482	22724	12572	16656

*For the primary crops, world averages have been calculated as the ratio of the global water use for the production of a crop to the global production volume. For processed products, the global averages have been calculated as the ratio of the global virtual water trade volume to the global product trade volume.

Table 2 Global average virtual water content of some selected products, per unit of product

Product	Virtual water content (litres)
1 glass of beer (250 ml)	75
1 glass of milk (200 ml)	200
1 cup of coffee (125 ml)	140
1 cup of tea (250 ml)	35
1 slice of bread (30 g)	40
1 slice of bread (30 g) with cheese(10 g)	90
1 potato (100 g)	25
1 apple (100 g)	70
1 cotton T-shirt (250 g)	2000
1 sheet of A4-paper (80 g/m^2)	10
1 glass of wine (125 ml)	120
1 glass of apple juice (200 ml)	190
1 glass of orange juice (200 ml)	170
1 bag of potato crisps (200 g)	185
1 egg (40 g)	135
1 hamburger (150 g)	2400
1 tomato (70 g)	13
1 orange (100 g)	50
1 pair of shoes (bovine leather)	8000
1 microchip (2 g)	32

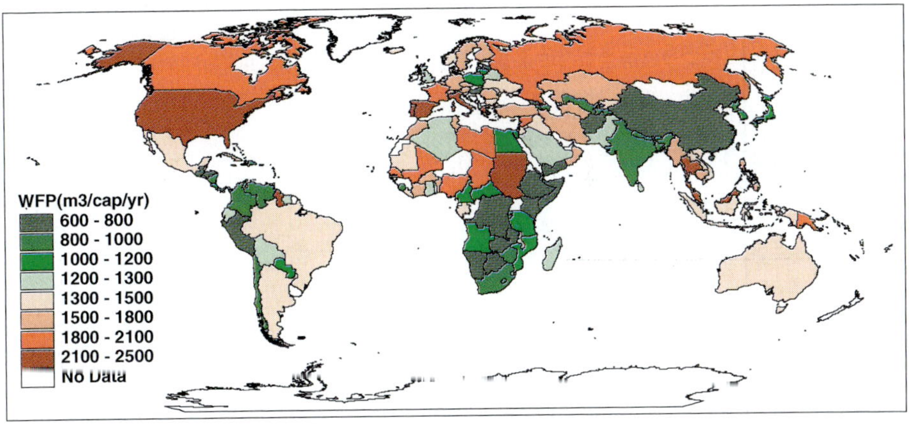

Fig. 2 Average national water footprint per capita (m^3/capita/yr). Green means that the nation's water footprint is equal to or smaller than global average. Countries with red have a water footprint beyond the global average

The size of the global water footprint is largely determined by the consumption of food and other agricultural products (Figure 3). The estimated contribution of agriculture to the total water use (6390 Gm^3/yr) is even bigger than suggested by earlier statistics due to the inclusion of green water use (use of soil water). If we include irrigation losses, which globally add up to about 1590 Gm^3/yr (Chapagain and Hoekstra, 2004), the total volume of water used in agriculture becomes 7980 Gm^3/yr. About one third of this amount is blue water withdrawn for irrigation; the remaining two thirds is green water (soil water).

The four major direct factors determining the water footprint of a country are: volume of consumption (related to the gross national income); consumption pattern (e.g. high versus

Table 3 Composition of the water footprint for some selected countries. Period: 1997–2001

		Use of domestic water resources					Use of foreign water resources					Water footprint by consumption category				
		Domestic water withdrawal	Crop evapotranspiration*		Industrial water withdrawal		For national consumption		For re-export of imported products	Water footprint		Domestic water	Agricultural goods		Industrial goods	
Country	Population	(Gm³/yr)	For national consumption (Gm³/yr)	For export (Gm³/yr)	For national consumption (Gm³/yr)	For export (Gm³/yr)	Agricultural goods (Gm³/yr)	Industrial goods (Gm³/yr)	(Gm³/yr)	Total (Gm³/yr)	Per capita (m³/cap/yr)	Internal water footprint (m³/cap/yr)	Internal water footprint (m³/cap/yr)	External water footprint (m³/cap/yr)	Internal water footprint (m³/cap/yr)	External water footprint (m³/cap/yr)
Australia	19071705	6.51	14.03	68.67	1.229	0.12	0.78	4.02	4.21	26.56	1393	341	736	41	64	211
Bangladesh	129942975	2.12	109.98	1.38	0.344	0.08	3.71	0.34	0.13	116.49	896	16	846	29	3	3
Brazil	169109675	11.76	195.29	61.01	8.666	1.63	14.76	3.11	5.20	233.59	1381	70	1155	87	51	18
Canada	30649675	8.55	30.22	52.34	11.211	20.36	7.74	5.07	22.62	62.80	2049	279	986	252	366	166
China	1257521250	33.32	711.10	21.55	81.531	45.73	49.99	7.45	5.69	883.39	702	26	565	40	65	6
Egypt	63375735	4.16	45.78	1.55	6.423	0.66	12.49	0.64	0.49	69.50	1097	66	722	197	101	10
France	58775400	6.16	47.84	34.63	15.094	12.80	30.40	10.69	31.07	110.19	1875	105	814	517	257	182
Germany	82169250	5.45	35.64	18.84	18.771	13.15	49.59	17.50	38.48	126.95	1545	66	434	604	228	213
India	1007369125	38.62	913.70	35.29	19.065	6.04	13.75	2.24	1.24	987.38	980	38	907	14	19	2
Indonesia	204920450	5.67	236.22	22.62	0.404	0.06	26.09	1.58	2.74	269.96	1317	28	1153	127	2	8
Italy	57718000	7.97	47.82	12.35	10.133	5.60	59.97	8.69	20.29	134.59	2332	138	829	1039	176	151
Japan	126741225	17.20	20.97	0.40	13.702	2.10	77.84	16.38	4.01	146.09	1153	136	165	614	108	129
Jordan	4813708	0.21	1.45	0.07	0.035	0.00	4.37	0.21	0.22	6.27	1303	44	301	908	7	43
Mexico	97291745	13.55	81.48	12.26	2.998	1.13	35.09	7.05	7.94	140.16	1441	139	837	361	31	72
Netherlands	15865250	0.44	0.50	2.51	2.562	2.20	9.30	6.61	52.84	19.40	1223	28	31	586	161	417
Pakistan	136475525	2.88	152.75	7.57	1.706	1.28	8.55	0.33	0.67	166.22	1218	21	1119	63	12	2
Russia	145878750	14.34	201.26	8.96	13.251	34.83	41.33	0.80	3.94	270.98	1858	98	1380	283	91	5
South Africa	42387403	2.43	27.32	6.05	1.123	0.40	7.18	1.42	2.10	39.47	931	57	644	169	26	33
Thailand	60487800	1.83	120.17	38.49	1.239	0.55	8.73	2.49	3.90	134.46	2223	30	1987	144	20	41
United Kingdom	58669403	2.21	12.79	3.38	6.673	1.46	34.73	16.67	12.83	73.07	1245	38	218	592	114	284
USA	280343325	60.80	334.24	138.96	170.777	44.72	74.91	55.29	45.62	696.01	2483	217	1192	267	609	197
Global total/avg.	5994251631	344	5434	957	476	240	957	240	427	7452	1243	57	907	160	79	40

* Includes both blue and green water use in agriculture

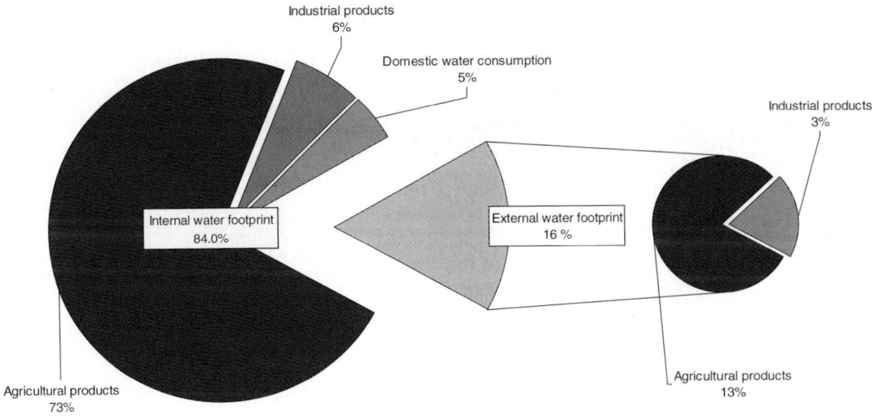

Fig. 3 Contribution of different consumption categories to the global water footprint, with a distinction between the internal and external footprint

low meat consumption); climate (growth conditions); and agricultural practice (water use efficiency). In rich countries, people generally consume more goods and services, which immediately translates into increased water footprints. But it is not consumption volume alone that determines the water demand of people. The composition of the consumption package is relevant too, because some goods in particular require a lot of water (bovine meat, rice). In many poor countries it is a combination of unfavourable climatic conditions (high evaporative demand) and bad agricultural practice (resulting in low water productivity) that contributes to a high water footprint. Underlying factors that contribute to bad agricultural practice and thus high water footprints are the lack of proper water pricing, the presence of subsidies, the use of water inefficient technology and lack of awareness of simple water saving measures among farmers.

The influence of the various determinants varies from country to country. The water footprint of the USA is high (2480 m^3/cap/yr) partly because of large meat consumption per capita and high consumption of industrial products. The water footprint of Iran is relatively high (1624 m^3/cap/yr) partly because of low yields in crop production and partly because of high evapotranspiration. In the USA the industrial component of the water footprint is 806 m^3/cap/yr whereas in Iran it is only 24 m^3/cap/yr.

The aggregated external water footprints of nations in the world constitute 16% of the total global water footprint (Figure 3). However, the share of the external water footprint strongly varies from country to country. Some African countries, such as Sudan, Mali, Nigeria, Ethiopia, Malawi and Chad have hardly any external water footprint, simply because they have little import. Some European countries on the other hand, e.g. Italy, Germany, the UK and the Netherlands have external water footprints contributing 50–80% to the total water footprint. The agricultural products that contribute most to the external water footprints of nations are: bovine meat, soybean, wheat, cocoa, rice, cotton and maize.

Eight countries – India, China, the USA, the Russian Federation, Indonesia, Nigeria, Brazil and Pakistan – together contribute fifty percent to the total global water footprint. India (13%), China (12%) and the USA (9%) are the largest consumers of the global water resources (Figure 4).

Both the size of the national water footprint and its composition differs between countries (Figure 5). On the one end we see China with a relatively low water footprint per capita, and on

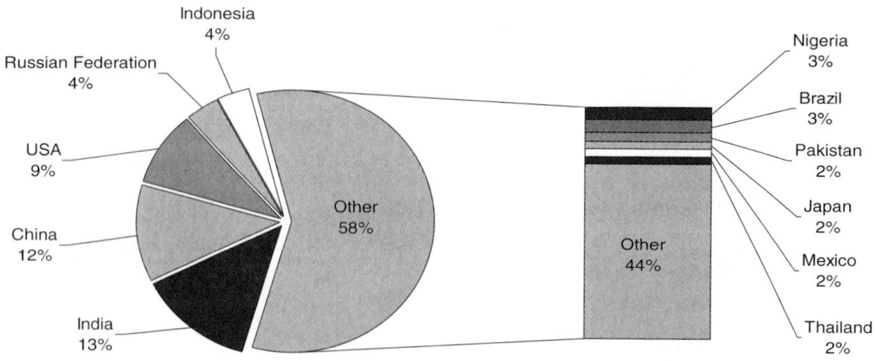

Fig. 4 Contribution of major consumers to the global water footprint

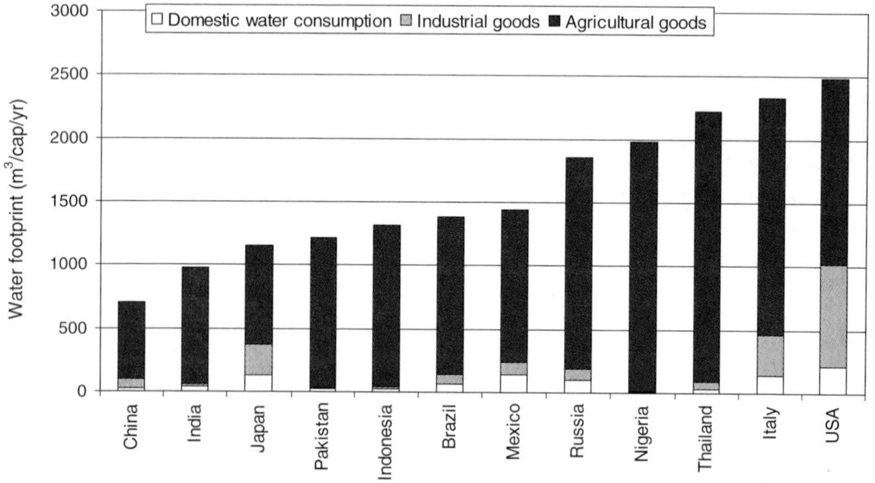

Fig. 5 The national water footprint per capita and the contribution of different consumption categories for some selected countries

the other end the USA. In the rich countries consumption of industrial goods has a relatively large contribution to the total water footprint if compared with developing countries. The water footprints of the USA, China, India and Japan are presented in more detail in Figure 6. The contribution of the external water footprint to the total water footprint is very large in Japan if compared to the other three countries. The consumption of industrial goods very significantly contributes to the total water footprint of the USA (32%), but not in India (2%).

Conclusion

The global water footprint is 7450 Gm^3/yr, which is in average 1240 m^3/cap/yr. The differences between countries are large: the USA has an average water footprint of 2480 m^3/cap/yr whereas China has an average water footprint of 700 m^3/cap/yr. There are four most important direct factors explaining high water footprints. A first factor is the total volume of

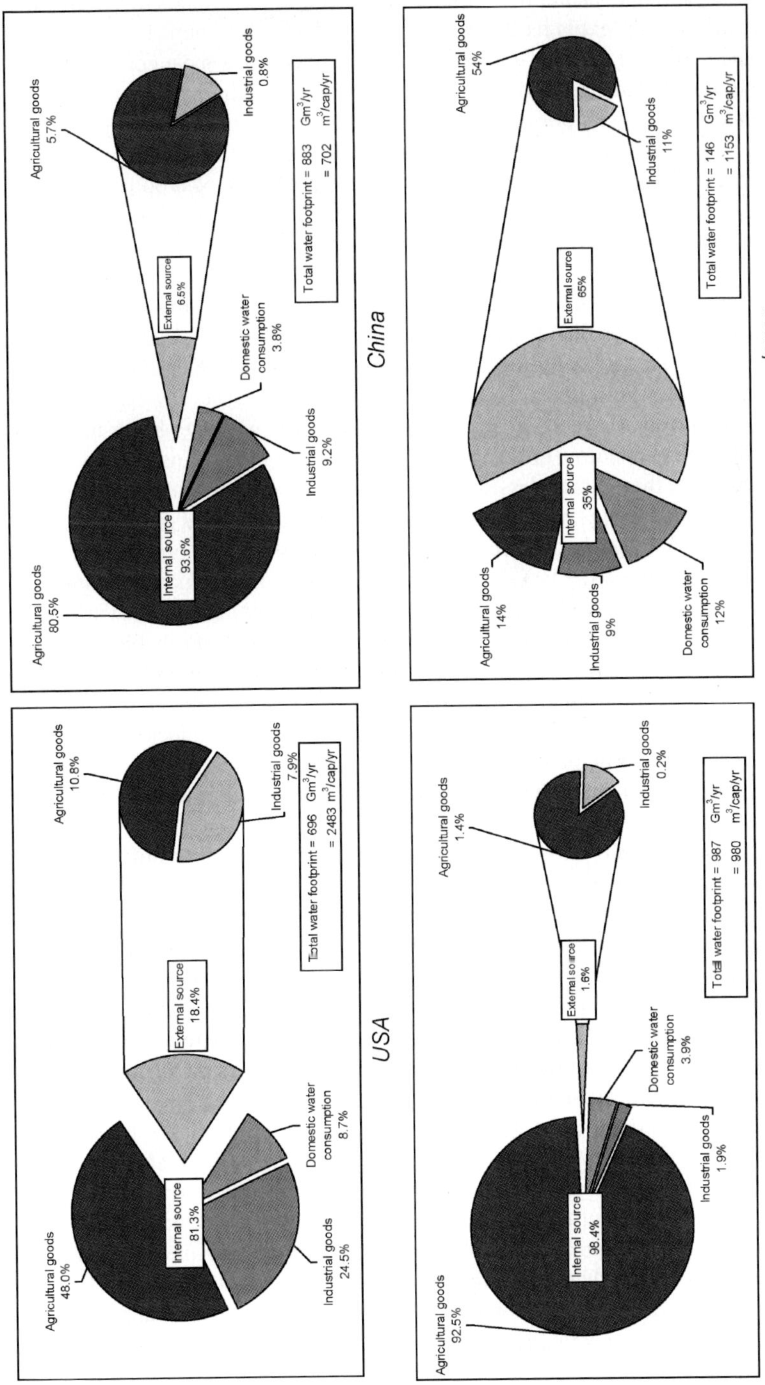

Fig. 6 Details of the water footprints of the USA, China India and Japan. Period: 1997–2001

consumption, which is generally related to gross national income of a country. This partially explains the high water footprints of for instance the USA, Italy and Switzerland. A second factor behind a high water footprint can be that people have a water-intensive consumption pattern. Particularly high consumption of meat significantly contributes to a high water footprint. This factor partially explains the high water footprints of countries such as the USA, Canada, France, Spain, Portugal, Italy and Greece. The average meat consumption in the United States is for instance 120 kg/yr, more than three times the world-average meat consumption. Next to meat consumption, high consumption of industrial goods significantly contributes to the total water footprints of rich countries. The third factor is climate. In regions with a high evaporative demand, the water requirement per unit of crop production is relatively large. This factor partially explains the high water footprints in countries such as Senegal, Mali, Sudan, Chad, Nigeria and Syria. A fourth factor that can explain high water footprints is water-inefficient agricultural practice, which means that water productivity in terms of output per drop of water is relatively low. This factor partly explains the high water footprints of countries such as Thailand, Cambodia, Turkmenistan, Sudan, Mali and Nigeria. In Thailand for instance, rice yields averaged 2.5 ton/ha in the period 1997–2001, while the global average in the same period was 3.9 ton/ha.

Reducing water footprints can be done in various ways. A first way is to break the seemingly obvious link between economic growth and increased water use, for instance by adopting production techniques that require less water per unit of product. Water productivity in agriculture can be improved for instance by applying advanced techniques of rainwater harvesting and supplementary irrigation. A second way of reducing water footprints is to shift to consumptions patterns that require less water, for instance by reducing meat consumption. However, it has been debated whether this is a feasible road to go, since the world-wide trend has been that meat consumption increases rather than decreases. Probably a broader and subtler approach will be needed, where consumption patterns are influenced by pricing, awareness raising, labelling of products or introduction of other incentives that make people change their consumption behaviour. Water costs are generally not well reflected in the price of products due to the subsidies in the water sector. Besides, the general public is – although often aware of energy requirements – hardly aware of the water requirements in producing their goods and services.

A third method that can be used – not yet broadly recognized as such – is to shift production from areas with low water-productivity to areas with high water productivity, thus increasing global water use efficiency (Chapagain *et al.*, 2005a). For instance, Jordan has successfully externalised its water footprint by importing wheat and rice products from the USA, which has higher water productivity than Jordan.

The water footprint of a nation is an indicator of water use in relation to the consumption volume and pattern of the people. As an aggregated indicator it shows the total water requirement of a nation, a rough measure of the impact of human consumption on the natural water environment. More information about the precise components and characteristics of the total water footprint will be needed, however, before one can make a more balanced assessment of the effects on the natural water systems. For instance, one has to look at what is blue versus green water use, because use of blue water often affects the environment more than green water use. Also it is relevant to consider the internal versus the external water footprint. Externalising the water footprint for instance means externalising the environmental impacts. Also one has to realise that some parts of the total water footprint concern use of water for which no alternative use is possible, while other parts relate to water that could have been used for other purposes with higher added value. There is a difference for instance between beef produced in extensively grazed grasslands of Botswana (use of green water without alternative use) and beef produced in an industrial livestock farm in the Netherlands (partially fed with imported irrigated feed crops).

The current study has focused on the quantification of consumptive water use, i.e. the volumes of water from groundwater, surface water and soil water that evaporate. The effect of water pollution was accounted for to a limited extent by including the (polluted) return flows in the domestic and industrial sector. The calculated water footprints thus consists of two components: consumptive water use and wastewater production. The effect of pollution has been underestimated however in the current calculations of the national water footprints, because one cubic metre of wastewater should not count for one, because it generally pollutes much more cubic metres of water after disposal (various authors have suggested a factor of ten to fifty). The impact of water pollution can be better assessed by quantifying the dilution water volumes required to dilute waste flows to such extent that the quality of the water remains below agreed water quality standards. We have shown this in a case study for the water footprints of nations related to cotton consumption (Chapagain *et al.*, 2005b).

International water dependencies are substantial and are likely to increase with continued global trade liberalisation. Today, 16% of global water use is not for producing products for domestic consumption but for making products for export. Considering this substantial percentage and the upward trend, we suggest that future national and regional water policy studies should include an analysis of international or interregional virtual water flows.

References

Allan JA (1993) Fortunately there are substitutes for water otherwise our hydro-political futures would be impossible. In: Priorities for water resources allocation and management, ODA, London, pp 13–26

Allan JA (1994) Overall perspectives on countries and regions. In: Rogers P, Lydon P (eds) Water in the Arab World: perspectives and prognoses. Harvard University Press, Cambridge, Massachusetts, pp 65–100

Allen RG, Pereira LS, Raes D, Smith M (1998) Crop evapotranspiration - Guidelines for computing crop water requirements – FAO Irrigation and Drainage Paper 56, FAO, Rome, Italy, http://www.fao.org/docrep/X0490E/x0490e00.htm

Chapagain AK, Hoekstra AY (2003) Virtual water flows between nations in relation to trade in livestock and livestock products. Value of Water Research Report Series No. 13, UNESCO-IHE, Delft, The Netherlands, http://www.waterfootprint.org/Reports/Report13.pdf

Chapagain AK, Hoekstra AY, Savenije HHG (2005a) Saving water through global trade. Value of Water Research Report Series No. 17, UNESCO-IHE, Delft, the Netherlands, http://www.waterfootprint.org/Reports/Report17.pdf

Chapagain AK, Hoekstra AY, Savenije HHG, Gautam R (2005b) The water footprint of cotton consumption. Value of Water Research Report Series No. 18, UNESCO-IHE, The Netherlands, http://www.waterfootprint.org/Reports/Report18.pdf

Chapagain AK, Hoekstra AY (2004) Water footprints of nations. Value of Water Research Report Series No. 16, UNESCO-IHE, Delft, The Netherlands, http://www.waterfootprint.org/Reports/Report16.pdf

FAO (2003) AQUASTAT 2003. Food and Agriculture Organization of the United Nations, Rome, Italy, ftp://ftp.fao.org/agl/aglw/aquastat/aquastat2003.xls

Gleick PH (ed) (1993) Water in crisis: A guide to the world's fresh water resources. Oxford University Press, Oxford, UK

Haddadin MJ (2003) Exogenous water: A conduit to globalization of water resources. In: Hoekstra AY (ed) Virtual water trade: Proceedings of the International Expert Meeting on Virtual Water Trade. Value of Water Research Report Series No. 12, UNESCO-IHE, Delft, The Netherlands, http://www.waterfootprint.org/Reports/Report12.pdf

Hoekstra AY, Hung PQ (2002) Virtual water trade: A quantification of virtual water flows between nations in relation to international crop trade. Value of Water Research Report Series No. 11, UNESCO-IHE Institute for Water Education, Delft, The Netherlands, http://www.waterfootprint.org/Reports/Report11.pdf

Hoekstra AY, Hung PQ (2005) Globalisation of water resources: International virtual water flows in relation to crop trade. Global Environmental Change 15(1):45–56

ITC (2004) PC-TAS version 1997–2001 in HS or SITC, CD-ROM. International Trade Centre, Geneva

Rees WE (1992) Ecological footprints and appropriated carrying capacity: what urban economics leaves out. Environ Urban 4(2):121–130

Shiklomanov IA (2000) Appraisal and assessment of world water resources. Water International 25(1):11–32

Wackernagel M, Onisto L, Linares AC, Falfan ISL, Garcia JM, Guerrero IS, Guerrero MGS (1997) Ecological footprints of nations: How much nature do they use? How much nature do they have?. Centre for Sustainability Studies, Universidad Anahuac de Xalapa, Mexico

Wackernagel M, Rees W (1996) Our ecological footprint: Reducing human impact on the Earth. New Society Publishers, Gabriola Island, BC, Canada

Water Resour Manage (2007) 21:49–62
DOI 10.1007/s11269-006-9040-4

ORIGINAL ARTICLE

Transitions towards adaptive management of water facing climate and global change

Claudia Pahl-Wostl

Received: 20 April 2006

Abstract Water management is facing major challenges due to increasing uncertainties caused by climate and global change and by fast changing socio-economic boundary conditions. More attention has to be devoted to understanding and managing the transition from current management regimes to more adaptive regimes that take into account environmental, technological, economic, institutional and cultural characteristics of river basins. This implies a paradigm shift in water management from a prediction and control to a management as learning approach. The change towards adaptive management could be defined as "learning to manage by managing to learn". Such change aims at increasing the adaptive capacity of river basins at different scales. The paper identifies major challenges for research and practice how to understand a transition in water management regimes. A conceptual framework is introduced how to characterize water management regimes and the dynamics of transition processes. The European project NeWater project is presented as one approach where new scientific methods and practical tools are developed for the participatory assessment and implementation of adaptive water management.

Keywords Adaptive water management · Social learning · Transition processes · Water management regimes · Global change

1. Introduction – Challenges for water management

Sustainable water management has become an issue of major concern over the past decade. It has become increasingly clear that the pressing problems in this field have to be tackled from an integrated perspective taking into account environmental, human and technological factors and in particular their interdependence. To emphasize the need for adopting an integrated approach the notion of "water system" is introduced encompassing all environmental factors of the resource base, technologies and human beings. The term water system has

C. Pahl-Wostl
Institute of Environmental Systems Research, University of Osnabrück, Barbarastrasse 12, D-49069 Osnabrück, Germany
e-mail: pahl@usf.uni-osnabrueck.de

been developed and defined for the global scale in the context of the global water system project (Box 1). The different "components" cannot be understood in isolation. Technical systems are perceived as part of the human component. Technologies are embedded in a network of social routines that link technologies to their function to achieve the overall management objectives. This area of research has not yet received sufficient attention since often technical systems have been studied and developed in isolation from their social context. Such negligence may lead to failures in the introduction of new technologies in water management when the influence of cultural factors and social relationships prevail. One needs to better understand the interdependence and co-evolutionary development of management objectives and paradigms, environmental characteristics, technologies and social routines.

Box 1. Definition of Water Systems in the GWSP Science Plan (Framing Committee, 2004)

As a working definition, we define the global water system as the global suite of water related human, physical, biological, and biogeochemical components and their interactions. These components include:

1. *Human components* – These are the sum of water-related organizations, engineering works, and water use sectors. Society is both a component of the global water system and a significant agent of change within the system.
2. *Physical components* – These are the physical attributes and processes of the traditional global hydrologic or water cycle, including runoff, geomorphology, and sediment processes.
3. *Biological and biogeochemical components* – This category includes the sum of aquatic and riparian organisms and their associated ecosystems and biodiversity. These organisms are also integral to the geochemical functioning of the global water system and not simply recipients of changes in the physico-chemical system. Hence we also include here the biogeochemistry of the global water system and water quality.

Water management is a purposeful activity with multiple and partly conflicting goals to maintain and improve the state of water resources. Water as a resource must be allocated among competing uses. In many areas the available water is polluted and hence cannot be used for many purposes or requires expensive treatment. An uncontrolled urbanization and fast industrialization in developing and threshold countries contribute to exacerbate the pressure on the water resource. Technologies and knowledge are in most cases available in general but the implementation into practical action is slow. The importance of environmental flows needs has received more attention over the past years. Environmental flows refer not only to the absolute amount of water available for ecosystem water needs but to the spatio-temporal distribution of quantity and quality of water flows that are of key importance to maintain the integrity of riverine ecosystems and preserve their ability to provide services valuable to humans (Dyson *et al.*, 2003).

Dealing with extremes has always been one of the major challenges for water management. Natural water supply varies over time and some variability can be compensated by the buffering capacity inherent in the water system such as natural storage or adaptation in water use patterns. Technical facilities have been used to extend the capacity of the water system to cope with extremes. Large-scale technical infrastructure has been implemented to shield human activities from the variability of the resource. Reservoirs, water diversions and artificial storage are used to enhance the resource base in case of droughts. Dikes and levees were built to protect settlements from floods in case of excess precipitation. However, the limitations of the ability to control extremes by technical means have become very clear during weather extremes occurring over the past years. The extreme flood of the Elbe in 2002 was followed by a century drought and heat wave in 2003 in Europe where dikes broke due to excess dryness of the building material. Such events have triggered an increasing awareness of water managers for the possible challenges posed by global and climate change. It becomes more and more difficult to predict probabilities for weather extremes which are fundamental for the current strategy to deal with risks. Improving our understanding of the likelihood of extreme events based on experience derived from historical records does not tell us much about the likelihood of extremes in the future given the uncertainties caused by climate change. It has become even more difficult to quantify potential damages caused by weather extremes. As a consequence of taking into account the true complexity of water systems at different scales and an increase in uncertainty, radical changes are needed in water management:

- Move from technical management to a true integration of the human dimension.
- Make management more adaptive and flexible to make it operational under fast changing socio-economic boundary conditions and climate change.

This poses considerable challenges to the tradition of water management characterized by a prediction-and-control approach and an emphasis on technical solutions. To face those challenges adaptive water management under uncertainty is advocated as a timely extension of water management and a requirement to really move towards IWRM (Pahl-Wostl *et al.*, 2004). The guiding principle of this paper is that water management has to become more adaptive but that the major obstacle is to understand and manage the transition process given the high inter-connectedness and complexity of riverine water systems. Hence the attention will be devoted to the processes of change. First adaptive water management as used in this paper will briefly be characterized. Then concepts are developed which are required to analyse and understand water management regimes and their transformation.

2. Adaptive water management

The idea of adaptive management has been discussed in ecosystem management for quite some time (Holling, 1978; Walters, 1986; Pahl-Wostl, 1995; Lee, 1999). It is based on the insight that the ability to predict future key drivers influencing an ecosystem, as well as system behaviour and responses, is inherently limited. Hence management must be adaptive and include the ability to change management practices based on new experience and insights. Adaptive management refers thus to a systematic process for continually improving management policies and practices by learning from the outcomes of implemented management strategies. The most effective form of adaptive management employs management programs that are designed to experimentally compare selected policies or practices, by evaluating alternative hypotheses about the system being managed (e.g. Gunderson, 1999; Kiker *et al.*, 2003; Richter *et al.*, 2003). This implies that hypotheses can be generated and that the outcomes of experiments allow distinguishing between different hypotheses. This is a very positivistic approach and the reality of complex socio-ecological systems may not allow such unambiguous conclusions. Learning may be also based on a more inductive approach where insights are derived from new information and dynamic hypotheses guide reasoning and structured argumentation. The generation of dynamic hypotheses embedded in a collective learning process should also sharpen the awareness to be prepared for the unexpected. As Bormann *et al.* (1994) defined it "Adaptive management is learning to manage by managing to learn". Adaptive management has as one target to *increase the adaptive capacity of the (water) system*. Adaptive capacity can be defined as the potential or capability of a system to adjust, via changes in its characteristics or behaviour, so as to cope better with existing and future stresses. More specifically, adaptive capacity refers to "the ability of a socio-ecological system to cope with novelty without losing options for the future" (Folke *et al.*, 2002) and "that reflects learning, flexibility to experiment and adopt novel solutions, and development of generalized responses to broad classes of challenges" (Walker *et al.*, 2002).

Increasing the adaptive capacity of water systems implies thus integrated system design which may range from the introduction of new socio-technical systems to building social capital in an actor network to restoring a multi-functional landscape. The problem to be tackled is to increase the ability of the whole system to respond to change rather than reacting to undesirable impacts of change. It is a pro-active management style that must be based on a sound understanding of what determines a basin's adaptive capacity and vulnerability. The difference between prevailing and adaptive management approaches can be illustrated for the example of dealing with extremes.

Socio-ecological systems can cope with a certain variability of climate (e.g. precipitation) on diurnal, seasonal and annual time scales without major detrimental impacts on function. In the current management approach human activities are shielded from environmental variability by technical precautions. The boundaries of maintaining system function despite environmental variability are pushed to their extremes. Reservoirs maintain the supply of water in case of droughts and dams protect settlement from flooding. Investments in protective measures are often based on a quantitative assessment of risks. The costs for an investment at present are compared to the expected costs for damage (probability of an extreme event*damage costs caused by the extreme event) over a certain time period in the future. How much people are willing to invest to protect themselves from damage depends on their risk aversion and on the rate they apply to discount the costs for future damage. The damage is not a natural given but depends on the ability of the current system to cope with extremes. The whole management approach is caught in a positive feedback loop that even increases the vulnerability of the water system to extreme events: given the fact that technical infrastructure is implemented to shield the water system from extremes, less precautions are taken elsewhere to cope with extremes (e.g. settlements in former floodplains, high population densities with large water demand in drought areas). The damage in case of technical failure increases which puts an even higher pressure on implementing even larger infrastructure. If extremes exceed the protective capability of the technical infrastructure the damage may be disastrous. Hence there is a sharp threshold separating complete protection from disaster.

The first step in the adaptive approach aims at increasing the ability of the water system to operate under a wider range of environmental variation. Technical infrastructure is not designed to entirely shield the system from environmental variability but socio-technical means are implemented to allow maintaining a water system's functionality despite its being exposed to environmental variation. This may imply for example that instead of building larger reservoirs to maintain supply in case of drought, management of demand is used to reduce and/or shift the requirements of certain water uses if supply is scarce. Regarding floods one may introduce multi-functional landscapes with restored floodplains and temporal flooding zones. In the long-term, adaptive management needs to establish the ability to change system structure – e.g. change to other types of crops and change life-styles or the allocation of water quota to certain uses. System design must aim at implementing water systems with

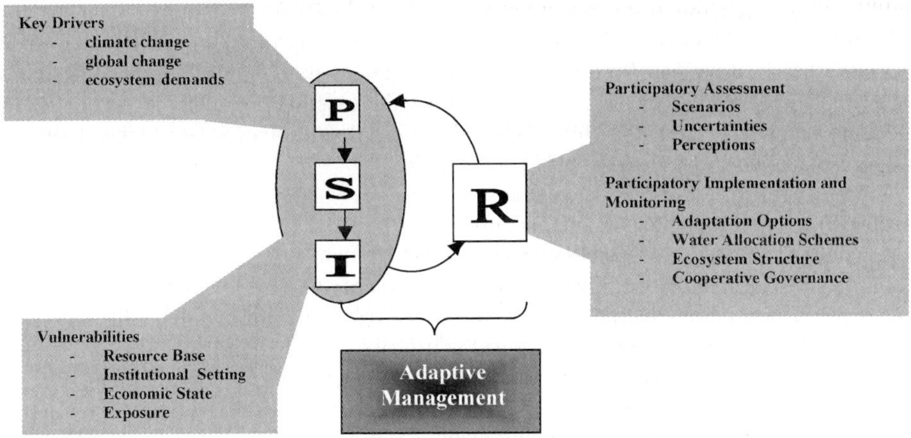

Fig. 1 Adaptive management represented in an extended PSIR (Pressure-State-Impact-Response) framework – system design to increase the ability of the system to cope with change – R as part of the autonomous response strategies. The whole process has to be perceived as being iterative and proceed in cycles in contrast to the quite linear and sequential approach that is often adopted when using the PSIR scheme. The boxes indicate the type of variables and processes that are of importance

inherent degrees of freedom to adapt at a range of spatial and temporal scales in iterative learning cycles. Figure 1 summarizes this schematically in the so-called PSIR framework. The figure emphasizes the cyclic nature of the process to make the distinction to the quite common linear approaches in using the PSIR approach.

What is the type of mechanisms one has to put in place to increase the ability of the system to "learn"? To answer this question one needs to develop a better base to understand what determines the ability of a water system to change and adapt.

3. Requirements for the ability of a "water system" to adapt

For a system to be able to adapt to change or to be prepared for uncertain future change the following two aspects are key requirements:

– new information must be available to the system and the system must be able to process this information,
– the system must have the ability to change based on processing new information.

It is assumed here that an improved understanding of water systems and their adaptive capacity has to be based on the notion of complex adaptive systems (CAS) which by their definition have a high adaptive capacity (Pahl-Wostl, 1995; in press):

A CAS is a complex, nonlinear, interactive system which has the ability to adapt to a changing environment. Such systems are characterized by the potential for self-organization, existing in a nonequilibrium environment. CAS derive their adaptive capacity from their multi-level, modular structure. The system elements are diverse in both form and capability and they adapt by changing their rules of interactions and, hence, behaviour, as they gain experience. Complex, adaptive systems evolve, meaning their past or history, i.e., their experience, is added onto them and determines their future trajectory.

Complex adaptive systems are characterized by self-organization and distributed control. Adaptation is not controlled by a central unit. Distributed control has proved to be efficient and flexible in allocating resources to many users (e.g. ecosystems, markets). Such adaptive behaviour has implications for the response to change and for the development of management and intervention strategies. To use such understanding as base for the improved design of water management regimes, one needs to identify the essential elements of water systems and their interaction. Here it is useful to first develop a definition of what is understood by a water management regime and its main components.

A management regime is here referred to as the whole complex of technologies, institutions,[1] environmental factors and paradigms that are highly interconnected and together form the base for the functioning of the management system targeted to fulfil a societal function. The different components of a management regime have co-evolved over time. Due to the resulting high interconnectedness and internal logic, it is assumed that individual elements of the regime cannot be exchanged arbitrarily.

A water management regime is characterized first of all by its societal function:

In general, the key objectives of water management are the allocation of water for multiple uses taking into account different requirements for water quality and the spatio-temporal variability of both supply and demand. Sustainable management implies that the pillars of social equity, economic efficiency and environmental sustainability are all taken into account. Water management has also to provide protection from water related risks.

Water management regimes are characterized by structural dimensions that form an integral part of a regime. One could conceive of more than one possible approach to categorize these dimensions. The following is one attempt to do so – its overall usefulness has yet to be proven in more practical applications:

- Management Paradigms of system approach and intervention strategies including risk management approaches.
- Governance:

  – Institutions (formal legal structures and informal norms), horizontal and vertical interplay and fit with physical boundaries,
  – Actor networks (role, linkages) and policy arenas.

- Scale(s) of operation – vertical (local, regional, national) and horizontal (sectoral) integration.
- Information management and sharing.
- Technological infrastructure – size, life-time, costs.
- Risk management.
- Environmental factors taken into consideration.

Table 1 gives an overview of the typical characteristics of current regimes and of what are considered to be typical characteristics of integrated, adaptive regimes.

The most widely spread water management regime can be described as a "prediction and control" regime (Pahl-Wostl, 2002; Moberg and Galaz, 2005). The system's approach is derived from a mechanistic thinking. System behaviour and response can be predicted and optimal control strategies can be designed. Decision making is shaped by regulatory frameworks including technical norms and legal prescriptions.

Information needs are defined by technical experts. The main data collection and regular monitoring focuses on the state of the environment. Knowledge is not shared and

[1] Institutions refer to the formal (e.g. laws) and informal (e.g. norms) rules and decision making practices that determine the behaviour and roles of actors. Institutions do not include the material entities that are referred to as organizations.

Table 1 Comparison between current and an integrated, adaptive regime water management regimes

	Prediction and control regime	Integrated, adaptive regime
Management paradigm	Prediction and control based on a mechanistic system's approach	Learning and self-organization based on a complex systems approach
Governance	Centralized, hierarchical, narrow stakeholder participation	Polycentric, horizontal, broad stakeholder participation
Sectoral integration	Sectors separately analysed resulting in policy conflicts and emergent chronic problems	Cross-sectoral analysis identifies emergent problems and integrates policy implementation
Scale of analysis and operation	Transboundary problems emerge when river sub-basins are the exclusive scale of analysis and management	Transboundary issues addressed by multiple scales of analysis and management
Information management	Understanding fragmented by gaps and lack of integration of information sources that are proprietary	Comprehensive understanding achieved by open, shared information sources that fill gaps and facilitate integration
Infrastructure	Massive, centralized infrastructure, single sources of design, power delivery	Appropriate scale, decentralized, diverse sources of design, power delivery
Finances and risk	Financial resources concentrated in structural protection (sunk costs)	Financial resources diversified using a broad set of private and public financial instruments
Environmental factors	Quantifiable variables such as BOD or nitrate concentrations that can be measured easily	Qualitative and quantitative indicators of whole ecosystem states and ecosystem services

communication with stakeholder and interest groups is mainly by passive channels (cf. reviews and case studies in Timmorman and Langaas, 2003). Technological infrastructure relies mainly on large-scale infrastructure (dams, reservoirs, centralized waste water treatment plants) supposed to guarantee long-term stability and reliability and a higher efficiency in implementation, maintenance and operation than de-centralized small-scale structures. Due to long-life times and large sunk costs, change is largely impossible once a certain infrastructure is in place. In current water management regimes responsibilities are fragmented with little interaction between areas such as flood protection, regional planning, waste water treatment, water supply or hydropower generation. This allows for a high specialization of dealing with a single problem but prevents integration and also prevents change and learning. The environmental factors taken into consideration refer mainly to quantifiable variables characterizing system states such as BOD or nitrate concentrations that can be easily and unambiguously measured.

It is evident that current management regimes include only very limited possibilities to introduce change based on new insights – high flexibility was not a design criterion. On one hand decision making structures are inflexible. On the other hand the type of long-lived, large-scale and expensive infrastructure leaves little opportunities for adjustments. In addition, criteria for getting the dimensions of technical infrastructure have often been based on designing capacity to deal with extremes. Large-scale water supply capacity has been designed to meet daily peak demand which implies the need to provide accurate forecast for demand extremes for decades (Tillman *et al.*, 2005).

To overcome the shortcomings of current approaches, Gleick (2003) advocated a "soft path" to build greater flexibility in water management regimes to address the rising uncertainty from global change: "*A transition is under way to a 'soft path' that complements centralized physical infrastructure with lower cost community-scale systems, decentralized and open decision-making, water markets and equitable pricing, application of efficient technology, and environmental protection.*"

Given the very plausible assumption that elements of a water management regime are closely linked within an internal logic, this claim by Gleick implies a fundamental shift in the water management paradigm: the paradigm of "management as control" has to be replaced by "management as learning". The right column in Table 1 outlines the expected characteristics of an adaptive and integrated water management regime. Further in depth investigations are required to provide a sound base to understand under which socio-economic, cultural, technological and environmental conditions these characteristics result really in a more adaptive regime. First analyses conducted within the context of the NeWater project provide clear evidence that in some countries in Europe a transition has already started (Huitema and Becker, 2005). But change is slow and will require many instances of social learning in the process of change and in adaptive management itself.

4. The importance of processes of social learning

4.1. What is social learning?

Social learning in river basin management refers to developing and sustaining the capacity of different authorities, experts, interest groups and the public to manage their river basins effectively. Collective action and the resolution of conflicts require that people recognize their interdependence and their differences and learn to deal with them constructively. The different groups need to learn and increase their awareness about their biophysical environment and about the complexity of social interactions.

4.2. Why is social learning needed in the transition to adaptive water management?

As pointed out previously, technical infrastructure (e.g. large technical infrastructure for flood protection), citizen behaviour (expectations regarding safety in floodplains, risk perception) and habits, and engineering rules of good practice are often mutually dependent and stabilize each other. In many cases they have co-evolved over a long period of time. Hence one observes so-called lock-in situations with the effect that changes towards new resource management schemes are blocked and require collective learning and decision making processes (Pahl-Wostl, 2002).

A new concept for social learning in river basin management has been developed in the context of the European project HarmoniCOP[2] (Harmonizing COllaborative Planning). The

[2] More information on the HarmoniCOP project is available on the webpage – www.harmonicop.info. The main objectives of HarmoniCOP have been to increase the understanding of participatory river basin management

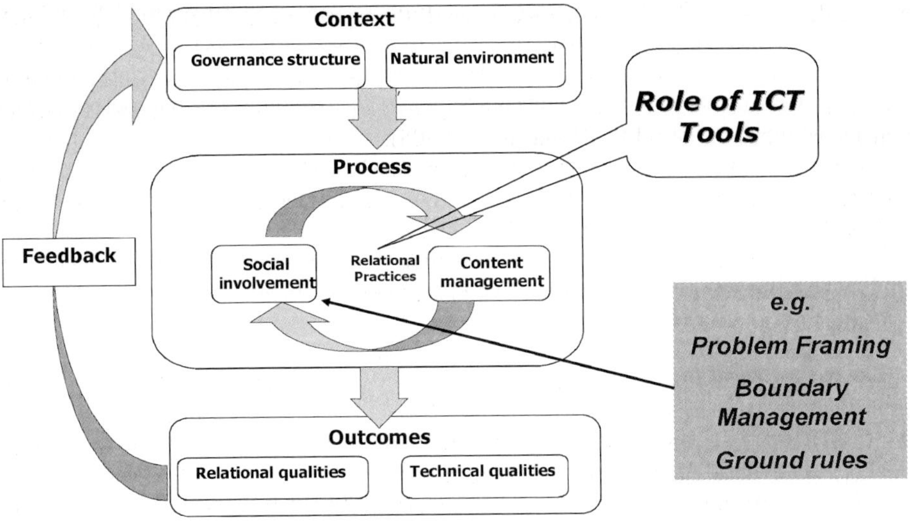

Fig. 2 Conceptual framework for social learning in resources management. Information and communication technology tools may play a decisive role in supporting and shaping relational practices that link social involvement and content management. This implies also a new role for simulation models in such processes

approach adopted by the HarmoniCOP project is characterized by a broad understanding of social learning that is rooted in the more interpretative strands of the social sciences. Figure 2 represents the framework for social learning developed to account for learning processes in water resources management (Bouwen and Taillieau, 2004, Craps *et al.*, 2003; Pahl-Wostl, 2002, in press). The framework is structured into context, process and outcomes and a feedback loop to account for change in a cyclic and iterative process. The context refers the governance structure and the natural environment in a river basin. To improve the state of the environment implies in practice implies most often a change in the governance structure. Social learning is assumed to occur at two levels – on short to medium timescales at the level processes between actors and on medium to long timescales at the level the change in the governance structure.

The process concept referring to multi-party interactions in actor networks has two pillars (Figure 2). They relate to the processing of factual information on a problem (content management) and engaging in processes of social exchange (social involvement). Social involvement refers to essential elements of social processes such as the framing of the problem, the management of the boundaries between different stakeholder groups, the type of ground rules and negotiation strategies chosen or the role of leadership in the process. This concept has as central hypothesis that the management of content and the social involvement are strongly interdependent and cannot be separated and that ICT tools play an important role. The overall process leads to both technical qualities such as the improvement of the state of the environment and to relational qualities such as an increase in the capacity of a stakeholder group to manage a problem and/or institutional change. This leads as well to a different interpretation of the role of information and the ability of an actor network to use new

in Europe, to generate practically useful information about and improve the scientific base of social learning and the role of ICT tools in river basin management and support the implementation of the European Water Framework Directive).

information in social learning processes and derive based on it collective action. Such learning environments are perceived to be crucial for the adaptive governance of socio-ecological systems (Folke *et al.*, 2005; Pahl-Wostl, 2005). Hence, an entirely new element of monitoring refers to the quality of the communication process in actor networks, the appropriateness of a chosen institutional setting. Social learning is assumed to be crucial for the transition towards and for sustaining adaptive management practices which supports the statement of Bormann *et al.* (1994) "Adaptive management is learning to manage by managing to learn". However, the understanding of processes of social learning has to be embedded in a wider context of understanding the overall dynamics of a transition towards adaptive management and to understand all transformation processes required for change.

5. The dynamics of the transition

The concept developed here makes use of recent work on socio-technical transitions, an active area of research building strongly on complex systems and evolutionary approaches. In order to understand change and transitions it is useful to distinguish the following three levels (macro-meso-micro) of a system (Pahl-Wostl, 1995; Geels, 2001; Rotmans *et al.*, 2002):

- The Landscape or macro-level with stabilizing factors that constitute the context for a water management regime. The landscape encompasses e.g. environmental variability, legal frameworks, deeply rooted societal norms and cultural values. The landscape provides the context and also the selection environment within which a management regime unfolds. The landscape level must not be entirely independent from the micro and meso level since feedback processes can operate bottom-up (e.g. diffusion of innovation) and top-down (e.g. selection of regime).
- The management regime or meso-level with stabilizing interdependencies between the elements as described in section three.
- The niches or micro-level where innovative approaches can develop in a locally protected environment (e.g. large scale research projects, subsidized pilot studies) and/or in new areas of application such as the restoration of riverine landscapes that has started to become an integral part of water resources management.

The distinction between macro-meso- and micro level is quite common for complex adaptive systems. The dynamics of transitions are assumed to follow the typical sigmoid-shaped curve of change between alternative stable states in feedback systems where after an initial slow phase where change has to be supported by external input, change becomes autonomous due to positive feedback until a new stable configuration is reached. The innovative contribution of recent advances in transition research is the attempt to make this approach operational for understanding socio-technical change in strongly interconnected socio-technical systems. Further research efforts are still needed to understand transitions in strongly interconnected human-technology-environment systems.

The understanding of the causes and dynamics of transition processes is still limited. Nevertheless one can identify some clear indications that a transition process has started in water management. The landscape in general provides a stabilizing context for a management regime but it may also impose a pressure on it if the landscape changes and existing regimes cannot adapt. The latter is currently the case for water management with respect to global and climate change and the overall increase in the dynamics of socio-economic developments. The increased awareness for the complexity of systems and for management as learning rather than control seems to be an overall trend in different fields (Senge, 1990; Pahl-Wostl, 1995,

2004; Levin, 1998; Hartvigsen *et al*., 1998; Berkes *et al*., 2002). On one hand the systems to be managed and the problems to be tackled have become indeed more complex. The pace of change in socio-economic conditions and technologies is tremendous. Uncertainties arising from global change in general and climate change in particular pose major challenges for the management of environmental resources. On the other hand the awareness for the need to take the complexity of problems fully into account has increased and the frame of analysis has partly changed.

The introduction of the European Water Framework Directive constitutes a major change in context for water resources management in Europe. However, in comparison to other areas, environmental resources management, in general, and water resources management, in particular, has been quite slow in adopting such changes. Galaz (2005) analysed, for example, the current realization of the WFD in Sweden and concluded that it might at worst reduce the resilience of nested social-ecological freshwater-systems, the capacity of freshwater systems to deal with change and perturbations. According to his analyses learning processes are not stimulated, water management institutions disregard complexity and uncertainty and water policy is poorly prepared to tackle global environmental change. One possible reason for this lack of innovation is the strong interdependence of the factors stabilizing current management regimes. One cannot, for example, move easily from top-down to participatory management practices without changing the whole approach to information and risk management. Hence, research is urgently needed to better understand the interdependence of key elements of water management regimes and the dynamics of transition processes in order to be able to compare and evaluate alternative management regimes and to implement and support transition processes if required.

6. The NeWater project

The challenges of understanding the transition towards adaptive water management is tackled in a European project: NeWater[3] (New methods for adaptive water management under uncertainty). The project develops a conceptual framework for understanding and a comprehensive methodology for analysing and implementing transitions to adaptive water management. Investigations are performed in a number of case studies in Europe, Africa and Central Asia. Emphasis is given to the assessment of key drivers of global change and the vulnerability of river basins.

To understand the "Transition to Adaptive Management" new concepts and methods are developed for understanding and implementing the transition from current practices to more adaptive management to increase the adaptive capacity of river basins and thereby reducing their vulnerability to global change. Currently, the hypothesis is tested that one can identify a finite number of typical water management regimes and typical transition pathways based on understanding the interdependence and the role the key factors – management paradigms, governance, information management, scale of operation and integration, technical infrastructure and environmental factors. Emphasis is given to investigate the influence of key drivers such as climate change or ecosystem water requirements on the vulnerabilities and adaptive capacity of river basins and to develop practical toolkits to set the baseline for understanding the priorities to be addressed by adaptive management strategies. Case study research emphasizes stakeholder processes and coordinates empirical research in the selected

[3] The NeWater project (www.newater.info) is an Integrated Project in FP 6 of the EU with 12 Mio Euro EU funding over 4 years (January 2005 – December 2008).

river basins to generate input to the development of new concepts and methodologies and to provide a test bed for their plausibility and applicability under different environmental and societal conditions. The case studies in the Rhine, Elbe, Tisza, Guadiana, Amudarya, Nile and Orange basins were chosen to provide a rich base of empirical knowledge covering different environmental, institutional, cultural and economic settings. Specific emphasis is given to the development of tools and guidance for practitioners based on new conceptual insights, experience collected in the basins and the needs from ongoing policy processes, in particular the European Water Framework Directive[4] and the European Water Initiative.[5] Research activities and developments of guidance for practitioners pay much attention to integrating results from previous and ongoing EU projects and to engaging in an intensive dialogue with the wider community of IWRM experts. International Platforms have been established as link to relevant European activities and ongoing policy processes. The platforms provide immediate feedback from and to policy processes.

NeWater devotes much attention to achieve a true integration between social, natural and engineering sciences and to bridge the science policy gap. Despite many efforts in recent years, there is still a huge gap between the social and the natural/engineering sciences. Whereas more formal approaches such as decision theory have started to be integrated, more qualitative approaches in the social sciences are still neglected. NeWater addresses the strong need to bridge the "hard" and "soft" approaches in systems analysis.

Soft systems approaches take into account that reality is partly socially constructed and that an understanding of subjective perceptions and the collective framing of a problem situation are essential to deal with complex environmental problems and management tasks (Checkland, 1999; Walker *et al.*, 2002; Pahl-Wostl, 2002a, 2004). Hard systems approaches emphasize the need for factual analysis and "objective" and "hard" decision criteria. It may be highly misleading and even detrimental to achieving sustainable resources management if one relies on hard systems approaches in situations where uncertainties in the factual knowledge base are high and conflicts about values and management objectives are substantial. This is where social learning comes into play.

The promise of applying systems science as a bridge between hard and soft systems approaches is realized as all stakeholders join to review technologies, policies, underlying assumptions and worldviews and re-assess the main goals and questions on which policies and practice are based. This allows participants recurring chances to correct the hypotheses, policies, action plans, and measuring tools (such as indicators) in a transparent and cyclic process. Systems methods help people see what they normally do not consciously think about or discuss in an open forum: feedback loops with complex interactions and delays that create long and mid-term impacts (e.g. unexpected effects of flood protection policies as outlined before), expectations they hold about other people's behaviour and framing of the context into which they embed the problem under investigation. Such a dialogue is greatly facilitated by qualitative and quantitative modeling. The transition to adaptive management requires that stakeholders grasp how the system behaviour emerges from structure and the underly-

[4] The European Water Framework Directive adopted October 2000 prescribes to all member states of the European Union to achieve a good water status of all European waters (art. 1 and 4 of the WFD) by 2015. It requires to develop and implement river basin management plans and to include stakeholders and the public into this process (europa.eu.int/comm/environment/ water/water-framework/index_en.html).

[5] The European Water Initiative supports the Johannesburg development goals to half the number of people without access to safe water and basic sanitation by 2015 and to generalise the adoption and practice of integrated river basin approaches based on knowledge and innovation (europa.eu.int/comm/research/ water-initiative/index_en.html).

ing worldviews. These methods help in that transition by exposing links between natural, economic and social processes that may sometimes be counterintuitive. Hence, NeWater aims as well at promoting innovation in research approaches in order to be able to analyse and understand human-technology-environment systems.

7. Conclusions and outlook

The paper summarizes arguments supporting the need for a change in current water management practices towards more adaptive and flexible approaches. The hypothesis is stated that change is impeded due to the strong interconnectedness of factors stabilizing current water management regimes. A couple of promising developments suggest that one can expect major breakthroughs in the understanding of water management regimes and the dynamics of transition processes over the next years which would provide tools for analysis and methods for assessing and implementing management regimes adapted to the environmental, technical, institutional, cultural and socio-economic context. Breakthroughs in both research and practical implementation require processes of social learning which include changes in the role of different stakeholder groups and in the framing of current water management problems.

The considerations in previous sections have been made without giving much attention to the scale of intervention. Whereas historical water management has been local, modern IWRM approaches are based on river catchments (watersheds) as scale for water management based on the boundaries of the hydrological unit. There are strong indications that one has even to adopt a global perspective to understand water management problems and to derive appropriate response strategies for their management (Vorosmarty *et al.*, 2004). The research approach outlined in this paper will provide a sound base for a comparative analysis of management regimes of river basins at global scale and will thus provide a base for understanding the adaptive capacity of the global water system, one of the key research questions to be addressed under the umbrella of the Global Water System Project (www.gwsp.org).

Acknowledgements The ideas developed here profited from intensive discussions with participants from the NeWater project and the framing committee of the Global Water Systems Project. The research was financially supported by the European Commission under the contract no 511179 for the Integrated Project NeWater.

References

Bandura A (1977) Social Learning Theory. Ehglewood Cliffs, Prentice-Hall, N-J

Berkes FL, Colding J, Folke C (eds.), (2002) Navigating Social-Ecological Systems: Building Resilience for Complexity and Change. Cambridge University Press, Cambridge, MA

Bormann BT, Cunningham PG, Brookes MH, Manning VW, Collopy MW (1994) Adaptive ecosystem management in the Pacific Northwest. USDA For. Serv. Gen. Tech. Rep. PNW-GTR-341. 22 pages

Bouwen B, Taillieu T (2004) Multi-party collaboration as social learning for interdependence: Developing relational knowing for sustainable natural resource management. J Community Appl Soc Psychol 14:137–153

Checkland P (1999) Soft Systems Methodology: A 30-year retrospective. Wiley and Sons, Chichester

Craps M (ed) (2003) Social Learning in River Basin Management. Report of workpackage 2 of the HarmoniCOP project (www.harmonicop.info)

Dyson M, Bergkamp G, Scanlon J (eds) Flow. The Essentials of Environmental Flows. IUCN, Gland, Switzerland and Cambridge, UK. xiv

Folke C, Carpenter S, Elmqvist T, Gunderson L, Holling CS, Walker B (2002) Resilience and sustainable development: Building adaptive capacity in a world of transformations. Ambio 31:437–440

Folke C, Hahn T, Olsson P, Norberg J (2005) Adaptive governance of social-ecological systems. Annu Rev Environ Resour 30:8.1–8.33

Framing Committee of the GWSP (2004) The Global Water System Project: Science Framework and Implementation Activities. Earth System Science Partnership. (Draft: February 6, 2004). http://www.gwsp.org

Galaz VR (2005) Does the EC Water Framework Directive build Resilience? Harnessing Socio-Ecological Complexity in European Water Management. Policy Paper I by "The Resilience and Freshwater Initiative", Swedish Water House, Stockholm

Geels FW (2002) Technological transitions as evolutionary reconfiguration processes: a multi-level perspective and a case-study. Research Policy 31:1257–1274

Gleick PH (2003) Global freshwater resources: Soft-Path solutions for the 21st century. Science 302:524–528

Gunderson L (1999) Resilience, flexibility and adaptive management – antidotes for spurious certitude? Conservation Ecology 3(1):7. [online] URL: http://www.consecol.org/vol3/iss1/art7

Hartvigsen G, Kinzig A, Peterson G (1998) Use and analysis of complex adaptive systems in ecosystem science: Overview of special section. Ecosystems 1:427–430

Holling CS (ed) (1978) Adaptive Environmental Assessment and Management. John Wiley and Sons, New York

Huberman BA (ed) (1988) The Ecology of Computation. Elsevier, Amsterdam

Huitema D, Becker G (2005) Governance, institutions and participation: a comparative assessment of current conditions in selected countries in the Rhine, Amu Darya and Orange river basins. NEWATER report series, No 7, www.newater.info

Kiker CF, Milon JW, Hodges AW (2001) Adaptive learning for science-based policy: the Everglades restoration. Ecological Economics 37:403–416

Lee KN (1999) Appraising adaptive management. Conservation Ecology 3:3–16

Levin SA (1998) Ecosystems and the biosphere as complex adaptive systems. Ecosystems 1:431–436

Moberg F, Galaz V (2005) Resilience: Going from Conventional to Adaptive Freshwater Management for Human and Ecosystem Compatibility. Swedish Water House Policy Brief Nr. 3, SIWI

Pahl-Wostl C (1995) The dynamic nature of ecosystems: Chaos and order entwined. Wiley and Sons, Chichester

Pahl-Wostl C (2002) Towards sustainability in the water sector: The importance of human actors and processes of social learning. Aquatic Sciences 64:394–411

Pahl-Wostl C, Hare M (2004) Processes of social learning in iuntegrated resources management. Journal of Applied and Community Psychology 14:193–206

Pahl-Wostl C (in press) The implications of complexity for integrated resources management. Environmental Modeling and Software, in press

Pahl-Wostl C (2006) The importance of social learning in restoring the multifunctionality of rivers and floodplains. Ecology and Society 11(1):10. [online] URL: http://www.ecologyandsociety.org/vo111/iss1/art10/

Pahl-Wostl C, Downing T, Kabat P, Magnuszewski P, Meigh J, Schlueter M, Sendzimir J, Werners S (2004) Transition to Adaptive Water Management; The NeWater project. NeWater Report Series, No. 1, www.newater.info

Richter BD, Mathews R, Harrison DL, Wigington R (2003) Ecologically sustainable water management: managing river flows for ecological integrity. Ecological Applications 13:206–224

Rotmans J, Kemp R, van Asselt M (2001) More evolution than revolution: Transition management in public policy. Foresight 03:15–31

Senge P (1990) The fifth discipline. Doubleday, New York

Walker B, Carpenter S, Anderies J, Abel N, Cummings G, Janssen M, Lebel L, Norberg J, Peterson GD, Pritchard R (2002) Resilience management in social-ecological systems: a working hypothesis for a participatory approach. Conservation Ecology 6(1):14 (online), http://www.consecol.org/vol6/iss1 /art14/main.html

Walters C (1986) Adaptive Management of Renewable Resources. Macmillan, New York

Tillman DE, Larsen T, Pahl-Wostl C, Gujer W (2005) Simulation for strategy development in water supply systems. Hydroinformatics 7/1

Timmerman JG, Langaas S (eds.) (2003) The role and use of environmental information in European transboundary river basin management. IWA Publishing, UK

Vörösmarty C, Lettenmaier D, Lévqêue C, Meybeck M, Pahl-Wostl C (2004) Humans transforming the global water system. EOS 85:513–516

Walker B, Carpenter S, Anderies J, Abel N, Cummings G, Janssen M, Lebel L, Norberg J, Peterson GD, Pritchard R (2002) Resilience Management in Social-ecological Systems: A Working Hypothesis for a Participatory Approach. Conservation Ecology 6(1):14 (online), http://www.consecol.org/vol6/iss1/art14/main.html

Water Resour Manage (2007) 21:63–77
DOI 10.1007/s11269-006-9041-3

ORIGINAL ARTICLE

Stakeholder-driven, enquiry-driven, or stakeholder-relevant, enquiry-driven science?

W. James Shuttleworth

Received: 20 April 2006

Abstract There is currently debate within the international hydrological community on whether hydrological science should give priority to providing measurements, knowledge, and understanding pre-determined as being needed by stakeholders, or priority to more basic enquiry-driven science that will stimulate the continued health and growth of hydrology as an important Earth science discipline. Two recent major international initiatives in hydrology reflect these two perspectives. One, the *Hydrology for the Environment, Life, and Policy* (HELP) program, is primarily fostered by UNESCO-IHP and is focused on stimulating the stakeholder-driven hydrological science required in specific catchments that have become members of a global network. The second, the decade on *Prediction in Ungauged Basins* (PUB), which is appropriately managed by IAHS, is primarily driven by scientific enquiry and is focused on creating new scientific methods and understanding, albeit with practical application ultimately in mind. This paper summarizes the nature, origins, growth, and progress of these two international programs but also describes the subtly different approach that has been adopted by the U.S. National Science Foundation's (NSF's) Center for *Sustainability of semi-Arid Hydrology and Riparian Areas* (SAHRA). NSF is a federal agency whose primary goal is to "enable the future" by stimulating novel science. Because SAHRA is a federally-funded entity supported by an agency with this goal, the Center clearly cannot operate in stakeholder-driven, response mode in competition with the already effective private U.S. consultancy industry. Nonetheless, SAHRA's mission is to create knowledge and build understanding that will enhance the prospects of sustainable water management in semi-arid regions, especially the southwestern U.S. To resolve this apparent conflict, SAHRA looks ahead to future stakeholder needs and builds its research agenda around selected critical stakeholder-relevant questions that require substantial and sustained investment in basic, multidisciplinary, enquiry-driven science. This paper describes SAHRA's approach and reports on associated research and outreach activities.

W. J. Shuttleworth
Department of Hydrology and Water Resources, University of Arizona, NSF Center for Sustainability of semi-Arid Hydrology and Riparian Areas, Tucson, AZ, USA

Keywords HELP · PUBS · SAHRA · Hydrological science · Stakeholder-driven hydrological science · Stakeholder-relevant hydrological science · Semi-arid hydrology

1. Introduction

Hydrology is inherently an applied subject because quantifying, modeling, and predicting surface and sub-surface flows has intrinsic practical value. But hydrology was recognized internationally as being a "science" as early as 1922, at the General Assembly of the International Union of Geodesy and Geophysics (IUGG) in Rome when a committee recommended that a "International Branch of Scientific Hydrology" be created to further hydrography. Although the International Association of Hydrological Sciences (IAHS) became established before World War II, hydrology as it is understood today was unknown to the general public and hydrology was not taught at universities except as components of such disciplines as physical geography, hydraulic engineering, geology, and hydraulics. Between 1948 and 1980, and particularly in the 1960s and 1970s, hydrology became better recognized as a subject of great significance for economic development. Subsequently hydrology then grew to become an important Earth System Science discipline critical to the wellbeing of mankind.

As engineers and scientists learned to measure components of the hydrological cycle and to build (initially often empirical or statistical) relationships between them, there have been times when progress paused long enough to define a then-current "accepted understanding". Practitioners used this understanding as the basis for planning, design, and management decisions and this stimulated and facilitated growth of the hydrological consultancy industry. Once established, the presence of accepted hydrological practice, especially when it becomes enshrined in law or policy, can provide a disincentive to investment in the acquisition of new hydrological knowledge and an inhibition on the application of any new hydrological understanding resulting from such investment. In part because of this, some hydrologists have become sensitive to the suggestion that enough is known to solve most practical problems and they argue for preserving the purity of hydrology as a scientific discipline motivated by scientific enquiry. However, others argue that the most able hydrologist scientists have lost practical relevance and that hydrological research should primarily be guided by the needs of stakeholders who have need of information to aid water management and water policy within a catchment.

Over the last five years, two major international initiatives have emerged that reflect these two perspectives on hydrology. One, the *Hydrology for the Environment, Life, and Policy* (HELP) program, which is primarily being fostered by the United Nations Educational, Scientific and Cultural Organization's (UNESCO's) International Hydrology Program (IHP), is focused on stimulating the basin-specific stakeholder-driven hydrological science required in catchments that have been selected to become members of a global network. The second, the decade on *Prediction in Ungauged Basins* (PUB), which is being implemented by IAHS, is primarily driven by scientific enquiry and is focused on creating new scientific methods and understanding, albeit with practical application ultimately in mind. The nature, origins, growth, and progress of these two international programs are described in Sections 2 and 3, respectively.

The U.S. National Science Foundation's (NSF's) Center for *Sustainability of semi-Arid Hydrology and Riparian Areas* (SAHRA) has adopted a subtly different approach. As an alternative to either the pure *stakeholder-driven* approach or the *enquiry-driven* approach, the SAHRA Center has adopted a "*stakeholder-relevant, enquiry-driven*" approach. This involves looking ahead to anticipate future stakeholder needs and then building a research agenda selected to address critical stakeholder-relevant questions that require substantial and sustained investment in basic, multidisciplinary, enquiry-driven science. Section 4 describes SAHRA's approach and provides examples of associated research and outreach activities.

2. The hydrology for environment life and policy (HELP) program

The essential philosophy around which the HELP program is built is that integrated catchment management is most effectively made by first creating a framework for water law and policy experts, water resource managers, and water scientists to work together on water-related problems. This network (or partnership) then defines and implements the required "stakeholder-driven" hydrological research agenda and specifies the water related physical (hydrological, climatological, ecological) and non-physical (technical, sociological, economics, administrative, law) observations that must be made to address the most critical policy and management issues. A catchment selected to become part of the international HELP network necessarily must accept this stakeholder-driven approach and demonstrate commitment to pursuing it, drawing advice, guidance, and support as it does so from other catchments in the HELP network through information exchange and shared experience.

HELP is a joint initiative of UNESCO and the World Meteorological Organization (WMO) and, following the recommendations of the 5th UNESCO/WMO International Conference on Hydrology (February 1999), HELP is led by UNESCO's International Hydrological Programme (IHP). The program was originally stimulated by several factors including a call for a UN Global Water Quality Initiative in 1996, and a request for UNESCO and WMO to consider the launching of a 2nd International Hydrological Decade by the British Hydrological Society in the so-called "Exeter Statement" in 1998 (BHS, 2005a). Important steps towards creating HELP include a UNESCO-sponsored expert group meeting in Wallingford, UK in 1998 where the stakeholder-driven framework was established; endorsement of the framework for the HELP program by the 5th Joint UNESCO/WMO Conference on International Hydrology in Geneva, Switzerland and later by the IHP Bureau in 1999; detailed specification of the first phase of HELP at a Task Force Meeting in Tucson, USA in 1999; and approval for HELP at the 14th Session of UNESCO-IHP in Paris and at 11th Session of WMO-CHy in Abuja, Nigeria in 2000.

Subsequent development of the HELP program has been rapid and impressive. In Dublin in 2000, the HELP Interim Management Committee (IMC) agreed on a two-stage process for program implementation, with activity initially focused on a network derived from the 25 basins that had at that time already submitted unsolicited statements of interest in HELP and a later formal solicitation for proposals for additional participating basins when the program had successfully undergone this initial proof of concept. Written proposals were then requested from the first 25 basins. These proposals were reviewed by sub-committee of the HELP IMC and, on the basis of its maturity in implementing the HELP philosophy, each basin was assigned a provisional classification into the following four categories:

- "Proposed" HELP basins (6 basins).
- "Evolving" HELP basins (14 basins);
- "Operational" HELP basins (4 basins); and
- "Demonstration" HELP basins (1 basin).

Consistent with the approach of encouraging information exchange and sharing experience between basins in the network, HELP then sponsored several conferences which included representation from among these initial HELP basins. Notable among these conferences

Table 1 Characteristics used to classify the basins selected for inclusion in the HELP basin network during Phase 2 of the program

Group P: *Proposed* HELP basins are characterized by the following:
•may need to provide more detail for various aspects described in the Proposal Document.; may not have yet achieved any initial operational activity.
•may not have yet begun full stakeholder involvement;
•may have identified too few or too narrow a range of the HELP key issues;
•may also need to provide further information about official endorsement, support and funding commitments.
Group E: *Evolving* HELP basin are characterized by the following:
•has demonstrated initial progress and commitment to develop the basin in accordance with HELP principles;
•has plans to involve stakeholder groups in regular meetings for HELP basin management;
•has budgetary and stakeholder commitment secured and is awaiting implementation;
•has a comprehensive project plan for proposed activities with timelines and milestones;
•has plans for workshops, regular reporting, publications and web site;
•may become operational following at least one year's implementation of the project plan;
•may solicit external support, if local resources do not suffice.
Group O: *Operational* HELP basin are characterized by the following:
•has implemented the HELP philosophy;
•has involved most HELP stakeholder groups in basin management;
•is substantially functioning across several HELP key issues in an integrated manner;
•demonstrates an active interface between science and water managers, and society;
•has established mechanisms for unrestricted information and data access and exchange;
•follows the WMO Resolution 25 on international exchange of hydrological and related data.
Group D: *World Demonstration* HELP basin are characterized by the following:
•has a high quality web site and a response facility for requests for information;
•is prepared to partner, cooperate with or engage in a twinning agreement with another HELP basin;
•is able to provide facilities for local seminars and/or visits from other HELP basins;
•is able to promote and attract sponsors for the HELP concept;
•addresses problems in all of the five HELP key issues areas;
•has a wide and varied range of stakeholders, including water resource managers, scientists, national and local government, private sector and NGOs.

were the 5th IHP/IAHS George Kovaks Colloquim in Paris in 2000 and a major conference entitled "Towards Integrated Catchment Management: Opening the Paradigm Locks between Hydrology, Ecology, and Policy Making" in Sweden in 2002.

Based on the success of the first phase of HELP, in October 2003 the formal call for new or continued participation in the HELP Basin network was made. The subsequent review resulted in 67 basins being selected and again classified into the four categories given above (UNESCO, 2004). The criteria characterizing these classifications are given in Table 1. Figure 1 shows the location of basins selected to be in the newly defined HELP Network. 29% of the basins are in Europe, 20% in Africa, 12% in North America, 11% in Latin America and the Caribbean, 7% in Australia and the Pacific, and 4% in the Middle East. In 2004 and 2005, HELP-related international conferences span the globe. They include the International Conference on "Integrated Water Resources Management in Vulnerable Ecotypes" in Xinjiang, China in August 2004; a HELP session at the 2nd International Symposium on "Transboundary Waters Management" in Tucson, Arizona in November 2004; a HELP Workshop on "The effects of mining on water resources quality" in Vanuatu Islands, S.W. Pacific in December 2004, the HELP International Conference on "North-South Analysis on

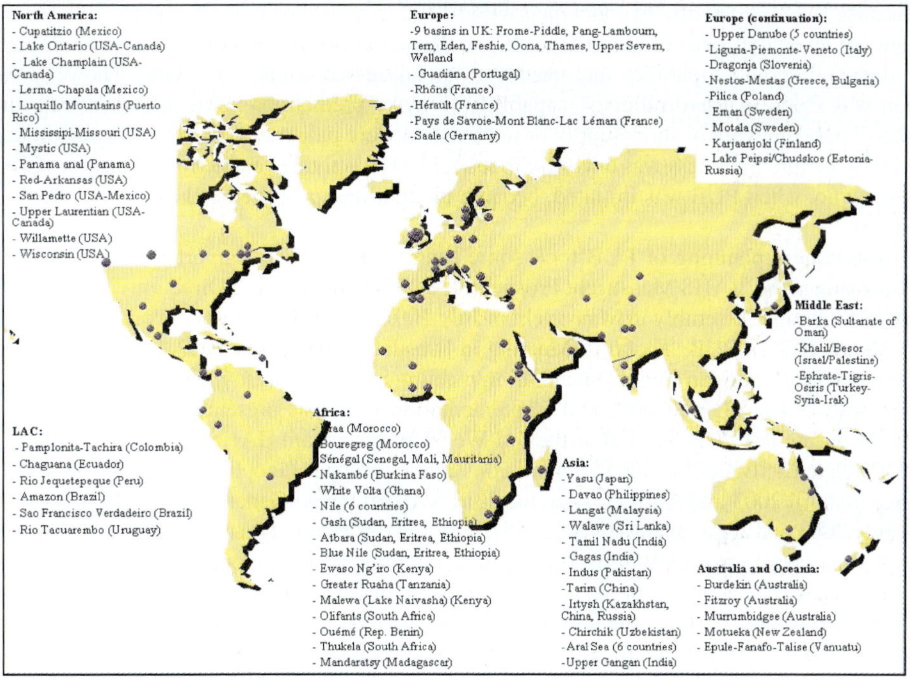

Fig. 1 Location of the network of participating basins in phase 2 of the HELP program

Global Change Impacts on Watersheds" in Bonn, Germany in February 2005, and a Joint HELP-IWMI-IGBP meeting on "The Science needed to address Water and Food" in South Africa later in 2005.

An important new development in the organisation of the HELP program is a movement toward organizing activity through Regional Coordinating Units (RCUs) which arguably better sample any regional or cultural distinctions that may influence Integrated Catchment Management (ICM) practice. At this time, four RCUs so far exist to coordinate HELP-related activity in Australasia and South East Asia; Latin America and the Caribbean; North America; and Europe. In summary, the HELP program remains vigorous and effective and it continues to grow both in terms of the level of participation and the effectiveness of information exchange between basins.

3. IAHS decade on predictions in ungauged basins (PUB)

IAHS's "decade" (2003–2012) on *Predictions in Ungauged Basins* was stimulated by two motivations, one institutional and one scientific. The program resulted from an internet-based discussion which was initiated in September 1999 by the then President-elect of IAHS, Kuni Tacheuchi. The discussion (IAHS, 2005a) sought to define strategic directions within hydrological science that IAHS might take leadership of for a decade or more. In part, the purpose of this debate was to build a renewed excitement within IAHS itself, the perception at that time being that the organization had become too "institutional" in nature and that it needed to strengthen its position as the "academic" leader of the international hydrological community.

Because IAHS is essentially concerned with fostering new hydrological science at the international level, the focus of the internet-based discussion was on major gaps in hydrological understanding and capability that needed to be addressed community-wide. The observation was made that hydrologists' capability to make predictions in basins where relevant traditional data were in short supply or nonexistent were palpably poor. The suggestion that addressing this weakness was a worthy focus for IAHS activities came from Murugesu Sivapalan who, when PUB was initiated, became the first leader of the PUB Scientific Steering Group.

Subsequent planning of PUB took some time to accomplish. Important steps included discussion at the "IAHS Maastricht Prospective Workshop on Future Directions" at the VIIth IAHS Scientific Assembly in Maastricht in July 2001, a PUB Preparatory Workshop in Kofu in March 2002, a PUB "Kickoff" Meeting in Brasilia in November 2002, and a PUB Open Forum at the American Geophysical Union meeting in San Franciso in December 2002. The first draft of the PUB Science and Implementation Plan was presented in March 2003 in Kyoto during the PUB Session at the 3rd World Water Forum (IAHS, 2005b). Subsequent PUB-related activities include major sessions at the European Geophysical Union meeting in Nice in April 2003 and April 2004, at the 23rd XXIIIrd IUGG General Assembly in Saporro in July 2003 and at the 8th IAHS Scientific Assembly in Foz do Iguaçu, Brazil in April 2005.

As an overall goal, PUB aims to formulate and implement appropriate science programs which engage and energize the scientific community, in a coordinated manner, towards achieving major advances in the capacity to make predictions in ungauged basins. PUB has set community wide objectives for the hydrological community, as follows:

- Advance the ability of hydrologists worldwide to predict the fluxes of water and associated constituents from ungauged basins, along with estimates of the uncertainty of predictions;
- Advance the knowledge and understanding of climatic and landscape controls on hydrologic processes occurring at all scales, in order to constrain the uncertainty in hydrologic predictions;
- Demonstrate the value of data for hydrologic predictions, and provide a rational basis for future data acquisitions, including alternative data sources, by quantifying the links between data and predictive uncertainty;
- Advance the scientific foundations of hydrology, and provide a scientific basis for sustainable river basin management; and
- Actively promote capacity building activities in the development of appropriate scientific knowledge and technology to areas and communities where it is needed.

The goals of PUB will be accomplished through the activities of Working Groups. In an attempt to establish PUB as a "bottom-up" program in which ideas and enthusiasm comes from the broad worldwide community of hydrological scientists (as opposed to the traditional "top-down" approach of coordinating science though high level IAHS Commissions), the process through which individual PUB Working Groups are established is currently little inhibited by a formal approval process. Groups of scientists interested in working together on PUB-relevant science can self-organize and propose to become a PUB working Group by applying through an internet-based application process. Although PUB is still in its infancy, at this writing ten PUB Working Groups have so far registered their intent to provide leadership in areas of science relevant to PUB (Table 2), in some cases seeking to lead regional or national contributions that cover several scientific topics.

Figure 2 illustrates the organizational infrastructure for PUB. In terms of governance, PUB operates as fully-fledged IAHS Working Group which reports directly to the IAHS Bureau through the Chair of the PUB Scientific Steering Group who is a member of the Bureau.

Table 2 Current working groups participating in the IAHS decade on Predictions in Ungauged Basins (PUB)

Working group No.	Scientific topic or regional or national focus of working group
WG1	Top-Down Modelling Working Group [http://www.stars.net.au/tdwg/]
WG2	MOPEX Working Group
WG3	Orographic Precipitation, Surface and Ground Water Interactions and their Impact on Water Resources
WG4	Japan Working Group [Suimon Adventure for Knowledge Evolution; SAKE]
	WG4.1 – Estimation of Extreme Events
	WG4.2 – Model Selection and Uncertainty Evaluation
	WG4.3 – Hydrologic and Landscape Diveristy
	WG4.4 – Global Hydrologic Modelling
	WG4.5 – Downscaling Global Hydrologic Information
WG5	Design Flows for Ungauged Basins
WG6	China Working Group
	WG6.1 – Hydrological Modelling and Water Resources Assessment under High Water-Stress
	WG6.2 – Evaluation and Prediction of the Groundwater
	WG6.3 – Flood Forecast and Damage Estimation
	WG6.4 – Prediction of Water Resources and its Consumption in the Arid Region
	WG6.5 – Study of the Ecologically Vulnerable Basins
	WG6.6 – Development of the Coupled Model for the Hydrologic Cycle and Water Quality in the Urbanized River Basins
	WG6.7 – Applications of New Technologies, Theories and Methods to the Hydrological Prediction in Ungauged Basins
WG7	Uncertainty Estimation for Hydrological Modelling
WG8	Remote Sensing and Data Assimilation
WG9	Mediterranean Climate Ungauged Basins
WG10	Drought and Flood Risk: Hydrology and Sediment Transport in Mountain Catchments

The Scientific Steering Group (SSG), provides the primary source of scientific guidance to PUB. Members of the SSG are appointed by invitation for 2 years with a possibility of renewal and, from January 2005, and will be drawn from the grassroots membership of the IAHS who are active in PUB Working Groups. For the first five years, the SSG supported by a PUB Strategic Advisory Group (SAG) to provide strategic advice, foster linkage to related programs, and to seek to secure funds, and lead capacity building activities. During the decade of PUB, activities and performance will be periodically reviewed by an ad hoc Review Panel of eminent scientists and representatives of user groups and linked programs. The next major review will be conducted in the first half of 2005.

4. NSF center for sustainability of semi-arid hydrology and riparian areas (SAHRA)

The Center for *Sustainability of semi-Arid Hydrology and Riparian Areas* (SAHRA) is funded by the U.S. National Science Foundation (NSF). Because NSF is a federal agency whose primary goal is to "enable the future" by stimulating novel, state-of-the-art science and because the SAHRA Center is a federally-funded entity supported by NSF, SAHRA clearly cannot operate in response mode to meet local stakeholder-selected demands in competition

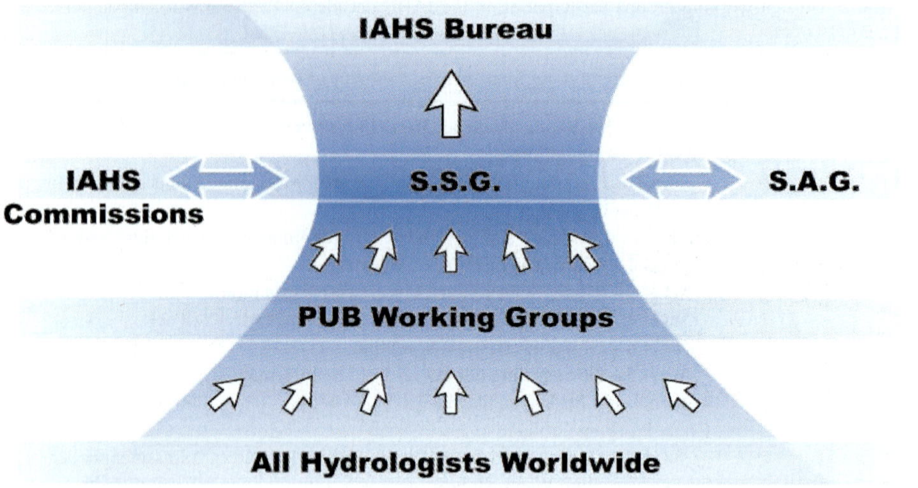

Fig. 2 Organizational diagram of the IAHS decade for Prediction in Ungauged Basins (PUB) within the IAHS bureaucratic structure

with an already effective private U.S. consultancy industry. Nonetheless, SAHRA's mission (SAHRA, 2005a) is to create knowledge and to build understanding that will enhance the prospects of sustainable water management in semi-arid regions, especially the southwestern U.S. Therefore, SAHRA must provide practical outputs that have direct relevance. To resolve this apparent conflict, the SAHRA Center has adopted the strategy of looking ahead to anticipate future stakeholder needs and then building a research agenda to address selected stakeholder-relevant questions which are already critical or soon to be critical, and which require substantial and sustained investment in basic, multidisciplinary, enquiry-driven science.

The mission of SAHRA, is "*to promote sustainable management of water resources by conducting water resources-related science in the context of critical water management issues of semiarid and arid regions*". However, the power to improve sustainability of water resources properly rests with elected officials, professional water managers, and legal experts at local, state, and national levels. Consequently, what is required is an effective mechanism for synthesis, integration, education, and outreach in support of better informed decision-making. Consequently, SAHRA's mission has a dual nature and involves both (a) conducting basin-focused multidisciplinary research that addresses critical and stakeholder-relevant knowledge gaps in the hydrological understanding of semiarid and arid regions; and (b) conveying what is known and what is being learned regarding arid and semiarid hydrology to improve water management and policy. SAHRA's approach is:

1. to anticipate the future need for critical hydrological understanding in arid and semiarid regions by defining stakeholder-relevant questions that cannot be addressed by individual-investigator research; then
2. to bring together appropriate multidisciplinary teams of researchers to address these questions effectively; and
3. simultaneously to build partnerships with a relevant spectrum of stakeholders (public agencies and private organizations) so that this new understanding will be effectively

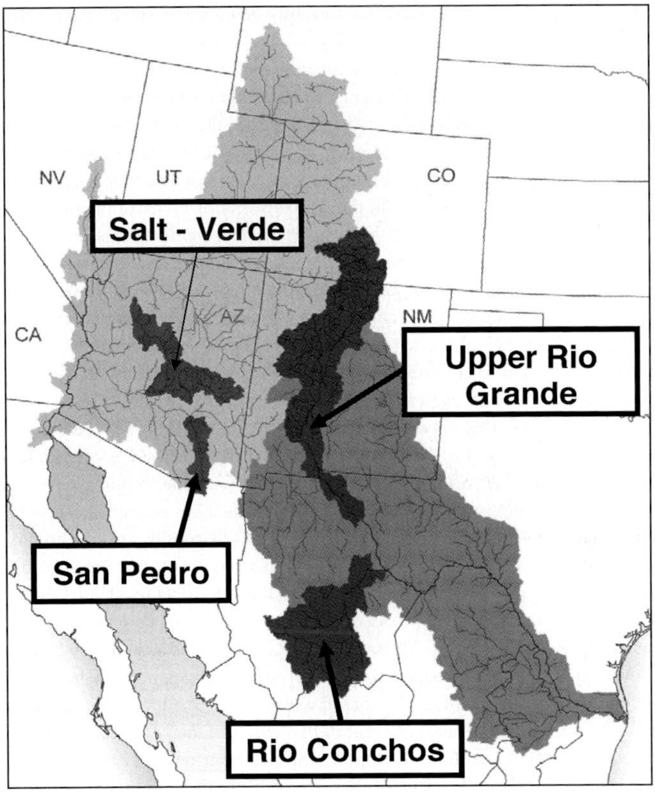

Fig. 3 River Basins in the Southwestern U.S. where the SAHRA Center's Research has Focus

applied to the efficient and equitable management of water resources and to the rational implementation of public policy.

At this writing, the three stakeholder-relevant integrating questions (Table 3) which are being used to focus SAHRA scientific research are broad-based and capable of engendering and crosscutting many related topics of inquiry and touch on scenarios that are of prime interest in this region: namely land use changes, population growth, and climate variability. SAHRA further frames its science and stakeholder activities in a river basin context because basin-focused science yields a synergy of activities that helps to drive science integration. Moreover, stakeholder issues are tied to river basins, consequently a river basin focus also helps to drive the application of research results. SAHRA's primary geographical focus is currently on two river basins (Figure 3): the Rio Grande/Rio Bravo and the Upper San Pedro River basins, although the Center maintains interest and some activity in the Salt-Verde River and Rio Conchos basins.

In the context of the present paper the important point is that, once the link to stakeholders has been made by defining stakeholder relevant questions and through on-going aggressive outreach efforts, the hydrological research that is then undertaken can be (and is) enquiry-driven and both the research and outreach activity can be novel and state-of-the-art. For example, research into SAHRA's the stakeholder-relevant questions "*What are the impacts of vegetation change on the basin-scale water balance*?" and "*What are the costs and benefits*

Table 3 The three integrating research questions currently being addressed by the SAHRA Center

Riparian question: *What are the costs and benefits of riparian restoration and preservation?*
In the semiarid Southwest, most human settlements, irrigated agriculture, and regional biodiversity are located in riparian corridors. These riparian systems integrate the hydrologic and biogeochemical processes that occur within a basin. Consequently, water resource management decisions may impact river systems not only through changes in streamflow, but also through changes in water quality, the socioeconomic value of the river system, and the structure and diversity of the riparian ecosystem. A complete evaluation of the costs and benefits of important management decisions regarding riparian preservation and restoration therefore requires an integrated, multidisciplinary understanding. SAHRA research consequently focuses on developing fundamental, process-level understanding in three areas: (1) determining the water balance of riparian systems, (2) evaluating ecosystem dynamics and values, and (3) understanding nutrient and solute sources and cycling. The resulting understanding will further the development of integrated river system models that stakeholders can use to evaluate costs and benefits of potential restoration or preservation efforts.
Water markets question: *Under what conditions are water markets and water banking feasible?*
In the Southwest, water markets and water banking are increasingly viewed as potentially effective mechanisms for allocating water resources, providing economic benefits and avoiding potential conflicts associated with water scarcity. For these mechanisms to be truly effective, detailed knowledge of the available water supply and the factors that affect water demand is critical. To this end, SAHRA is developing products to better estimate precipitation rates and snow-pack volumes at the basin scale. SAHRA is also improving understanding of the factors that determine residential, industrial, and agricultural demand for water, using approaches such as experimental economics and water use micro-logging to disaggregate demand. These products and knowledge will then be integrated into a model that allows water resource managers to consider the trading of water rights and third party impacts in evaluating the potential of market-based mechanisms to allocate water resources effectively.
Vegetation question: *What are the impacts of vegetation change on the basin-scale water balance?*
Vegetation change is a common feature of the Southwestern landscape. Over the last several decades this has occurred in the form of shrub invasion of grasslands, expansion of pinyon-juniper and mesquite, thickening of ponderosa pine forests, and anthropogenic land-use changes. More recently drought related fires and bark beetle infestations are resulting in large-scale vegetation change. While a widespread perception exists that such changes have reduced water resources available for human use, research that documents the actual changes on the basin-scale water balance is lacking. SAHRA seeks to understand the role of vegetation type and structure in the partitioning of rain and snow into evaporation/sublimation, runoff, and infiltration, and how moisture stored in the soil is shared between transpiration, recharge, and streamflow. SAHRA's approach involves: (1) intensive field measurements at selected plot- to hillslope-scale sites to investigate vegetation controls on partitioning and guide development of methods to model and scale these processes; (2) exploring the use of remotely sensed data to determine key hydrologic variables across basins; and (3) integrated modeling to evaluate the effects of vegetation change.

of riparian restoration and preservation?" necessarily requires the development of new approaches to quantifying the stream flow in semi-arid catchments (much of which originates as winter snow) and new understanding of the way different species control water and carbon exchanges with the atmosphere. Investigation of the value of assimilating remotely sensed data into a model of snow melt in the Tokopah River basin in California (Molotch *et al.*, 2005) is an example of SAHRA's stakeholder-relevant but enquiry-driven research. Figure 4 compares the measured river discharge for this basin with that predicted using the standard approach (I.e., the US Army Core of Engineers standard snowmelt model (UASCE) and assumed snow distribution extrapolated from SNOTEL sites), and shows how prediction is improved when remotely sensed albedo and solar radiation data are used in the snow melt modeling. Another example of stakeholder-relevant but enquiry-driven research is the

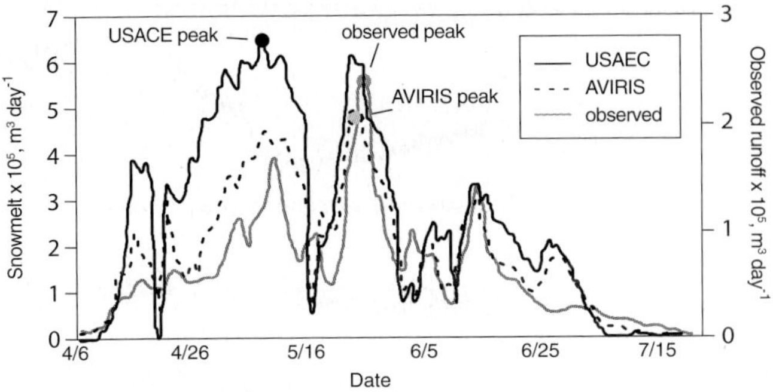

Fig. 4 Measured discharge for the Tokopah river basin in California compared with that predicted using the US Army Core of Engineers standard snowmelt model (UASCE) with snow distribution extrapolated from SNOTEL sites and with that when using remotely sensed albedo and solar radiation data (AVIRIS)

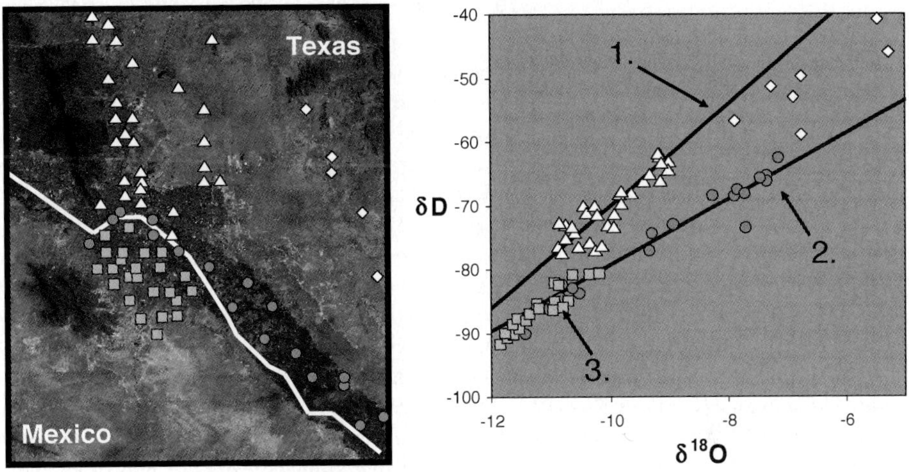

Fig. 5 Isotopic signatures if water samples from wells around the town of Juárez in Mexico indicating water sources as (1) recent precipitation, (2) present day Rio Grande river water, after evaporation from reservoirs along the river, and (3) original Rio Grande river water prior to the building of dams on the river

development and use of state-of-the-art isotope identification techniques to identify the source of groundwater in semi-arid aquifers. Figure 5 illustrates an example of this (Hibbs *et al.*, 2003) in which the relative abundancy of isotopes in the ground water in wells around the city of Juárez in Mexico reveals that the local groundwater beneath the town was recharged prior to the building of dams on the Rio Grande River.

Other interesting new enquiry-driven research findings which have stakeholder relevance include the fact that some of the more successful plants practice a form of "water banking" in semi-arid environments (Hultine *et al.*, 2004). Water is drawn in by surface roots and transferred downwards by deep roots into the soil when there is plentiful precipitation, later to be withdrawn and used when required by the plants (Figure 6). This mechanism may well be one of the reasons why some woody plants, such as mesquite, have been so successful in competition with other species. Similarly, Figure 7 shows the measured evapotranspiration and carbon exchange for woodland, grassland, and shrubland stands in the San Pedro river basin for the 2003 growing season and shows the very substantial difference in these (Scott

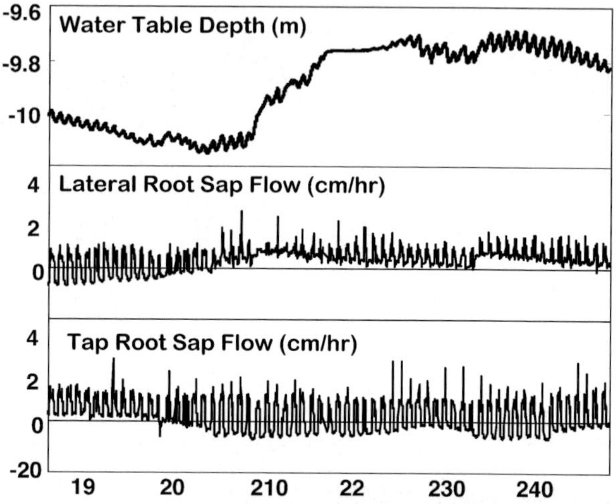

Fig. 6 (a) Water table depth below a mesquite stand and measured sap flow in (b) lateral and (c) tap roots

et al., 2005). All these vegetation covers have some access to groundwater, but the woodland stand is particularly effective in tapping into groundwater to obtain 473 mm of the total 639 mm evapotranspiration for the growing season. The net carbon uptake by woodlands and, to a less extent, shrubland is much greater than for grassland but so is the loss by respiration, especially after periods of rainfall when microbial activity is greatest.

If the results of stakeholder-relevant, enquiry-driven research are to find application, it is essential also to be imaginative in finding effective methods for transferring knowledge to stakeholders and educating students and the general public. The SAHRA Center is particularly aggressive in doing this. Among the many activities of the SAHRA Knowledge Transfer team is a web-based *Global News Watch* information system (SAHRA, 2005b) which searches for and distributes water-related news stories translated from eight languages; a trade publication, *Southwest Hydrology* (SAHRA, 2005c) which is published six times each year to inform and connect the water communities of the semi-arid and arid Southwest; the SAHRA Hydroarchive (SAHRA, 2005d), which is a system that allows creators of hydrologic software to share their work with researchers and practitioners in the hydrological community together with updates on research and forecasts related to droughts, fire risks, monsoons, and snowpack made via media briefings (SAHRA, 2005e) and the use of kiosks, displays, Web sites and DVDs at selected, much frequented sites, such as Kartchner Caverns (SAHRA, 2005f), to raise awareness of the hydrologic cycle among the general public.

In addition, international outreach efforts within SAHRA include focused studies on the Mexico boarder, including a joint US-Mexico project which is being carried out with the Instituto Mexicana de Technologia del Agua (IMTA) and the Sandia National Laboratory, to build a model of the Lower Rio Bravo/Rio Grande and associated demands (agriculture, urban, environmental, etc.) from Fort Quitman to the mouth, to assess the impacts of alternative land use and water resource management strategies. SAHRA's other noteworthy international outreach activities include an "alternative futures" study for the town of La Paz, Baja, Mexico, which is being carried out in collaboration with Centro Interdisciplinario Ciencias Marinas (La Plaz) and Harvard University, to investigate how economic performance, demographic changes, private and public investments, and public policy choices could influence urban growth and land use change over the next 20 years.

SAHRA's education program is also aggressive and includes the distribution of water education kits (SAHRA, 2005g) to provide classroom support for teachers, education of

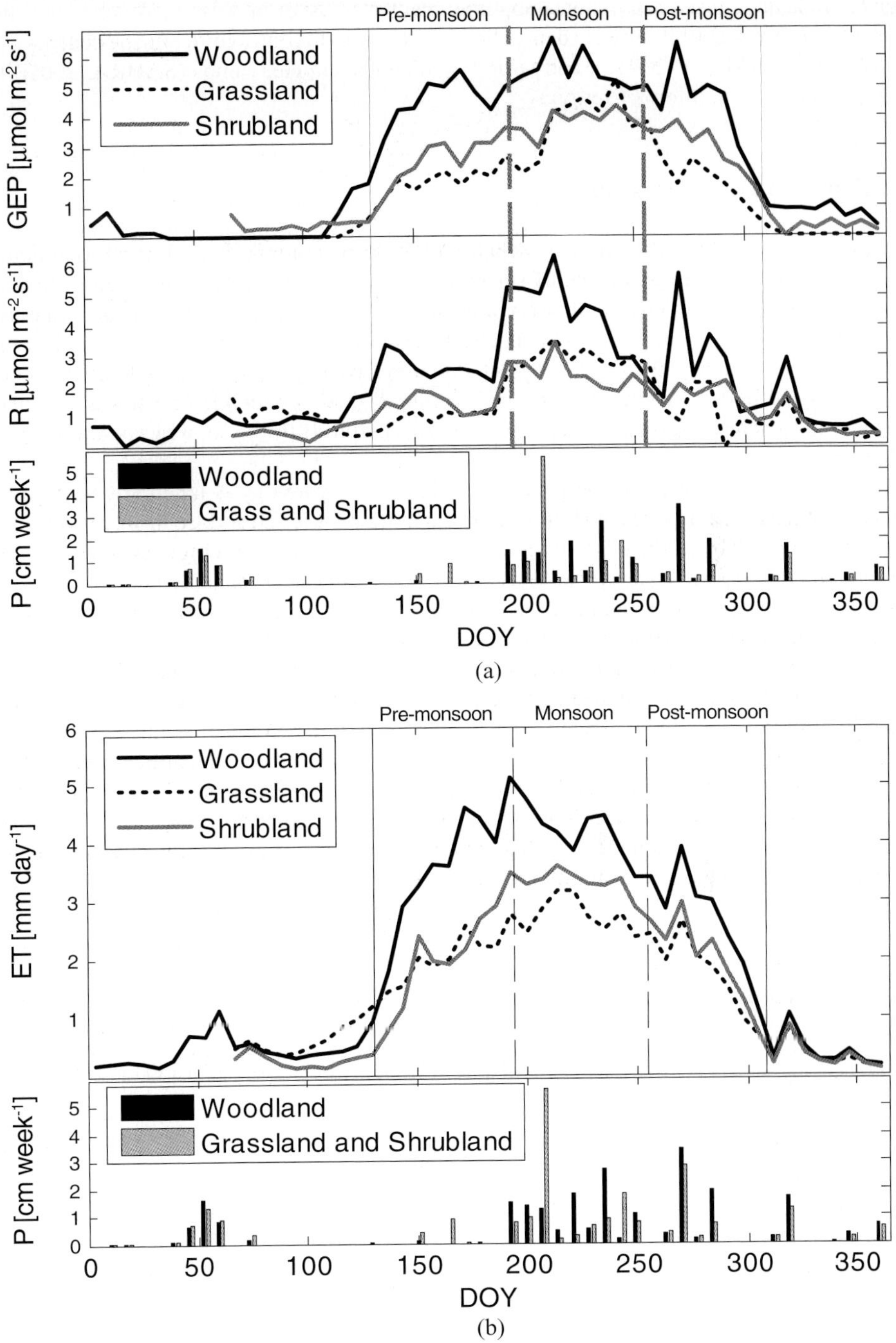

Fig. 7 Measure evapotranspiration, gross ecosystem production, and respiration measured at woodland, grassland, and shrubland covered sites in the San Pedro river basin through the 2003 growing season

undergraduate non-science majors about regional water issues in Arizona Water Issues; the involvement of school children in data collection networks to contribute to much needed water quality data (SAHRA, 2005h); and hydrology-themed summer camps (SAHRA, 2005i) to support learning in informal settings.

5. Summary and conclusions

The present paper draws attention to a debate within the international hydrological community regarding the relative importance of stakeholder-driven hydrological science and enquiry-driven hydrological science and overviews the two recent major international initiatives in hydrology, HELP and PUB, which reflect these two perspectives. It also describes the approach of defining stakeholder-relevant, enquiry-driven hydrological research which has been adopted by the SAHRA Center. In general, all three are legitimate ways to organize hydrological science and all three are proving successful. If this is so, why should there be debate?

As stated at the outset of the paper, the fact is that hydrology is inherently an applied subject because quantifying, modeling, and predicting surface and surface flows has intrinsic practical value. The truth is also that over the last thirty years hydrology has come of age as an important Earth System Science discipline of significance for economic development and critical to the future wellbeing of mankind and the planet. Surely, then, there is no need for such debate. Hydrology no longer needs to fret about its status as a "science" and it can easily accommodate a whole spectrum of interests and perspectives on priority within the hydrological community. In fact, many of the "movers and shakers" in both HELP and PUB (and SAHRA) are actually the same people. In the coming decades, the need to seek sustainable provision of safe water to satisfy human and ecological needs and allow further economic development is the single greatest challenge facing mankind. To meet this challenge, it is necessary both to provide new hydrological knowledge and capability and, at the same time, to build understanding among stakeholders of what is already known and what is being learned. These should not be not competing objectives, they must be pursued together.

Acknowledgements This work was supported by SAHRA (Sustainability of semi-Arid Hydrology and Riparian Areas) under the STC Program of the National Science Foundation, Agreement No. EAR-9876800 and NSF award DEB – 0415977. The author gratefully acknowledges the inputs from his co-participants in the HELP and PUB programs and from many colleagues in the SAHRA Center and the assistance of Marisa Rodriguez in final preparation of the manuscript.

References

BHS (2005a) available via http://ewre-www.cv.ic.ac.uk/bhs/exeter.html

Hibbs B, Phillips F, Hogan J, Eastoe C, Hawley J, Granados A, Hutchison B (2003) Hydrogeologic and isotopic study of the groundwater resources of the Hueco Bolson aquifer, El Paso, Texas/Juarez, Mexico area. Hydrological Science and Technology 19(1–4):109–119

Hultine KR, Scott RL, Cable WL, Williams DG (2004) Hydraulic redistribution by a dominant, warm desert phreatophyte: seasonal patterns and response to precipitation pulses. Functional Ecology 18:530–538

IAHS (2005a) Discussion documented at http://www.cig.ensmp.fr/~iahs/ via the *PUB Corner* and then *IAHS Strategic Science Discussion* links

IAHS (2005b) Final draft of PUB Science Plan downloadable via http://www.cig.ensmp.fr/~iahs/

Molotch NP, Painter TH, Bales RC, Dozier J (2004) Incorporating remotely-sensed snow albedo into a spatially-distributed snowmelt model. Geophysical Research Letters 21:L03501–L03505

SAHRA (2005a) see http://www.sahra.arizona.edu/

SAHRA (2005b) www.sahra.arizona.edu/newswatch/

SAHRA (2005c) www.sahra.swhydro.arizona.edu

SAHRA (2005d) www.sahra.arizona.edu/software/index_main.html

SAHRA (2005e) www.sahra.arizona.edu/kt/media

SAHRA (2005f) www.sahra.arizona.edu/kartchner/index.html

SAHRA (2005g) www.sahra.arizona.edu/water/index.html

SAHRA (2005h) www.sahra.arizona.edu/education/teachers/rio_grande.html

SAHRA (2005i) www.sahra.arizona.edu/education/k-12/camp.html

Scott RL, Huxman TE, Williams DG, Goodrich DC (2005) Ecohydrological impacts of woody plant encroachment: seasonal patterns of water and carbon dioxide exchange within a semiarid riparian environment

UNESCO (2005) Evaluation of the 2003–2004 global call for HELP proposals (http://portal.unesco.org/sc_nat/ev.php?URL_ID=4324&URL_DO=DO_TOPIC&URL_SECTION=201 &reload=1107306140)

Water Resour Manage (2007) 21:79–95
DOI 10.1007/s11269-006-9042-2

ORIGINAL ARTICLE

Learning Alliances for the broad implementation of an integrated approach to multiple sources, multiple uses and multiple users of water

Frits W. T. Penning de Vries*

Received: January 2005 / Accepted: October 2005

Abstract 'Multiple use systems' are systems that allow efficient and effective supply of water from different sources to communities for their domestic and for their productive purposes and that allow interaction with providers of water related services. Such systems are probably highly desirable from the perspective of using scare water efficiently and also from the perspectives of gender equity and improving livelihoods. It is therefore useful to carry out scientific research to validate this statement about a water-innovation. The mode of research must be 'action research'.

The specific form and management of multiple use systems depends on local biophysical and socio-economic factors, as well as on local institutions and legislation. Eleven 'cornerstones' need to be in place to realize a full multiple use system. Since a blue print cannot be made and many parties are involved, 'learning alliances' are to be set up in specific geographic areas and at national level to identify how much of these cornerstones of multiple use systems are still lacking, and to work together to create or implement these. Guidelines for setting up Learning Alliances and for actually implementing systems of multiple water use are needed.

Keywords Learning alliance · Action research · Integrated water resource management · Upscaling · Domestic water · Productive water · Implementation · Multiple use systems · Innovation

---

Manuscript for the International Conference on 'Integrated Assessment of Water Resources and Global Change: A North-South Analysis', 23–25 February 2005, Bonn, Germany (http://www.zef.de/watershed2005). Submitted after review and revision 21 October 2005.

---

*The paper is based on extensive discussions and notes on multiple use systems shared among the members of the MUS project and others in November 2004, including E. Boelee, J. Butterworth, T. Cousins, A. Duran, J. Hagmann, B. Mintesinot, S. Morardet, P. Moriarty, I. Restrepo, S.Ruaysoongnern, S. Suryarwanshi, S. Smits, B. van Koppen and B. Yoder.

F. W. T. Penning de Vries
International Water Management Institute, South Africa
e-mail: f.penningdevries@cgiar.org
Current Address: IPO MAIRS 40 Huayanli, IAP-CAS Chaoyang District, P.O. Box 9804, Beijing 100029, China

Introduction

Why does 'integration' need special attention?

Communities at the scale of villages or settlements in catchments use water for multiple purposes: 'domestic' water for drinking, cooking, cleaning, sanitation, and 'productive' water to earn an income such as by gardening, rearing livestock, brick making, growing of field crops (Yoder, 1983; Van der Hoek *et al.*, 2001). They source water from wells, canals, rivers, reservoirs, and so on (Meinzen-Dick and Van der Hoek, 2001). In small traditional communities integration around water is obtained through informal insight, experience and discussion. Achieving equity among gender and wealth groups in the communities is often not a local priority (Van Koppen, 2001). In larger communities, groups of stakeholders can have representatives in water user associations (Faysse, 2004).

As communities increase in size and wealth and demand more and/or better water, government organizations respond. Since governments are usually structured such that domestic water and productive water are dealt with by different departments, their single-target responses tend to be uncoordinated. Such interventions achieve sub-optimal efficiency and, more importantly, are often not sustainable (Moriarty *et al.*, 2004). Hence, we do need to give particular attention to integration, or perhaps better: to re-integration.

What should be integrated?

Water is crucial for a wide range of activities of every person and in every ecosystem and environment. It is available in limited quantities, and can be used, wasted, lost, polluted, stored, transported, re-used and recycled. Optimum use by an individual user or a single process can be achieved by focussing at a single objective. Optimum use for societies and ecosystems, however, means meeting multiple objectives, and integrated approaches to water use and supply are required. Biswas (2004) argues that while integrated water resource management is a good principle, there is no guideline to operationalize it. We believe that operationalization is feasible by following a gradual, multipartner joint learning approach. That will allow the handling three dimensions of 'integrated approaches to water': (i) *spatial* dimension: across what area to integrate? (note that water catchments are logical units for integration of water management, but communities integrate also social and financial needs that are generally not constraint by catchment boundaries); (ii) *temporal* dimension: over how much time to integrate and find the optimum solution?, and (iii) the *social* dimension: recognize the water sources and needs from different stakeholder groups and aim at acceptable compromises in water usage.

How to integrate users and users?

There are no blueprints for multiple water use systems for at least four reasons: (1) 'optimal' is judged by end users and not by planners, (2) user's criteria and perspectives for 'optimal' change over time and with development, (3) water sources may change (climate change), and (4) reality is too complex for general solutions. Multiple processes of optimization proceed in parallel, continuously and locally. Capacity building of people and organizations is necessary to allow them to identify their own multiple use optimum solutions.

In the 1990's, participatory approaches were introduced to involve rural stakeholders in formal management of resources through participatory approaches and with a view to facilitate and accelerate uptake of innovations (Engel, 1995). These involve NARES as sources of information and targets of capacity building, and later also farmers and farmer organizations (Penning de Vries *et al.*, 2000). Other categories of stakeholders, however, need to be involved for an effective and speedy adaptation and adoption process. The project TRANSCOL (Technology Transfer program on Water Supply Treatment) in Bolivia successfully promoted Inter-Institutional Regional Working Groups (IRWG's) to bring stakeholders together and learn how best to adopt the innovation of multi-stage filtration (Visscher *et al.*, 2005). CIAT (Centre for International Tropical Agriculture) and CARE started in S. America a 'Learning Alliance' on agro-enterprises, and after that with the Catholic Relief Services (CRS) in Africa. In their words: a Learning Alliance is 'a process undertaken jointly by research and development agencies through which research outputs are shared, adapted, used and innovated upon to strengthen and create local capacities, improve the research outputs, generate and document development outcomes and identify future research needs and potential areas of collaboration.' (Lundy, 2002). We expand the definition to include the end users of the innovation and define a Learning Alliance as 'a group of organizations and individuals in a particular area with a shared interest in an innovation and the scaling-up of that innovation'. A Learning Alliance follows a flexible but structured, re-iterative path to progress.

Multiple water use systems are little researched as such. How people tap 'productive water' (often from irrigation supplies) for household use, or 'domestic water' (generally piped) for small productive purposes (e.g. gardens) has been well documented. This literature confirms the demand for holistic views. But studies that address multiple use systems from the outset, analyze practices how these are achieved, with a view of recommending improved planning method are rare (judged by our global literature collection of the subject (MUS, 2005)). Hence, positive results obtained in some countries are difficult to access and to repeat in other places.

This paper describes briefly how the MUS-project is adopting Action Research to evaluate multiple use systems and learn lessons to do that effectively, and starts using the a Learning Alliance approach for significantly outscaling and upscaling of integrated systems for multiple use of water. We have no new facts yet, but process documentation to acquire them has been initiated in five river basins. A first update to this paper is already being prepared (Van Koppen *et al.*, 2005).

The MUS project

The project Multiple-Use Systems (MUS) was created in response to a call for research proposals by the CGIAR-Challenge Program for Water and Food (CPWF, 2005). Its first phase is funded by the CPWF from 2004 till 2008, and new partners and donors are joining. The CPWF aims at stimulating innovative ways of doing water-research, at broad partnerships, and particularly at impact.

In most countries, the water sector is fragmented into sub-sectors, the most important ones from a the point of view of resource management and multiple use services being those dealing with domestic water supply, sewerage and waste-water, irrigation, and other productive uses of water in small and medium enterprises. Other relevant water sectors include local government, rural development, social welfare and health, livestock, fisheries, forestry and environment. In the past, centralised planning has made it difficult to achieve the flexibility needed to bring these, typically, governmental stakeholders together to work effectively at the local level. In other cases, it is the inadequate capacity of local planners that promotes

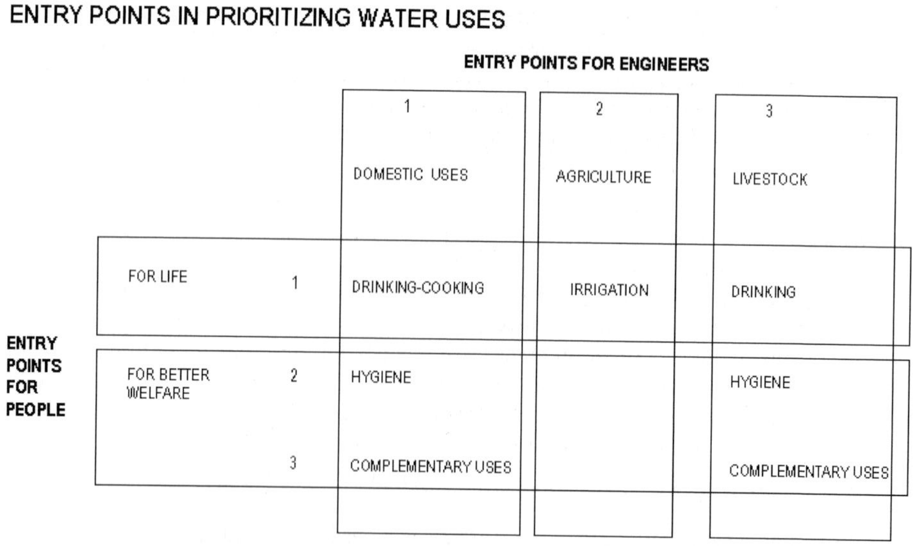

Fig. 1 Entry points into multiple use systems from different angles (source: CINARA)

fragmentation. The problem is illustrated in Figure 1. The current trends towards increased decentralization and capacity building provide an opportunity to bring back flexibility.

The ultimate objective of MUS-project is to '*enhance land and water productivity, improve rural livelihoods and promote gender equity*' (see http://www.musproject.net/index.php/ intranet). MUS does this through promotion of multiple-use systems, in particular by designing, testing and promoting models, guidelines and tools for the upgrading of existing systems to systems where sources, users and users are effectively integrated. We also formulated two objectives:

Capacity building: To engage, inform, prepare and strengthen the capacity of project partners and of other participants of Learning Alliances, including professionals and policy makers from the domestic and productive water sectors in NGOs, government, financing institutions, private sector, and development organizations, to jointly promote a 100-fold wider implementation of multiple-use water supply systems after this project.

New knowledge: To generate new knowledge and synthesize existing knowledge into innovative models, guidelines, and tools that can be used to produce quantifiable positive impacts on the food security, income, work load, health and well-being of the poor, particularly of women and children, HIV/AIDS victims and child headed households.

The first objective has platforms of stakeholders in its focus and the Learning Alliance is its key approach. The second objective supports the first; it focuses at end users of water and uses Action Research as it main approach. But the two are interwoven: doing effective action research has its own elements of a learning alliance, and the alliances cannot perform well without some action research.

In MUS, Action Research and Learning Alliance will be carried out at various locations in five CPWF-basins: Indus-Ganges (India), Limpopo (Southern Africa), Mekong (S.E. Asia), Nile (NE Africa), and the virtual Andes (S. America) basin. Sites have been selected with local partners. There are quite different experiences with m.u.s. in each of these basins that

Table 1 Lead partners in the MUS-project

Name	Key expertise	Key geographic area
International Development Enterprises (IDE), USA	Implementation small scale water technologies	S. Asia
International Water and Sanitation Centre (IRC), Netherlands	Domestic water policies and dissemination	W. Asia, L. America
International Water Management Institute (IWMI), S. Africa, Ethiopia, India, Thailand	Management agricultural water, policies	Asia, Africa
Khon Kaen University (KKU), Thailand	Land and water management	Thailand
Mekelle University, (MKU), Ethiopia	Land and water management	Ethiopia
CEMAGREF, Unité Mixte Recherche Eau, France	Water pricing	Europe

Table 2 Associate partners for implementation at the basin level (and the list is growing)

Name	Key expertise	Key geographic area
Association for Water and Rural Development	Community mobilization and development	S. Africa
Catholic Relief Services	Community development, micro-finance, sanitation	Africa
Centro-Agua	Participatory water research	Bolivia
CINARA	Participatory water research	Colombia
IDE-Nepal and the Smallholder Irrigation and Marketing Initiative (SIMI)	Community development, marketing	Nepal
Local Wisdom farmer organization	Community mobilization and development, policy advocacy	Thailand
Population and Community Development Centre (PDA)	Community development	Thailand
Mvula Trust	Community mobilization and development, water	S. Africa
Population and community Development Association (PDA)	community development; self financing	S.E. Asia
World Vision S. Africa	Child care and welfare, community development	S. Africa

can be relevant for the other basins. The project started to gather data, analyze these, compare experiences and produce an integrated overview.

MUS has a wide range of expertise, skills and contacts in science and in rural development. At the moment, it has six lead partners (Table 1) that complement each other and has a growing list of associated partners (Table 2).

To operationalize the concept of 'multiple use systems', partners must share a conceptual framework. We conceived the framework based on experiences and insights of all partners (Boelee *et al.*, 2004) using the LearningWheel method (Hagmann, 2005) and defined its 'cornerstones': elements that must be in place in order to have a full multiple use system, be it at the end user, the district or the national level. We identified eleven cornerstones (Figure 2, Box 1). We use these to focus attention, to see in what areas international exchange to be promoted, and to measure our progress in research and capacity building. For each cornerstone, MUS will establish the base line information. Research and implementation actions are derived from these cornerstones, some of which do apply and can be realized for all sites. The framework also allows integration of the results into the bigger picture. The framework will guide the iterative process of action, reflection and joint re-planning, and help to integrate conceptual development, planning and knowledge management (Van Koppen *et al.*, 2005).

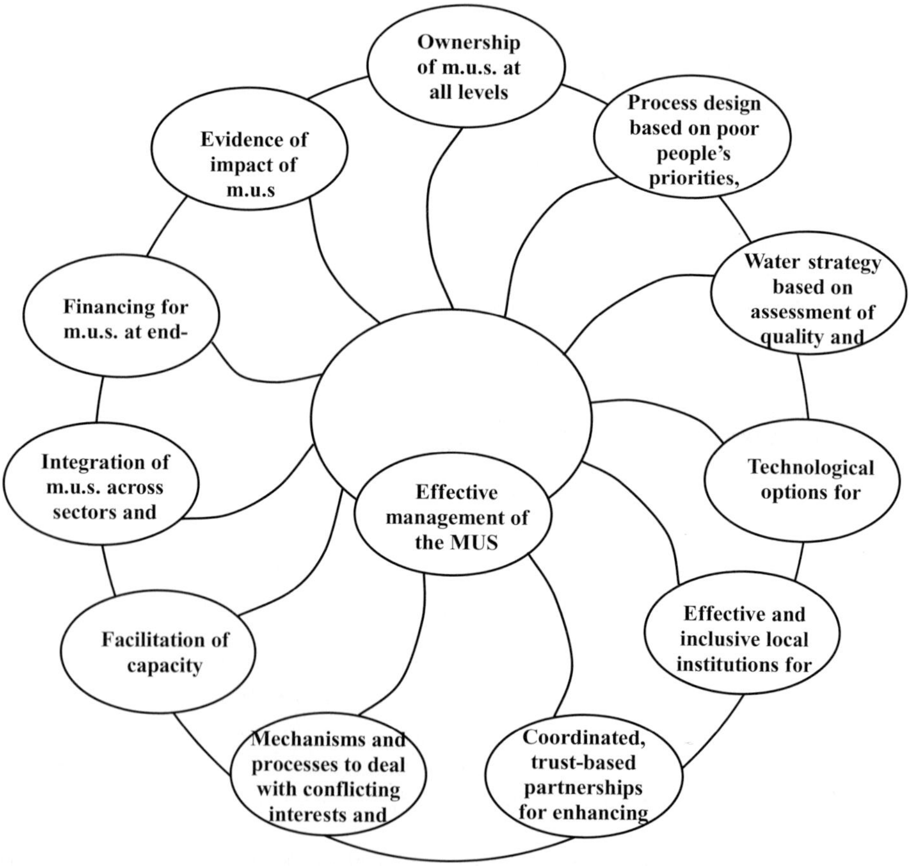

Fig. 2 The eleven cornerstones that must be in place in order to realize a full multiple use system and services (Boelee *et al.*, 2004). The connections between the cornerstones indicate that all interdependent but not that there is any particular sequence to be followed. The central element in the diagram is about management of the MUS-project and does refer to the concept of multiple use system

Box 1. Eleven cornerstones for multiple use systems with some key words to characterize them (Boelee *et al.*, 2004):

1. Ownership of 'multiple use systems' at all levels
   understanding of the concept at local, district and national levels.
2. Process design based on poor people's priorities, problems, perspectives.
   interventions should fit livelihood strategies.
   participation in decision making.
   involve local wisdom.
3. Water strategy based on assessment of quality and quantity
   relate quality to purpose of use.
   consider upstream and downstream uses.
4. Technological options
   appropriate and affordable technologies.
   tariff structures that reflect diverse users to promote equity and empowerment.
5. Effective and inclusive local institutions to manage multiple use systems
   need for champions, leaders.
   gender sensitive.
   promote social capital.
6. Coordinated partnerships for enhancing multiple use systems
   trust based partnerships.
   facilitation, coordination, learning alliance.
   influence policy levels.
7. Mechanisms for conflict resolution.
   with willing pasrtners
   with unwilling partners
8. Facilitation for upscaling of multiple use systems
   aim for practical results, not academic.
   often limited capacity for implementation.
9. Integration across sectors and levels
   break down horizontal and vertical barriers.
   local government may have narrow mandates, incentives, capacity.
   cooperation 'moves mountains'.
   institutionalize new knowledge.
10. Financing for multiple use systems at end user level
    resources from public and from private sectors are used
11. Evidence of superiority of multiple use systems
    proof is more convincing than theory.
    better basis for upscaling.

The Learning Alliance approach for multiple water use systems and services

The first objective of MUS is to build capacity to achieve a 100-fold wider implementation of multiple-use water systems than what the project itself can do (3000 households). With this ambitious aim we hope to contribute to achieving the Millennium Development Goals (World Bank Group, 2002). To have a realistic chance of achieving this, we start with it from the beginning. But how?

Researchers typically come into a community, do their research (participatory or otherwise), produce a report and academic papers, do a 'dissemination workshop' and, urged by the sponsor, move on to the next project. Often, there is no consolidation of lessons learned, no true sharing of results, and no development of local ownership. Uptake and scaling-up is left to ill-defined processes of 'dissemination' and 'advocacy'. In addition, many research projects do not allow for building the capacity of staff working in the relevant institutions such as local government, private sector, NGOs or extension services. This approach clearly does not prepare for major upscaling.

Implementers have also innovated: rope pumps, community gardens, family ponds and community small-dams are well-known innovations coming from the field. Many of such innovations are well integrated into the local water use system, but still have often failed to go to scale. One of the reasons for the failure is that scaling up is not carried out within an institutional, organisational, economic, physical and/or environmental setting that is realistic to that country or region. Examples of practices that are handicaps for uptake include: input subsidy for farmers, paying for people's participation, the use of subsidized facilitators to overcome bottlenecks, creation of parallel structures to bypass 'failing' government, use of highly motivated project teams that cannot be replicated, unrealistic levels of resources for base line studies, demonstrations, vehicles etc.

Connectivity to the local situation, continuity and learning must all be ensured to take innovations to scale. What is needed is 'a group of individuals or organisations with a shared interest in innovation and the scaling-up of innovation', or for short: a Learning Alliance. A Learning Alliance is a structured platform of a range of partners in a particular geographic area with different concerns (those of the various end users) and capabilities (implementation, regulation, policy and legislation, research, learning, documentation and dissemination). It breaks down barriers to sharing of information and creates a means for negotiation, and thus to speed up the process of identification, adaptation, and uptake of new innovation. Working together in implementation and research within an alliance of practitioners, researchers, policy makers and activists will lead to greater impact and more potential to go to scale through development of (i) capacity of Learning Alliance members, (ii) ownership of the concepts and process, and (iii) locally appropriate innovations.

Good results with a Learning Alliance approach for natural resources management by communities have already been registered in ten Asian countries with respect to adoption and impact (Gonsalves and Mendoza, 2003); the term they used to describe the movement is 'learning by doing' and 'learning spiral'. The gradual process builds capacity and implements solutions to jointly felt problems. Visscher *et al.* (2005) report that the TRANSCOL project with a learning alliance approach 'avant la lettre' resulted in significant outscaling (repetitions) and upscaling (adoption by authorities) and appeared sustainable (in as far as this shows in one decade).

Probably the most important element of a successful Learning Alliance is a shared understanding of the problem to be solved and a set of common objectives. All organisations participating in a Learning Alliance should have a vested interested in dealing with a specific issue and in the innovation to overcome it. The first step of establishing a Learning Alliance is therefore scoping and defining the area of intervention, and identifying the stakeholders to be involved in it. The MUS-basin core team will begin the process with a 'vision' in mind. In the early steps of setting up the Learning Alliance they will work in 'advocacy' mode: selling the idea to potential partners. But relatively quickly it must shift into a 'facilitator' mode: helping the new partners to understand, adapt, and own their own vision and objectives that will undoubtedly diverge from the starting point!

Learning Alliances are always context specific. Hard rules about who should be involved and in what manner cannot be formulated. It may depend on the specific topic of work, organisations available and interested, resources available, etc. The key points are to have a common vision as to what the objectives of the alliance are, and to include stakeholders from implementation, policy, regulation, learning and dissemination backgrounds. Deciding who is to be involved in a Learning Alliance is critical to the immediate success of sharing the results of action research, and to the likelihood of successful scaling-up. The choice should be based on a thorough process of stakeholder analysis and a clear view of the role that partners will take in further uptake and scaling-up. Table 3 provides an example of the result of stakeholder mapping; Box 2 shows factors to consider in partner selection.

Box 2. Factors to take into account in the selection of participants for a Learning Alliance:

ongoing work that is relevant to the issue or innovation;
personal interest;
ability to commit and take decisions;
ability to provide resources (financial, human);
potential to take up findings (become a champion);
ability to block or impede the project (local politicians).

Effective communication between members of the Learning Alliance is crucial. Elements include: identifying and understanding different perspectives, constant checking that there is still common understanding, sharing results and experiences horizontally and vertically, shared experimentation and learning within the boundaries of existing or realistically attainable institutions and policies. The methods used for communication draw on approaches from a range of disciplines: adult education and capacity building, action research, process documentation, dissemination and sharing, and process facilitation.

When there are two or even three levels of Learning Alliances involved, information flows in all directions are critical to ensure that ownership of (and responsibility for working with) the findings of pilot activities is achieved. The flow of appropriate information between different levels is enhanced by having an alliance, but it may also need professional facilitation.

Towards a guideline for Learning Alliances

A Learning Alliance follows a structured yet flexible and re-iterative path to progress. Tables 4 and 5 present a draft of the main and generic ‘steps’ needed in the process of establishing and working with a Learning Alliance at the National level and at the District level. They guide our processes of establishing Learning Alliances at different levels, but clearly should not be followed mechanistically. The processes to promote Learning Alliances needs to be dynamic, flexible, and fuzzy (as in ‘fuzzy logic’: i.e. not very precise, responsive to the actual situation, gradually improving). The ‘steps’ in the tables are like markers or waypoints on a journey that may start from several different points and follow several different routes, but in which most of these markers will have to be visited at least once. For example, in the beginning the activities may be initiated at the national level and then go to the local level. But it is equally valid to start at the district or at community level if, for example, an implementing partner is already involved in work there. What is important is that, wherever we start, we end up with a proper alliance that carries within it the necessary elements to allow for fast scaling-up.

Table 3 Example of matrix for mapping stakeholders for a multiple use systems Learning Alliance

Category	Stakeholder	Role in Learning Alliance	Strength	Weakness
Regulation/policy making	Ministry of Water	Review norms and standards	Capacity to scale up	Politicised
	Ministry of agric.	Create enabling policies	Capacity to scale up	Politicised
Innovation	National/local University	Research on new methodology	Strong in content	Often isolated; academic
	Government Research	Research on adoption, impact	Access to sites	Under resourced
Planning	Local government	Adopt MUS approach in planning	Capacity to adopt and support uptake	Politicised, under-staffed
Implementation	Dept. for Domestic Water	Scale up through implementation	Big reach; continuous presence	Politicised, under-staffed
	Private sector actors	Scale up through implementation	Sustainable, flexibility	Unaccountable; profit oriented
	International NGO	Scale up through implementation	Strong capacity	No long term continuity
	Dept. of irrigation	Investments and extension support	Strong extension officers	Sectoral bias; lack of flexibility
Dissemination/advocacy	Association of Municipalities	Mobilise other district councils	Big reach; credibility	Little content expertise
	National resource centre (NGO)	Document and disseminate lessons learned	Strong capacity	Isolated, under-resourced
	Local University	Research and documentation	Formulate messages	Isolated; under-resourced
Service provider	Community based organizations with district council	Manage the m.u.s. after project completion	Local level, relatively well skilled	Lack of empowered communities
	Local private sector	Day to day operations and maintenance; spare parts	Local level; flexible	Lack of skills, profit driven

Table 4 Creating a Learning Alliance (LA) at the National level; the process is iterative particularly in steps 4 and 5

Step	Objective	Activities	Tools	Outputs	Remarks
Step 1: Scoping	Come to an agreement as to the boundaries of the issue	Discussion within partnership	Discussion	Short (1–2 page) description of issue for use with stakeholders in steps 2 and 3	• Discussions at district level normally start from innovative work that people may want to scale up. • At national level: identify the 'innovation' to be introduced.
Step 2: Mapping stakeholders	Know who is somehow engaged with the issue defined earlier	Initial stakeholder mapping exercise (likely to be repeated in next step)	• Functional matrix • RAAKS tools (Rapid Appraisal Agricultural Knowledge Systems)	An initial list of likely stakeholders who may be approached to join the national LA	• Different types of functions are to be represented • Normally, one targets the line ministries and national organisations from whom the district organisations depend.
Step 3: Creating interest in a national LA, and formalizing it.	• Reach agreement on the common objective • Consolidate commitments • Review the m.u.s. cornerstones • Identify roles	• Stakeholder workshop • Institutional SWOT • Meetings with key stakeholders • Make TOR LA explicit	• Strength weakness, opportunities, threat analysis • Sector scan of tools	• Terms of reference for the LA • Determine the degree to which cornerstones are in place and which ones are missing or need work.	• Initial contacts can take place before the workshop. • Good facilitation will be essential, particularly when there are blockages or good opportunities to take innovations forward. • It may be necessary to create a National Steering Committee with its own secretariat.

(*Continued on next page*)

Table 4 *(Continued)*

Step	Objective	Activities	Tools	Outputs	Remarks
Step 4: Planning and design	• Scoping of national process finalized • Agreed structure and scope of learning and implementation process	As above	Project cycle management tools.	Work plan for the LA: clear plans for planning, design of interventions, implementation, monitoring and evaluation	• For each step roles and responsibilities need to be defined between the member organisations. • Some activities, especially around policy development, may not have proper project cycles. Interactions with the LA-workplan are still needed
Step 5a: First year: Identify pilots	• Pilot areas agreed • roles and tasks partners defined	In district level discussions with stakeholders ascertain interest, suitability	• Stakeholder mapping • Resource assessment	Reports on • Institutional 'readiness' • Water resource assessment • experiences elsewhere	• Criteria may include: commitment of organisations in district, presence of representatives of national stakeholders in district, ongoing initiatives, etc. • The criteria for institutional linkages are crucial.
Step 5b: Later on: Carrying out activities in pilots	Get a larger part of the missing m.u.s. cornerstones in place	field work, lobby, information campaign, discussion sessions	as relevant to the cornerstone	Progress reports on improving the cornerstones	The actual activities and results may be quite diverse among sites and countries. Communication vertical and horizontal, as well as between MUS project members in other countries, is crucial.

Table 5 Creating a Learning Alliance (LA) at the District level. May features are similar to those of the National level LA. Particularly steps 4 and 5 are re-iterated after one or two years

Step	Objective	Activities	Tools	Outputs	Remarks
Step 1: Scoping	Agree on the boundaries of the issue	Discussion within partnership	Discussion	Short (1–2 page) description of issue for use with stakeholders in steps 2 and 3	Discussions start from innovative work that people may want to scale up.
Step 2: Mapping stakeholders	Know who is somehow engaged	Initial stakeholder mapping exercise	Functional matrix	An initial list of stakeholders who may be approached to join the district LA	Different types of functions (water users, water providers) are to be represented
Step 3: Set up District LA	• Establish a District LA • Reach agreement on objectives • Create commitment • Set up a forum for implementation	• District level stakeholder workshop • District level institutional SWOT	Workshop		The criteria may include: geographical conditions, ongoing initiatives, presence of stakeholders in the village, etc.
Step 4: Planning and Project cycle at district level	Structure the learning and implementation process	Interaction between the District and the National LA (possibly with pilot villages)	Frameworks for planning, design of interventions, implementation, monitoring and evaluation	• Pilot villages can take part, so that Steps 3 and 4 coincide. • For each step roles and responsibilities need to be defined between the member organisations	Ensure that new planning and implementation approaches tailor as closely as possible with existing ones–and that where changes are necessary these are designed and are acceptable at both district and national level.

(*Continued on next page*)

Table 5 *(Continued)*

Step	Objective	Activities	Tools	Outputs	Remarks
Step 5a: Identify pilot villages	• Identify potential pilot villages/ communities • Have pilots as members of LA			Pilot sites identified with reports on readiness and the five capitals	A 'champion' for m.u.s. in the district or village is very helpful.
Step 5b: Later on: Carrying out activities in pilots	Get a larger part of the missing m.u.s. cornerstones in place	Field work, lobby, information campaigns, discussion sessions	As relevant to the cornerstone	Progress reports on improving the cornerstones	Actual activities and results may be quite diverse among sites and countries. Communication vertical and horizontal is crucial.

In MUS we will further elaborate on these principles and practices to promote Learning Alliances and summarize these in the form of a guideline. What this guideline for upscaling through Learning Alliances may look like by the end of a project? Ideally it will be a sort of 'toolbox', electronic or printed, in which the table provides the framework, and in which the tools, outputs, activities and objectives have been validated and updated. It should be accompanied by case studies from projects (in the different basins, for instance), based on the process documentation; successes, failures, lessons learned etc. An electronic version will allow users to click on, for example, a tool, and then pull up a fact sheet that talks them through how to use that tool, with the fact sheet in turn linked to a case study in which it has been used.

Action research on multiple use systems

The second objective of MUS is to gain new knowledge with respect to multiple use systems and its water services. Research on multiple use systems and services is much needed, since the benefits and cost under various conditions are not fully worked out, and hard scientific proof of the value of the concept is required before donors will be convinced to support major activities in upscaling. Research is also needed to prepare and evaluate guidelines. We envisage that two sets of guidelines are needed: one for implementers of multiple use systems for end users, and one for development of Learning Alliances.

Extensive debates about participatory approaches have shown the importance of involving stakeholders in the analysis of development problems and in the design of possible solutions for communities. This will help to identify the *relevant* aspects of the problems, create *ownership* of these problems and the solution, and build the required *skills and capacities* to tackle similar future problems and manage the solution in a sustainable manner. Research with the stakeholders as active partners is called 'action research'.

Action Research needs also to be designed to reflect the realities of research and support agencies. This implies procedures with of (annual) cycles of problem identification, solution identification, action, reflection, learning lessons, identification/modification of new solutions. Such cycles must also be linked to the planning processes of the partner organisations. The challenge is to realize a workable harmonization without excessive bureaucracy and rigidity. Action Research implies the need to manage the field activities adaptively, in contrast to following strict procedures or logframes, since blueprints to common developmental problems do not exist, and intermediate level actors need to be empowered to be responsive. Hence another challenge for Action Research is to allow for management flexibility without compromising scientific rigor.

The framework of Cornerstones (Figure 2, Box 1) guides Action Research in four ways:

- As a tool for common understanding and vision. It helps to learn together and to recognise the complexity and get a grasp of how to handle it. In particular it leads to a solid research framework that all partners can contribute to without going back to the research leader too often.
- As a frame to design new activities. In setting up new activities, the context can be analysed together with the main stakeholders and the main areas of interventions of the projects can be defined on the basis of the joint analysis.
- As a frame to monitor and evaluate on-going activities in basins in a strategic way. Basin teams can use the frame to reflect on their intervention and analyse the state of the art for each cornerstone. This helps them to reach a common perspective on where they are, what they consider success and what the knowledge and design gaps are in their existing intervention in an iterative way.

- As a knowledge management tool. The lessons and experiences and methodologies/tools used to enhance each of these cornerstones can be collected, synthesized across programmes and put back into the framework. This way, the frame will build up and enhance a rigorous and systematic learning in institutions/networks. Increasing operational knowledge from different actors on how to manage successful m.u.s. can be integrated in the common frame.

The steps in Action Research that follow from the framework are:

- Document the functioning of systems that were designed for multiple use and of systems that were not designed as such (with the eleven m.u.s. cornerstones in mind) and compare them in terms of water use efficiency, contributions to livelihoods, income and gender; for the documentation. MUS has a liaison scientist in each of the basins to collect information with local partners.
- Assemble documents on m.u.s. as well as relevant documents from other sources for internal and external use (MUS, 2005).
- Analyze these documents to determine costs and benefits for both cases (designed as m.u.s. and not-designed as such) to determine impact of effective multiple use systems on the five capitals for development (human, social, financial, physical, natural, see Merrey *et al.*, 2005).
- Draft a guideline for implementation of multiple use systems, and draft a guideline for the creation of learning alliances (Tables 3–5 are preliminary versions).
- Implement at sites in all countries systems that are multiple use by design, usually by upgrading existing systems, and evaluate and improve the draft guidelines; this is particularly the 'action' part in the research and involves much collaboration with NGOs and (other) national partners.
- Document the process of formation of learning alliances and analyze the results from all project sites to test the guidelines and improve them; this involves much 'action' with local partners and end users of water.
- An annual review is made of progress in view of the m.u.s. cornerstones, of understanding of the costs and benefits of the multiple use systems, and the impact they have on livelihoods, income and equity.

While research in MUS will have CPWF-funding until 2008, it is expected that it will require several more years of time to follow developments at pilot sites and to determine the full impact (e.g. Moriarty *et al.*, 2005), and also to improve the contents and functionality of the guidelines (languages, medium, tools, etc). For that, we will seek additional funding, and collaborate with partners among the NGO's and farmer organizations that implement projects that will last 10–15 years.

Discussion

Even though the arguments that integrated multiple use systems are better than un-integrated systems and that learning alliances are a better way forward, proof of these statement cannot yet be provided. Not only because little research has been so far but also because the successful cases have hardly been exposed to sustainability tests with respect to environment and institutions. While time will tell, we do need to improve our insights, research concepts and methods to be able to provide conclusive answers and shorten the time to reach many people in need of better water solutions.

Water users will often not optimize their water use *perse*, or that of other natural resources, but their household income or livelihood. In that optimization, water is only one of the elements and may be even a minor issue (until it runs out). In other words: 'integration' is a process where water and other resources are continuously and interactively optimized by different groups and at different scales. Integrated use of water is an example of fuzziness *par excellence*. But that does not mean that the concept is 'useless' or cannot be made operational. It does imply that in some learning alliances 'water' will not be the centre but has a secondary role.

References

Biswas AK (2004) Integrated water resources management: a re-assessment. Water International 29(2):248–256

Boelee ES, Morardet J, Butterworth J (2004) Report of the MUS basin workplan workshop, Nov. 2004. IWMI, Pretoria, S. Africa, pp 90

CPWF (2005) CGIAR Challenge Program Water for Food. www.waterforfood.org

Engel P (1995) Facilitating innovation: an action-oriented approach and participatory methodology to improve innovative social practice in agriculture. PhD-thesis. Wageningen University, Wageningen, the Netherlands

Faysse N (2004) An assessment of small scale user's inclusion in large scale water users associations of South Africa. Research Report 84, IWMI, Pretoria, pp 36

Gonsalves JF, Mendoza LC (2003) External review of IDRC's community-based natural resource management program initiative in Asia. Final Report, IDRC, Ottawa, Canada, pp 60

Hagmann J (2005) LearningWheel – Creating common frameworks for joint learning and knowledge management: a methodology for conceptualising experiences in multi-stakeholder fora. (forthcoming, for information: jhagmann@aol.com)

Lundi M (2002) Learning Alliances with development partners. CIAT, 2002. Consulted on internet.

Merrey DJ, Drechsel P, Penning de Vries FWT, Sally H (2005) Integrating 'livelihoods' into integrated water resources management: taking the integration paradigm to its logical next step for developing countries. Regional Environmental Change (on line journal, DOI10/1007/s10013-004-0088-5)

Moriarty PB, Butterworth JA, van Koppen B (eds) (2004) Beyond domestic. case studies on poverty and productive uses of water at the household level. IRC Technical Papers Series 41. Delft: IRC, NRI, and IWMI pp 350

Moriarty PB, Fonseca C, Smits S, Schouten T (2005) Learning Alliances for scaling up innovative approaches in the water and sanitation sector. In: Smits S, Fonseca C, Pels J (eds) Proceedings Symposium on Learning Alliances for scaling up innovative approaches in the water and sanitation sector. Delft, the Netherlands, June 2005, www.irc.nl/page/16676

MUS (2005) A literature base on publications about multiple use systems. http://www.iwmi.cgiar.org/ multipleuses/files/multiuseqry.htm

Penning de Vries, FWT, Niamskul C, Bechstedt HD, Maglinao A, Sajjapongse A (2000) Participatory approaches for sustainable use of sloping land in Asia, pp 203–215. In: Roetter R *et al.* (eds) Systems research for optimizing future land use. IRRI, Los Banos, Philippines pp 266

Ruaysoongnern S, Penning de Vries FWT (2005) Learning Alliances development for scaling up of multipurpose farm ponds in a semi-arid region of the Mekong basin. In: Smits S, Fonseca C, Pels J (eds) Proceedings of the symposium on Learning Alliances for scaling up innovative approaches in the water and sanitation sector. Delft, the Netherlands, June 2005, www.irc.nl/page/16676

Van der Hoek W, Konradsen F, Ensink JHJ, Mudasser M, Jensen PK (2001) Irrigation water as a source of drinking water: is safe use possible? Tropical Medicine and International Health 6:46–55

Van Koppen B (2001) Gender in integrated water management: an analysis of variation. Natural Resources Forum 25 (2001) 299–312. Elsevier Science Ltd, United Nations

Van Koppen B *et al.* (2005) Multiple use systems: an assessment. IWMI, Pretoria, S Africa. (in prep)

Visscher JT, Garcia M, Roling N (2005) Multi-stakeholder learning in Colombia. In: Smits S, Fonseca C, Pels J (eds) Proceedings Symposium on Learning Alliances for scaling up innovative approaches in the water and sanitation sector. Delft, the Netherlands, June 2005. www.irc.nl/page/16676

World Bank Group (2002) Millennium Development Goals. http://www.developmentgoals.org/

Yoder R (1983) Non-agricultural uses of irrigation systems: past experience and implications for planning and design. ODI Network paper 7, Overseas Development Institute, London

Water Resour Manage (2007) 21:97–110
DOI 10.1007/s11269-006-9043-1

ORIGINAL ARTICLE

Possibilities and problems with the use of models as a communication tool in water resource management

Johanna Alkan Olsson · Lotta Andersson

Received: 20 April 2006

Abstract Politicians and policy-makers, as well as modellers, often nurses an expectation that model derived results is an objective source of information that can be used to support decisions. However, several prerequisites have to be dealt with in order to ensure that models can be used as legitimate and efficient tools in water resource management. Based on empirical material from recent studies on the use of models in stakeholder dialogues, mainly focusing on catchment nutrient transport, two central problems are identified: (a) Models are laden with choices and thus depend on assumptions and priorities of modellers. (b) There are several factors that influence ability and willingness of stakeholders (as information recovers) to criticize or accept results of the modelling exercise. Recognized factors likely to influence stakeholders' acceptance of model derived results include issues at stake, stakeholders' ability to criticize model derived information, and their trust in the institutions that have developed or applied the used models. Identified prerequisites for successful use of models in integrated water resource management include: consideration of user relevance, awareness of and preparedness to handle constraints linked to communication of model-based results, transparency of used models and data and of involved uncertainties, mutual respect between experts and stakeholders and between involved stakeholder groups, a robust institutional network, and sufficient time for dialogues. Development and use of strategies for participatory modelling, based on a continuous dialogue between experts and stakeholders is recommended as a way to facilitate that the prerequisites for a successful use of models in water resource management are fulfilled.

Keywords Participatory modelling · Catchment models · Scenarios · Water management · Communication · Understanding of science

J. A. Olsson
Lund University Centre for Sustainability Studies, (LUCSUS), Box 170, S-221 00, Lund, Sweden

L. Andersson
Swedish Meteorological and Hydrological Institute (SMHI), S-601 76 Norrköping, Sweden

Introduction

With the help of models it is possible to quantify how different sources have an impact on water availability and pollution levels at specified sites within the landscape. Models can also be used to predict the impact of various scenarios. Such results can either serve as a basis for discussions about different strategies to solve an environmental problem (e.g., eutrophication), or as a basis for discussions of how to adapt to future changes that a local community has little power to influence (e.g., less water availability due to a changing climate).

Model based information is used more and more frequently as a decision support in different types of management situations (Shackley, 1997; Rosa, 1999; Sarewitz *et al.*, 2000; Lindley, 2001), including the water resource management area (e.g., Andersson and Moody, 2004; Arheimer *et al.*, 2004). One reason is the increasing demand to visualise the responsibility of the individual polluter (polluter's pays principle). Another rationale is the increasing demand to involve stakeholders in the decision process, which has call forth the development of pedagogical tools to facilitate dialogues concerning today's environmental conditions and the probable impact of various remedies (Beierle, 1998; Johnson *et al.*, 2001; Blomqvist, 2004; Barreteau *et al.*, 2001, 2004; Cash *et al.*, 2003).

The power of models consists of their capacity to describe complex, interrelated, relationships and to handle large quantities of data (Sarewitz *et al.*, 2000; Alkan Olsson, 2003; Kasemir *et al.*, 2003a). However, the availability of data is always incomplete and relevant process descriptions are simplified and sometimes missing (Edwards, 1999). To make a model useful in water planning it is therefore crucial that it can be applied despite limited availability of data, at the same time as it produces results that are acceptable in relation to observed conditions (Andersson, 2004).

A model produces a manageable description of a multitude of complex processes in the environment including the impact of human actions. Using models has been proven successful in increasing awareness among lay people and decision-makers, regarding causes to environmental problems and the potential impact of different remedies (Dürrenberger *et al.*, 1997; Darier *et al.*, 1999b; Dahinden *et al.*, 2000). A model-based dialogue has also been shown to facilitate the creation of a common view on the problems at stake as well as a platform to test the possible environmental impact of various suggested remedies (Schulze *et al.*, 2004). A common view of the causes of the prevailing environmental problems and the possible impact of various remedies is an important starting point for a changed behaviour and action (Jaeger *et al.*, 1999). Model-based discussions may also lead to an increased understanding among stakeholders of why and how different groups act or think as they do. Model based dialogues may also increase the possibility for public participation in the decision process as the model or scenario serves as a pedagogical tool (Ravetz, 2003). At the local scale where the credibility of national environmental goals has to be anchored, participation has proven especially important (Andersson, 2004; Wittgren *et al.*, 2004).

Use of model-derived information in water resource management is connected to two scientific dilemmas (Schackley and Dariér, 1998; Pahl-Wostl *et al.*, 2000). Firstly, the fact that the results are produced with the aim to serve as a basis for decision-making (taking stands) contradicts the general view of science as being objective (Irwin, 1995; Darier *et al.*, 1999a). Secondly when models are used to predict impacts of various scenarios, the results can be validated until far after decisions based on them have been carried out.

In specific model applications it is often challenged if the used model is based on up-to-date knowledge, and if it provides realistic output when tested against available time series of monitored data. Sometimes the question of the expert's pedagogical skills to communicate their results is also touched upon. But is this enough?

Aim

The increased use of computer-based catchment models as a tool in water resource management makes it imperative to examine the conditions for their practical application. What prerequisites are there to ensure that models can be used as valuable tools in water resource management? When could models be useful and what can they not help us with?

This paper is based on experiences from case studies with participation by the authors, complemented by a synthesis of others experiences of the use of computer-based models as a communication tool and basis for decision in environmental management. Factors influencing potential receiver's perceptions and acceptance of model based information are discussed, followed by an examination of some characteristics of model-based information that may give rise to problems and advice on how these may be overcome.

Starting point: Models as tools for better decision-making

Politicians and policy-makers, as well as modellers, often nurses an expectation that model derived results is an objective source of information that can be used to legitimise decisions (Jasanoff and Wynne, 1998; Oreskes, 2000; Pahl-Wostl *et al.*, Sarewitz, 2000). This expectation is based on the perception that access to objective, scientifically-based knowledge influence people's willingness to change their behaviour or to carry out remedies (Jäger, 1998). Among decision-makers it is therefore often anticipated that access to model-generated information based on expert knowledge, can assist in solving conflicts and political controversies, which have their origin in subjective unscientific judgements (Bailey, 1997; Yearly, 1999).

Yet this expectation is problematic for two kinds of reasons: (a) Models are laden with choices and thus depend on assumptions and priorities of modellers. (b) There are several factors that influence ability and willingness of stakeholders (as information recovers) to criticize or accept results of the modelling exercise.

Choices likely to influence model outputs

We will never have access to landscape information with a spatial solution needed for a "perfect" description of reality. Availability of data is often the main determinant for model complexity and it determines which temporal and spatial scale the model can be set up for. Some models have their main deficiencies in their description of processes whereas others are constrained by the availability of databases with relevant information in relevant temporal and spatial scales. These differences are partly due to that different models are developed for different purposes. Several models describing water and nutrient transport were originally developed from purely scientific criteria. Such a purpose may for example have been to verify a theory against observed data. The producers of these types of models prioritise that the model gives a physically correct description of relevant processes. Other models have been developed for operational use, e.g., to fill the demands from the hydropower sector. Modellers that develop models with the aim to be used operationally, are usually more concerned with that the models can produce results that are in accordance with measured time series, than with the fact that they do it for the right reason. The choice of model is thus not objective, but based on the individual modellers preference and earlier experience of various types of models, weighted against what is perceived as the needs of the user and what is judged to be realistic with regard to data availability and costs involved. The choice of model or modeller can therefore influence the outcome and lead to different recommendations.

With regard to uncertainty in model-derived results, some is due to limited information about processes in nature. A substantial part of the uncertainty is, however, linked to the impossibilities to correctly describe the heterogeneity of the landscape mosaic out from available databases. Even if a reasonable good fit is found between, e.g., monitored and modelled water flow or concentrations of nutrient in a water course, this is no proof for that the model adequately have provided the "right answer" for the "right reason". In some cases, different parameterisations of a model can provide different explanations of the integrated flow in the river that have similar fit to an available monitored time series. Consequently, depending on the selection of model parameterisation, we might get a different source appointment, or different distributions of water that follows different flow pathways, which means that both the contribution from different polluters can vary, and the suggested measures for an individual polluter can vary. If relating this to the political ambition to make the polluter responsible for the pollution (i.e. polluter's pays principle) the use of model derived results is thus not always straightforward. Coping with uncertainty become especially problematic when it not is clear, which effect these uncertainties may have on the involved stakeholders, and if they will influence different stakeholder groups in a different way (Irwin *et al.*, 1996).

Factors likely to influence stakeholders' acceptance of model derived results

The perception that models are efficient tools for optimisation of strategies for, for example in water-management, is not unproblematic (Wynne, 1992; Irwin, 1996; Yearley, 1999a). If model-derived results are accepted and considered in following actions depends not only on the scientific quality of the results or on the expert's pedagogical skills. It also depends on how the receiver interprets it, and in which wider context the information is placed (Douglas, 1982; Latour, 1982; Yearley, 1999b). Public perceptions of model-derived, well as other types of scientific information is influenced by a mix of political, economical and social factors (Shackley, 1997; Dahinden, 2000; Dahinden, 2003; Kasemir, 2003a) (Figure 1).

Own interests influenced by model output and/or the issues at stake

Stakeholders understanding and acceptance of information depends on how personally influenced he ore she is by the information and if, from his or her perspective it is "good" or "bad" news (Yearly, 1999b; Alkan Olsson and Berg, 2005). Stakeholders that are pointed out as large contributors of pollution are likely to be more sceptical and unwilling to accept model results as a basis for decision-support. This distrust will most likely increase if suggested measures are going to cost them money or time.

Another factor to consider is that a self-employed person with constrained economy is more likely to stress economical considerations more than environmental considerations. A dialogue not taking into consideration economical issues would probably in that case not be constructive (Alkan Olsson and Berg, 2005).

For most local stakeholders fairness and justice within the local community is an important factor to take into consideration. It is generally perceived that everybody has to contribute to solve the problem and that nobody should be able to free ride. In this context

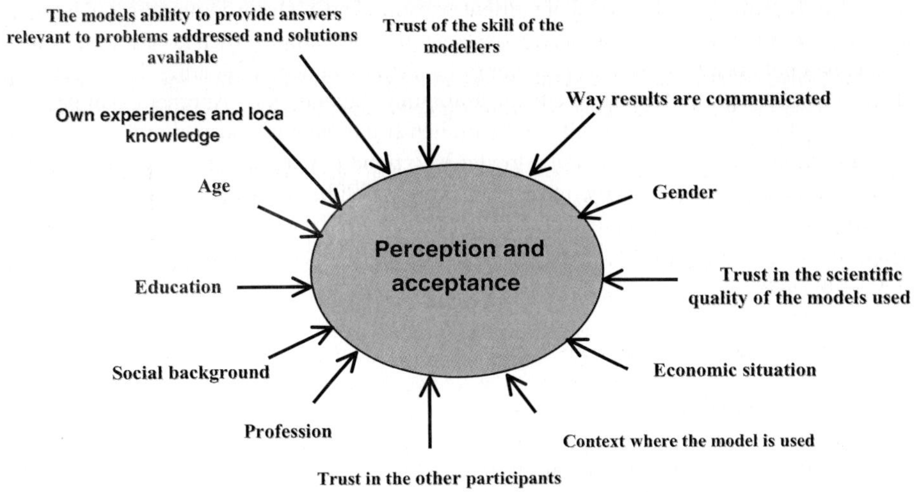

Fig. 1 Examples of factor influencing stakeholder's perception and acceptance of model-derived results

economic efficiency is frequently seen as less important than the assurance of everybody's equal contribution to the solution (Alkan Olsson and Berg, 2005). If the suggested actions based on model-generated information are seen as unfair, they are unlikely to be accepted.

Social-educational and economic background

Social, economical as well as educational background influences how used a stakeholder is to question model-based information (Schulze, 2001; Gough *et al.*, 2003). Knowledge about local conditions is another important factor. A farmer, in addition to having issues at stake, may be more critical towards model-output compared to person living in a town because the farmer has local knowledge of how nutrient leaches from his or her fields. Local knowledge may consequently on some occasions serve as a basis for criticism of model output. Criticism from stakeholders can, in addition to attention of bad compliance with observations of processes in the landscape (e.g., streambank erosion or surface runoff), consists of comments that already taken measures not have been considered on the models or neglect of specific point sources (Andersson, 2004; Alkan Olsson and Berg, 2005).

Trust in the institutions and ways of communicating

Acceptance of model-derived results does also depend on stakeholder's trust in the institutions that have developed the models, and the authorities that apply them for water resource management. This trust has been formed over time. Earlier positive or negative experiences of involved authorities or water experts, or by these groups in general will influence a later "communication situation". Several participants in the so-called Rönneå Study (Jöborn and Danielsson, 2005) had through earlier experiences with experts gained the view that "*they are doing their thing, they come here measure and then they leave again*" (Alkan Olsson and Berg, 2005). The fact that these measurements for example later could be used in the local set-up of and verification of a model of nutrient transport had not been communicated between the experts and the local stakeholders. Consequently, the earlier presence of researchers in the area had not been perceived as relevant to local stakeholders, indicating that experts did not leave something of interest to the local community behind them. Another example is the acceptance of a model and its result can be limited if the there is a suspicion from the local community that the purpose of using a model is to point out certain groups of stakeholders and increase their burden of taken remedies (Yearly, 1999b; Yearly, 2000). Earlier negative experiences may hence lock endeavours to create a model-based dialogue that truly aims to respect and consider all involved stakeholders opinions in the local implementation of water resource management.

Participatory modelling

In order to ensure that model-derived information is perceived as relevant and trustworthy, it is crucial to create a continuous dialogue between experts and stakeholders. It is hypothesised that such dialogue both will improve the model results and the possibility that they will be used in water resource management and that they will change behaviour in a way that contributes to improved environmental conditions. Various strategies to obtain this goal have been presented (Kasemir *et al.*, 2003b). We have chosen to use the term "participatory modelling" for this strategy due to its focus on the actual participation of local stakeholders in the modelling process. It is in line with suggestions by Cash *et al.* (2003), who emphasize that efforts to mobilize science and technology for sustainability are more likely to be effective when they manage boundaries between knowledge and action in ways that simultaneously enhance the salience, credibility, and legitimacy of the information, they produce.

One example of participatory modeling is when, in support of a local decentralization policy, multi-agent system (MAS) models, linked to geographic information systems (GIS) and other cartographic tools have been used together with role plays (Barretau *et al.*, 2001). The aim of these projects has been to help local rural authorities and the people under their jurisdiction to improve their empowerment on planning decisions about sustainable land use management (agriculture, animal production, the environment, etc). This combination of MAS and role playing games has proved to be an effective discussion support tool (Barreteau *et al.*, 2004).

Another example of participatory modelling is a recently started project "DEMO", where such a process may look is given in Figure 2. The principal basis of "participatory modelling" is that the users of the model-generated information, e.g., decision-makers and stakeholders, in a dialogue with modellers provides input to the set-up of models, based on legitimisation of used data-bases, own observations and other local knowledge. They are also involved in a continuous dialogue with regard to selection of e.g. remedies or mitigation strategies that are adopted into various modelled scenarios. Involved stakeholders could for example include representatives for farmers, forestry, industries, waste water treatment, and from summer houses and others that not are connected to waste water treatment plants.

The following section will give some advises of what is important to take into consideration in a stakeholder based dialogue based on applications of models. The examples are mainly taken from the Swedish Water Management Research Programme (VASTRA), but do also include examples from studies within the Hydrology for the Environment, Life and Policy

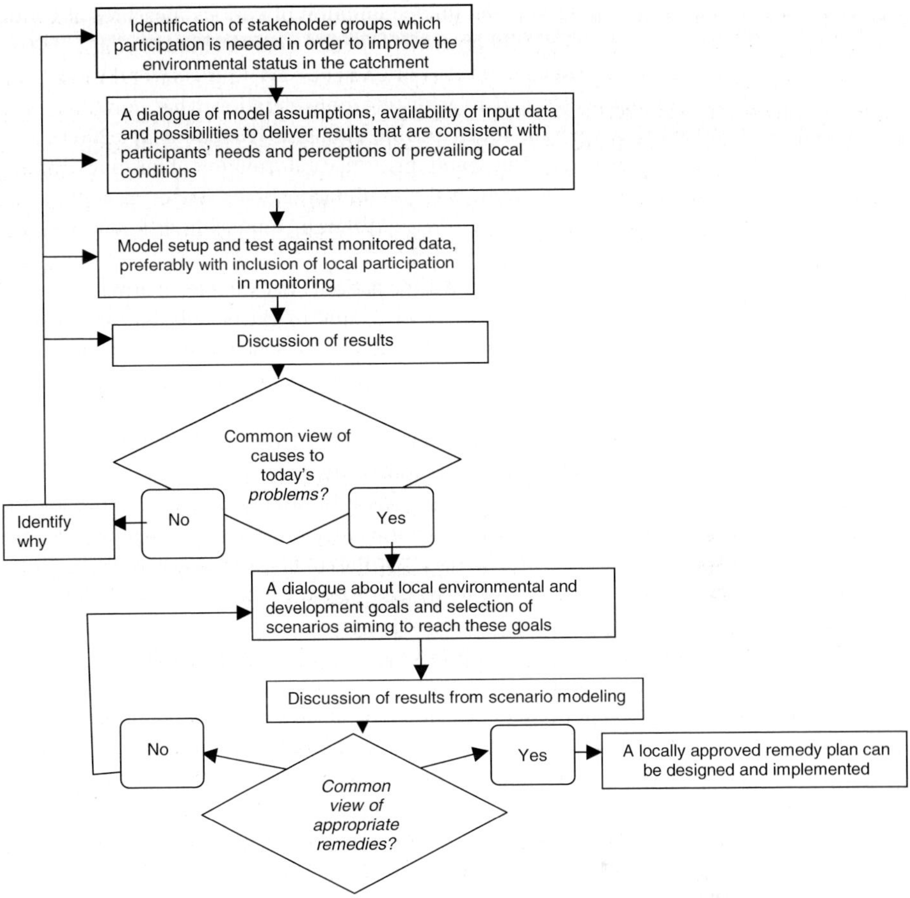

Fig. 2 Example of steps to be taken in a participatory modelling process

(HELP) programme, as well as other research projects studying the communication of model-based information.

Conditions for a successful use of models as a communication tool in environmental management

User relevance and user friendliness

Several studies have indicated that model-generated information often fails to contribute to answering questions that are central for local stakeholders.

The major reason is that experts are used to present their results for a scientific audience. For such purposes, it is often satisfactory to only show primary data, e.g., in the form of matrixes showing time series of water and nutrients flow for a soil profile, a field or a sub-catchment. Policy makers have other need of information and primary data has to be further analysed and often combined with other sources of information. Moreover policymakers are not used to handle complex tables and to be useful data should preferably be presented graphically, e.g. as maps indicating how the implementation of various measures at various sites in the catchment scenarios may fulfil an environmental goal (Dahinden *et al.*, 2000).

To create policies for water management it is often necessary to use several models that designed for different spatial scales and to take into consideration what combination of models that is most useful or relevant for a specific situation (Brandt *et al.*, 2004). With regard to eutrophication, a farmer will probably prioritise information about how different measures may influence leakage at the farm level. Simultaneously the farmer, as well as other stakeholders, needs to get an overall picture of how different sources in different parts of a catchment contribute to the overall pollution level.

Moreover stakeholders often need integrated information. For example, in provided model-derived reports, estimations of costs for taking measures might be missing or generated information is only is linked to one environmental parameter, whereas decisions need to be based on the impact of a remedy on several environmental parameters (Laird, 1992; Pahl-Wostl, 2000; Alkan Olsson and Berg, 2005).

Stakeholders are also interested in further derivations of the model output, e.g., how changed nutrient levels will influence plant and animal life.

Consequently, to create user relevant and user friendly information it is important that modellers and other experts within several scientific disciplines, including hydrologists, oceanographers, soil scientists and limnologists, environmental economists, as well as social scientists cooperate in between them selves as well as with the concerned stakeholders.

Awareness of and preparedness to handle constraints of using models in a dialogue

Decision-makers using models in stakeholder dialogues have to be aware that the dialogue can become an event per se that may stimulate several different and sometimes unexpected reactions (Yearley, 1999b). In some cases it may stimulate local stakeholders to act and solve the discussed problem. In other cases there might be a contrary reaction, as a result of that certain stakeholder groups feel that they are singled out as responsible for the observed problems, or because presented information is contradictory to their belief or their knowledge about local conditions (Kasemir *et al.*, 2003). It is therefore crucial that the facilitators of stakeholder dialogues are well aware of these possible problems and are responsive and flexible in handling situations that may arise.

When choosing between possible measures and comparing their effects it is important to be able to highlight the limitations of possible positive action. It is important to clarify the limits of their room of action. Decisions relevant to minimising eutrophication in Sweden are for example highly dependent on any developments in the Common Agricultural Policy (CAP) as well as of national Swedish agri-environmental politics.

Since the choice of model and of modellers might have an impact on the derived results and recommendations, the aim of using model generated information, as a tool in stakeholder dialogues is not primarily to find consensus about an optimised "best solution". The role of a model should rather be seen as a way to find a platform for a discussion where the relative importance of various sources and remedies can be demonstrated, and where different point of views of possible solutions may be considered and incorporated in new model runs. It is therefore important that stakeholders not are presented with ready-made information about sources and the impact of various remedies, but that the dialogue is able to, take into consideration stockholder's knowledge about for example local conditions (Andersson, 2004).

Another problem with model generated information forwarded by local stakeholders is that information often is limited to single environmental parameters (Alkan Olsson and

Berg, 2005). It is therefore not recognised that a solution to one problem may have a negative impact on another one. For example a decision of when to spread manure might be optimal from an eutrophication perspective, whereas it may have a negative impact on the loss of ammonia to the atmosphere which is negative from climatic change perspective. Since rational decision-making needs to consider more than one environmental problem at the time, it is therefore preferable not to limit a dialogue to only one environmental issue.

Consequently it is important to be aware of the constraints in using models as a tool in environmental management and never forget that the communication process is an important goal in itself.

Transparency as a quality of a dialogue

Several studies indicate that it is important that models are transparent (Nilsson, 1998; Yearley, 1999b). Used generalisations and assumptions have to be declared and openly explained if stakeholders feel it is required. It is also important that experts explain how different parameters and variables have been selected, in order to enable stakeholders to assess if they are representative for local conditions (Andersson, 2004; Alkan Olsson and Berg, 2005). The same type of transparency should exist in relation to assumptions made as a basis for scenarios. One way to ensure this transparency is to assure the presence of experts that are able to answer different types of technical questions when the model is used in a stakeholder dialogue (Guimarães Pereira *et al.*, 1999; Palerm, 2000).

Studies have showed that to increase the credibility it is important to show that the models have been verified against monitored data (Yearley, 1999b; Andersson, 2004; Alkan Olsson and Berg, 2005). This issue is especially complex when it comes to modelling of measures. Initially users of model-generated information might accept that there is a delay between taken measures and measurable effect in the watercourse. However, when measures have been taken local stakeholders will demand some kind of proof that measures will improve environmental conditions, and get some indication of expected time lag between action and response (Dahinden, 2003; Alkan Olsson and Berg, 2005). If no field studies of proven impacts can be presented, it is not probable that model-based results will increase the motivation to carry out measures.

The value of scenario modelling lays mainly in the possibility to compare the effect of different scenarios, not in providing exact quantifications. In several cases, it has been indicated that presentations of uncertainties have given rise to more confusion than clarification (Double, 1995; Kasemir, 2000; Andersson, 2004). However, it is the responsibility of the experts not to present model-generated information as exact but rather as the average from an interval wherein a plausible truth may exist. In this context, it is often more important to indicate which uncertainties that exist than to provide quantifications (Andersson, 2004). Presentations should thus not hide important uncertainties neither include very detailed quantifications.

Farmers are an example of a stakeholder group that in their professional life are used to manage uncertainties. Consequently, this stakeholder group is well prepared to handle and understand the presentation of uncertainties, especially if related to the climatic and landscape heterogeneity within their own surroundings (Alkan Olsson and Berg, 2005). Hence, forms of presentation of uncertainty need to be linked to for whom they are to be presented.

It is also important that the purpose of including models in a specific water resource management application is clearly stated and then let this purpose guide the work (Swollow *et al.*, 2001). If the purpose is to provide a decision support for the implementation of expensive measures, such as investments in constructions of wetlands or in connecting urban households to waste water treatment plants, assessments of costs and uncertainties in impact estimates will be central. If the purpose is to serve as an information base for local stakeholders about their environmental condition or to initiate a dialogue between local stakeholders and decision-makers, it is important to ensure that information and assumptions used for scenarios has gained local approval, which, e.g., can be obtained by involving local stakeholders in the collection and certification of used databases.

Mutual respect as a quality of a dialogue

Development of environmental models is based on advanced expertise and usually such expertise, combined with access to powerful computers and large digital databases is needed to set up the models, and sometimes also to run them (Edwards, 1999; Lam and Swayne, 2001). On the other hand, local knowledge might be important both for parameterisation and validation of a model (Andersson, 2004). Consequently, a mutual acknowledgement and definition of the roles of experts and users of model results in various phases of the process is needed in order to ensure a successful use of a model in water resource management. Without a dialogue between modellers and user's of model-derived results, there is a risk that certain groups uncritically may accept model-generated information where as others may dismiss models as something theoretical having little to do with reality (Bailey *et al.*, 1999; Chess and Purchell, 1999; Jaeger *et al.*, 1999).

Even when simpler, user-friendly models are used in stakeholder dialogues, the aid of experts is required in order to ensure a successful outcome of the dialogue. These models are often including databases that are constructed from results produced by more complex models for specified combinations of, e.g., soils, land-use and land-management. Expert knowledge is needed to clarify used assumptions, limitations and uncertainties involved, and to ensure that the models are adjusted for local conditions. Thereafter, lay people themselves, may test the impact of various scenarios, and then use this information in decisions related to and where various remedies should be implemented. (Collentine *et al.*, 2005)

When using model-derived results in communication between various stakeholder groups it is always important that mutual respect is created between those involved, for example it has to be avoided that some stakeholders perceive themselves as winners and others as losers. One proven way to create a good dialogue of mutual respect is to let moderators lead the dialogue (Mc Granahan and Gerger, 1999; Andersson, 2004; Jonsson, 2004). The moderator may also play a role as an independent judge between local authorities and local stakeholders. It is important that the moderator or other water experts participating in the dialogues understand that their role is to provide information, not educate local stakeholders to take the "right" decisions. The role of the experts is to assist the dialogue not to dominate it. It is important that the experts or involved decision-makers do not feel threatened if models or model-based results are criticised. One strategy to handle this problem is to see stakeholder comments as a resource to improvements of the model-generated result, with regard to their representatively for local conditions, as well as to improve presentation techniques.

Fears that there may be hidden motives behind the choice of a specific model, data and which scenarios that should be produced can only be avoided by a process based on mutual respect.

It is important that the moderators and experts recognise that dialogue participants have different perspectives, rhythms, social, cultural and economical resources and priorities. Where ever in the world a stakeholder dialogue is taking place it is crucial to give all people having a stake in the problem a possibility to participate in the process. "Silent" groups are often poor people with low education and it is frequently these groups that are suffering the most form environmental problems (for example eroded soil and lack of water) (Schultze *et al.*, 2004). In relation to eutrophication we know that the farming sector is responsible for a large part of the pollution. To create a constructive dialogue about the problem it is important that farmers are not pointed out as the black sheep. Even if it is the land managed by farmers that is leaching, we are all consuming the food produced. All involved has to show respect for the limits of what different groups can do to solve the problem.

Robust institutional network as a prerequisite of dialogue

It is important that the dialogue is made in a context, which is legitimate to all involved. The dialogue must guarantee transparency and be open to evaluation. Several studies have indicated that the existence of well functioning political institutions, which not are hampered by a misfit between hydroecological boundaries and administrative boundaries, and where different groups of actors (e.g., farmers, industries, and environmental NGO:s) have a possibility to influence decisions, is crucial for a successful environmental management (Wynn, 1984; Galaz, 2005). A robust institutional network is crucial in order to create a legitimate dialogue leading to action and changed environmental behaviour.

Time as a prerequisite of a dialogue

Time and effort put into the dialogue has been shown to be central for how local stakeholders understand and accept information (Andersson, 2004; Alkan Olsson and Berg, 2005). If stakeholders not are given sufficient time to understand and accept the produced information there is little use in employing this type of rather costly information. Repeated meetings including time for revisions of the model-generated information based on stakeholder discussions are important (Dahinden *et al.*, 2000). Environmental problems are complex and time for discussion is needed to avoid misunderstandings and allow exchange of views. The need of time is especially pronounced if the aim is to create action plans that are acceptable for all involved. If the need for time not is recognised, discussion may instead lead to enhanced conflicts and less probability to find acceptable solutions than was the case before the meeting(s). Even if the use of model-generated information in stakeholder dialogues not always lead to immediate actions, several studies have indicated that they might facilitate local cooperation in water resource management at a later stage. This is due to that discussions between different stakeholders, based on model-generated information, have increased the commitment to the problems at stake and increased understanding of how other stakeholder groups act and think which improves the possibilities for future cooperation (Alkan Olsson and Berg, 2005).

Conclusion

We are living at a dynamic planet with dynamic ecosystems and political processes. From a dialogue perspective, models serve as a tool, which is able to generate a systematisation of how different sources contribution to the overall pollution as well as the predicted impact of various scenarios, including the impact of suggested remedies. However, these tools are only useful if they are understood accepted and perceived as legitimate by those concerned.

One way to ensure legitimacy is to involve stakeholders at an early stage in both the modelling as well as in the decision process. Stakeholder participation could involve consultation about databases, assessment whether the produced model-generated information is in accordance with local knowledge and that the information is relevant for the problem at stake and discussion on which scenarios that are most likely to fulfil the goals defined for the environment. To ensure a legitimisation of the use of models in stakeholder dialogues, it is important to find a methodology for this process. Two important features of this methodology are user friendliness and the recognition that the communication process is an important goal in itself.

To ensure that dialogues are followed by actions, it is crucial to create a balance between model-generated information and other information and considerations. Models will never give an answer by themselves of the "best solution" of an environmental problem; only provide input to a decision in the form of indications of which sources that are important or the probable magnitude of impacts of a suggested measure. However, combined with other types of information, model-based information has a big potential to facilitate stakeholder dialogues and decision processes related to complex environmental problems.

Acknowledgements The authors acknowledge funding support from MISTRA through the VASTRA programme, from Formas through the DEMO (Participatory Modelling) project, and from the Swedish Meteorological and Hydrological Institute (SMHI). We also wish to thank C. Kotz and one anonymous referee for useful comments.

References

Alkan Olsson J, Berg K (2005) Local stakeholders' acceptance of model-generated data used as a communication tool in water management: the Rönneå study. Ambio 34:507–512

Alkan Olsson J (2003) Setting limits in nature, the case of the critical load concept. Department of Thematic studies, Linköping University, Linköping

Andersson L (2004) Riverine nutrient models in stakeholder dialogues. Int J Water Resour Dev 20:399–425

Andersson L, Moody D (2004) Foreword. Int J Water Resour Dev 20:267–274

Arheimer B, Torstensson G, Wittgren H (2004) Landscape planning to reduce coastal eutrophication. Agricultural practices and constructed wetlands. Landscape Urban Plann 67:205–215

Bailey P (1997) Integrated environmental assessment: a new methodology for environmental policy? Environ Impact Assess Rev 17:221–226

Bailey P, Yearley S, Forrester J (1999) Involving the public in local air pollution assessment: a citizen participation case study. Int J Environ Pollut 11:290–303

Barreteau O, Bousquet F, Attonaty JM (2001) Role-playing games for opening the black box of multi-agent systems: method and lessons of its application to Senegal River Valley irrigated systems. J Artificial Societies and Social Simulation 4(2)

Barreteau O, Bousquet F, Millier C, Weber J (2004) Suitability of mulit-agent simulations to study irrigated system viability: application to case studies in the Senegal River Valley. Agricult Syst 80:255–275

Beierle T (1998) Public participation in environmental decisions an evaluation framework using social goals. Resources for the Future, Discussion Paper 99–06, Washington, DC

Blomqvist A (2004) How can stakeholders participation improve European watershed management? the water framework directive, watercourse groups and Swedish contributions to Baltic Sea Eutrophication. Water Policy 6:39–52

Brandt M, Grahn G, Årnfelt E, Bäckman N (2004) Anpassning av TRK-systemet från nationell till regional nivå samt scenarioberäkningar för kväve – Tester för Motala Ström. Swedish Meteorological and Hydrological Institute, SMHI Report 94, Norrköping (in Swedish)

Cash D, Clark WC, Alcock F, Dickson NM, Noelle E, Guston DH, Jäger J, Mitchell RB (2003) Knowledge systems for sustainable development. Proc Natl Acad Sci USA 100:8086–8091

Chess C, Purcell K (1999) Public Participation and the environment: do we know what works? Environ Sci Technol 33:2685–2692

Dahinden U, Querol C, Jäger J, Nilsson M (2003) Citizen interaction with computer models. In: Kasemir B, Jäger J, Jaeger CC, Gardner MT (eds) Public Participation in Sustainability Science A Handbook. Cambridge University Press, Cambridge

Dahinden U, Querol C, Jäger J, Nilsson M (2000) Exploring the use of computer models in participatory integrated assessment- experiences and recommendations for further steps. Integr Assess 1:253–266

Dariér E, Gough C, De Marchi B, Funtowicz S, Grove-White R, Kitchener D, Pereira A, Wynne B (1999) Between democracy and expertise? citizens' participation and environmental Integrated Assessment in Venice (Italy) and St. Helens (UK). J Environ Policy Plann 1:103–121

Darier E, Shackely S, Wynne B (1999) Towards a "Folk integrated assessment" of climate change? Int J Environ Pollut 11:351–372

Double J (1995) Public opinion about issues characterized by technological complexity and scientific uncertainty. Public Understand Sci 4:95–118

Douglas M (ed) (1982) Essays in the Sociology of perception. Routledge and Kegan Paul, London

Dürrenberger G, Behringer J, Dahinden U, Gerger Å, Kasemir B, Querol C, Schüle R, Tàbara D, Toth F, van Asselt M, Vassilarou D, Willi N, Jaeger C (1997) Focus Groups in Integrated Asessment: A Manual for a Participatory tool. Ulysses WP-97-2. Centre for Interdisciplinary Studies in Technology, Darmstadt University of Technology, Darmstadt

Edwards P (1999) Global climate science, uncertainty and politics: data-laden models, model-filtered data. Sci Cult 8:437–472

Galaz V (2005) Social- ecological resilience and social conflict: institutions and strategic adaptation in Swedish water management. Ambio 7:567–572

Gough C, Darier É, De Marchi B, Funtowicz S, Grove-White R, Guimarães Pereira Ã, Schackley S, Wynne B (2003) Contexts of citizen participation. In: Kasemir B, Jäger J, Jaeger CC, Gardner MT (eds) Public Participation in Sustainability Science A Handbook. Cambridge University Press, Cambridge

Guimarães Pereira Ã, Gough C, Marchi B (1999) Computers citizens and climate change – the art of communicating technical issues. Int J Environ Pollut 11:266–289

Irwin A (1995) Citizens science: A study of people, expertise and sustainable development. Routledge, London

Irwin A, Dale A, Smith D (1996) Science and Hell's Kitchen: the local understanding of Hazard Issues. In: Irwin A, Wynne B (eds) Misunderstanding Science? The public reconstruction of science and technology. Cambridge University Press, Cambridge, pp 47–64

Irwin A, Wynn B (1996) Misunderstanding Science? The public reconstruction of science and technology. Cambridge University Press, Cambridge

Jaeger C, Schule R, Kasemir B (1999) Focus groups in integrated assessment: a micro-cosmos for reflexive modernization. Innovation 12:195–219

Jäger J (1998) Current thinking on using scientific findings in environmental policymaking. Environ Model Assess 3:145–153

Jasanoff S, Wynne B (1998) Science and decision-making. In: Rayner S, Malone E (eds) Human Choice and Climate Change, Vol. 1: The societal framework. Batelle Press, Cambridge, MA

Jöborn A, Danielsson I, Arheimer B, Jonsson A, Larsson MH, Lundqvist LJ, Löwgren M, Tonderski K (2005) Integrated water management for eutrophication control: public participation, pricing policy and catchment modelling. Ambio 7:482–488

Johnson N, Munk Ravnborg H, Westerman O, Probst K (2001) User participation in watershed management and research. Water Policy 3:507–520

Jonsson A (2004) Metoder för ökat deltagande. In: Lundqvist L, Jonsson A, Galaz V, Löwgren M, Alkan Olsson J (eds) Hållbar vattenförvaltning – organisering, deltagande, inflytande, ekonomi. VASTRA Rapport 5., Göteborg. (in Swedish)

Kasemir B, Schibli D, Stoll S, Jaeger C (2000) Involving the public in climate and energy decisions. Environment 12:32 12

Kasemir B, Jäger J, Jaeger CC, Gardner MT (eds) (2003a) Public Participation in Sustainability Science A Handbook. Cambridge University Press, Cambridge

Kasemir B, Jaeger C, Jäger J (2003b) Citizens participation in sustainability assessment. In: Kasemir B, Jäger J, Jaeger CC, Gardner MT (eds) Public Participation in Sustainability Science A Handbook. Cambridge University Press, Cambridge

Laird F (1993) Participatory analysis, democracy and technological decision making. Sci Technol Hum Val 18:341–361

Lam D, Swayne D (2001) Issues of EIS software design: some lessons learned in the past decade. Environ Model Software 16:419–425

Latour B (1986) Visualization and cognition: thinking with eyes and hands. Knowledge and Society: Studies in the Sociology of Culture, Past and Present 6:1–40

Latour B (1990) Drawing things together. In: Lynch M, Woolgar S (eds) Representation in scientific practice. MIT Press, London

Lindley J (2001) Virtual tools for Complex problems: an overview of the AtlasNW Regional Interactive Sustainability atlas for planning for sustainable development. Impact Assessment Proj Appraisal 19:141–151

Mc Granahan G, Gerger Å (1999) Participation and environmental assessment in northern and southern cities, with examples form Stockholm and Jakarta. Int J Environ Pollut 11:373–394
Nilsson M (1998) Computer Tool Experiences in Ulysses. Stockholm Environmental Institute, Stockholm
Oreskes N (2000) Why believe a computer? Models, measures and meaning in the natural world. In: Scheiderman, JS (ed) The earth around us: Maintaining a liveable planet. W.H. Freeman, San Francisco
Pahl-Wostl C, Schlumpf C, Büssenshütt M, Schönborn A (2000) Models at the interface between science and society: impacts and options. Integr Assess 1:267–280
Palerm J (2000) An empirical-theoretical analysis framework for public participation in environmental impact assessment. J Environ Plan Manage 43:581–600
Ravetz J (1999) Developing principles of good practice in integrated environmental asessment. Int J Environ Pollut 11:1–23
Ravetz J (2003) Models as metaphors. In: Kasemir B, Jäger J, Jaeger CC, Gardner MT (eds) Public Participation in Sustainability Science A Handbook. Cambridge University Press, Cambridge
Rosa E (1999) The quest to understand society and nature: looking back but mostly forward. Soc Nat Resour 12:371–376
Sarewitz D, Pielke RA, Byerly R (eds) (2000) Prediction science and the future of nature. Island Press, Washington DC
Schackley S, Darier É (1998) Seduction of the sirens: global climate change and modelling. Sci Public Policy 25:171–184
Shackley S (1997) Trust in models? The mediating and transformative role of computer models in environmental discourse. In: Redclift M, Woodgate G (eds) The International Handbook of Environmental Sociology. Edward Elgar, Cheltenham Glos, pp 237–260
Schulze R (2001) Managing water as a resource in Africa: are we asking the right questions? In: Gash J, Odada E, Oyebande L, Schulze R (eds) Freshwater resources in Africa. BAHC International Project Office PIK, Potsdam, Chapter 2: 9–14
Schulze R, Horan M, Seetal A, Schmidt E (2004) Roles and perspectives of the policy-maker, affected water sector and scientists in integrated water resource management: a case study from Africa. Int J Water Resour Dev 20:325–344
Swallow BM, Johnson N, Meinzen-Dick RS (2001) Editorial: Working with people for watershed management. Water Policy 3:449–455
Wittgren HB, Castensson R, Gippert L, Joelsson A, Jonason L, Pettersson A, Thunvik R, Torstensson G (2004) An actor game on implementation of environmental quality standards for nitrogen in a Swedish agricultural catchment. *Ambio* (in press)
Wynne B (1984) The institutional context of science, models, and policy: The IIASA Energy Study. Policy Sci 17:277–319
Wynne B (1992) Misunderstood misunderstanding: social identities and public uptake of science. Public Understanding Sci 1:281–394
Yearly S (1999a) What do we mean by "Science" in the public understanding of science. In: Dierkes M, von Gröte C (eds) between understanding and trust: The Public, Science and Technology. Harwood Academic, Reading
Yearly S (1999b) Computer models and the public's understanding of science: a case-study analysis. Soc Studies Sci 29:845–866
Yearley S (2000) Making systematic sense of public discontents with expert knowledge: two analytical approaches and a case study. Public Understanding Sci 9:105–122

Water Resour Manage (2007) 21:111–128
DOI 10.1007/s11269-006-9044-0

ORIGINAL ARTICLE

Integration of the biophysical and social sciences using an indicator approach: Addressing water problems at different scales

Caroline A. Sullivan · Jeremy Meigh†

Received: July 2005 / Accepted: January 2006

Abstract To be operationally sustainable, any system of environmental management needs to be based on a truly holistic assessment of all of the relevant factors influencing it. This is of course a daunting task, demanding as it does detailed and reliable data, not only from both the physical and social sciences, but also incorporating some representation of that part of knowledge which could be described as non-scientific. This could be said to include the uncertainties of market forces and political will, as well as traditional knowledge systems, and artistic representation. Recognising the limitations of our own knowledge system is important if we are to make progress in the achievement of sustainability. The development of less deterministic models is a step forward in that direction.

This paper provides some discussion on the challenges associated with the integration of data from different disciplines, and the application of that data at different scales. Alternative approaches to the assessment of water resources for policy making are highlighted, and the validity of using such assessments at different scales is discussed. Using the Water Poverty Index as illustration, examples are provided of how an integrated assessment framework can be used to provide consistency and transparency in decision-making, and how this can, in practice, be applied at a variety of scales.

Keywords Integration · Indicators · Water resources · Water Poverty Index (WPI) · Water management · Spatial scale · Poverty

Introduction

It has long been recognised that there is a link between human welfare and the environment. Some have argued (Kuznets, 1973; Stern *et al.*, 1996) that this relationship follows a predictable trajectory, where the development process is seen to impact at first negatively

†It is with regret that we announce that Dr. Jeremy Meigh died in February 2006.

C. A. Sullivan (✉) · J. Meigh
Centre for Ecology and Hydrology, Wallingford, Oxfordshire OX10 8BB, United Kingdom
e-mail: csu@ceh.ac.uk

on the environment, followed by a levelling off before a more positive relationship is then established. While there are a number of examples of countries in the later positive phase (New Zealand, Finland, Switzerland), the majority can be described as still being in that development phase where environmental impacts are by and large, negative. There is an urgent need for this negative human impact to be reversed, and this requires political action at the national and municipal levels (Huby, 2001).

Links between poverty and the environment have been demonstrated in many parts of the world, including both rural (Samal *et al.*, 2003) and urban zones (Satterthwaite, 2003). Feedbacks between environmental conditions and human fertility have been highlighted, (Crenshaw *et al.*, 2000), and where populations remain in poverty, higher reproductive rates continue to exacerbate pressure on the environment. This link has been examined further, focussing specifically on the relationships between poverty and water (Sullivan, 2002; Ahmad, 2003; Soussan, 2004) The pressing need to examine this relationship further is highlighted by the vast numbers of people still not properly provided for by water services, a challenge for the Millennium Development Goals (MDGs) as outlined at the World Summit on Sustainable Development in Johannesburg (WSSD, 2002). This is evidence of both a knowledge and policy failure, and lack of infrastructure, especially in Africa, where the problem was highlighted more than 15 years ago by both Falkenmark (1989) and Sen (1985). As a result of this failure, there are still severe water shortages occurring (or likely to occur) in many places on that continent (Meigh *et al.*, 1999; Alcamo *et al.*, 1997; Seckler *et al.*, 1998).

As Sen has pointed out more recently (1999), development requires a multi-faceted approach to achieve real poverty reduction, and others (Escamilla *et al.*, 2003) have emphasised the need for this to be embedded in a participatory process. Attempts to achieve this have been made, and one example where water issues and their relationship to both humans and the environment have been formalised is provided by the new water law of South Africa (DWAF, 2002). On a smaller scale, other attempts have been made to develop water management tools that facilitate a more sustainable approach to water management, by addressing more specifically the livelihood needs of local populations (Campbell *et al.*, 2002; Acreman *et al.*, 2000; Jackson *et al.*, 2004; Sullivan *et al.*, 2001).

Challenges of integration

Approaches to the application of integrated modelling for natural resource policy-making have been discussed in many contexts, including agriculture (Ruben *et al.*, 1998), the provision of ecosystem services (Jewitt, 2002), and the management of land use change (Lambin, 1997). The importance of such integration has been emphasized (Harris, 2002; Wallace *et al.*, 2003), and the use of indicators generated from these integrated approaches have been widely discussed (Bradbury, 1996; Turner *et al.*, 2000; Kammerbauer *et al.*, 2001; Sullivan, 2002; Yoffe and Ward, 1999). In particular, criticism has been made of integrated indicators in the form of the Human Development Index (HDI) (Streeten, 1994 1996; Srinivasan, 1994), but it has subsequently been clearly shown that such indicators have had a very significant impact on the way development progress has been assessed. In the case of water assessment, we argue that this approach, in spite of its imperfections, does provide a useful and easy to use tool for policy-making and management in the water sector.

The international commitment to the MDGs has increased the need to develop integrated tools to assess the development process, and many international agencies have called for the use of integrated indicators (UNESCO, 2003; IISD, 1999). The importance of linking these to livelihoods has been stressed (Hoon *et al.*, 1997), and an attempt to design and develop an integrated indicator for water resource assessment has been made by Sullivan *et al.* (2002, 2003), in the creation of the *Water Poverty Index* (WPI). As with all novel approaches, this has drawn some attention in both scientific literature (Salamah, 2000; Feitelson and Chenoweth, 2002; Molle and Mollinga, 2003) and the press (Hoffman and Hoffman, 2002). Other issues on the same subject have been raised (Saleth *et al.*, 2003; Schulze and Dlamini, 2002), and this demonstrates the fact that the development of a complex tool such as this must be regarded as an iterative process. As part of this iteration, the application of the WPI has being further investigated in a number of countries, and at a variety of scales. In this paper, we discuss some of the applications of the Water Poverty Index at these different scales, and it is our intention here to provide an overview of progress which has been made to date.

Types of knowledge

Any attempt to generate effective solutions for today's environmental problems will benefit from the inclusion of knowledge from a wide variety of sources. The traditional approach is to search for scientific understandings of processes, and from these, suggest best practice options. More recently, systems approaches take this further (Greiner, 2004; Gill, 1996) and the most innovative of these will try to incorporate relevant traditional knowledge and cultural values in the creation of a fully integrated systems model, from which integrated solutions can be found. In the case of water resource management, there are many examples of traditional approaches to both groundwater (for example in Pakistan) and surface water (for example in Bali), from which modern systems approaches could benefit. Ways to capture and reflect different types of knowledge have been discussed, and much interest in the use of integrated indicators has been generated (Hammond *et al.*, 1995; Rees, 1996).

The issue of scale

Water is highly variable, both on a spatial and temporal scale. The scale at which various types of knowledge can be applied to water management also varies widely. At a global level, scales most in use for the construction of global climate models tend to be based on grids of about 200 km by 300 km (IPCC, 2001). Assessment of water resources tends to be based on smaller grids (Meigh *et al.*, 1999; Schulze, 1999; Midgeley *et al.*, 1994; Arnell and King, 1998; Alcamo *et al.*, 1997, 2003; Seckler *et al.*, 1998), covering an area of some 2500 km^2 (i.e. about 50 km × 50 km). Most recently, attempts have been made to develop models at even finer resolutions, and recent increases in computing power have made hydrological modelling possible at grid scales of just 1 km^2. Another dimension of knowledge which has arisen as a result of advances in information technology is provided by the use of Geographical Information Systems. Numerous ways in which this can be used at a variety of scales to improve water management have been provided (van Tongeran *et al.*, 2001; Mallawaarachchi *et al.*, 1996; Vanvoris *et al.*, 1993; Gurnell and Montgomery, 1999).

In terms of water quality, it is important to recognize that there is major spatial and temporal variation in the scale of impacts of both point and diffuse sources of pollution (Meybeck, 2003; Nur *et al.*, 2001). These variable impacts have implications both at various ecological scales (Postel and Richter, 2003; Reynolds, 1998; Ricciardi and Rasmussen, 1999) and also at the human scale (Naiman and Decamps, 1997; Clarke, 1997; WCD, 2000), and innovative approaches are needed to take these into account for policy making (Sengupta and Bennett, 2003; Naiman *et al.*, 2002; Pahl-Wostl, 2002; Tharme, 2003; Richter *et al.*, 1997).

At the socio-economic and political levels, the scale relevant to policy making can range from the household to the nation. Governance and institutional temporal and spatial preferences are not always compatible with those of individuals or communities, or the maintenance of ecosystems. While altruistic and socially responsible values may influence policy decisions in some situations, there is often more likely to be a problem of power concentration, vested interests and lack of participation in the decision-making process. On the temporal scale there also is likely to be a mis-match between the time preference rates of politicians, scientists and the public at large, and in general, these are often totally out of keeping with what is relevant at ecological and hydrological scales.

On the temporal scale, water management is important to humans to secure adequate storage to ensure access to water when it is needed. Temporal variability of resources is more difficult to deal with than spatial variability, subject as it is to higher levels of uncertainty. At the other end of the spectrum from the low flows problem, more effective integrated water management is also needed to anticipate and mitigate against the impacts of flooding. Both high and low flows have impacts on habitats and biodiversity, and it is particularly in this respect that the maintenance of natural flow regimes is crucial for river health. Assessment of temporal variability is also important on an inter-annual basis, in order to develop a more effective understanding of the impacts of climate and other sources of global change.

In an attempt to bring all these issues together, and to generate more effective targeting for poverty reduction, efforts have been made to address these problems using geographical tools (Baker and Grosch, 1994). Addressing poverty linkages to water provision, through the development of an integrated indicator approach, the Water Poverty Index (Sullivan, 2002; Sullivan *et al.*, 2003) aims to target political and financial attention towards those most in need.

The structure and rationale of the Water Poverty Index (WPI)

The original concept of the Water Poverty Index was created by Sullivan (2001, 2002), building on the hydrological modelling work of Meigh *et al.* (1999). It was designed as a holistic tool, to capture the links between water availability and livelihoods (Scoones, 1998; Carney, 1998), while at the same time addressing the need to maintain ecological integrity. By using an integrated index structure, it was designed to be a water management tool which was accessible to water decision-makers at various levels. The overall structure and the component variables of the Water Poverty Index were identified through participatory consultation with scientists, water managers and stakeholders. The resulting structure is believed to capture a more comprehensive picture of the water management challenge (Sullivan *et al.*, 2002, 2003; Sullivan and Meigh, 2003). The components are:

- *Resources* – how much water is there available, taking account of seasonal and inter-annual variability and water quality;
- *Access* – A measure of how well provisioned the population currently is, including for domestic use and irrigation;
- *Capacity* – to manage water resources, based on education, health and access to finance;
- *Use* – this captures the use we make of the water, and its contribution to the wider economy;
- *Environment* – this tries to capture the environmental impact of water management, attempting to ensure long term ecological integrity.

The individual variables for each component need to be selected through a stakeholder process to ensure that they are available and they reflect the stakeholder's main concerns.

This was done in the examples presented here, and some minor changes were made in some locations to take account of cultural differences which existed in the pilot study cases. These different components of the WPI are combined within a composite index structure, as outlined in Box 1.

Box 1. Combining variables in the WPI

The WPI is derived from the weighted average of the five components *Resource* (R), *Access* (A), *Capacity* (C), *Use* (U), and *Environment* (E).:

$$\text{WPI} = \frac{wrR + waA + wcC + wuU + weE}{wr + wa + wc + wu + we}$$

where WPI is the Water Poverty Index value for a particular location, and w is the weight applied to each of the components. Each of the five main components is made up of a number of sub-components, and a weighting can be applied to indicate the importance of each variable. components are standardised to fall in the range 0 to 100; giving a final WPI value between 0 and 100. The highest value, 100, is taken to be the best situation, that is, the lowest possible level of water poverty, while 0 is the worst. To avoid problems of subjectivity, a baseline value of the WPI should be first calculated with these weightings set equally.

Source. Sullivan *et al.* (2002)

Using the WPI to investigate water resources at the community level

During the process of the development of the Water Poverty Index, this composite index methodology was tested in a total of 12 pilot sites in Tanzania, Sri Lanka and South Africa (Sullivan *et al.*, 2002, 2003). These represented both urban and rural communities, and it was found that the tool was able to identify strengths and weaknesses in the water sector at each location, as well as being able to discriminate between different locations. This comparison can be made based on the final indicator values, and also through an examination of the individual component values to identify the source of the problem in particular places. For the purpose of testing the Water Poverty Index methodology, data was collected from a total of over 1500 households, and the variables used to compile the WPI values at that scale included information and issues relevant at the household scale, as shown in Table 1. It must be noted however that one of the over-arching objectives of the WPI is to try to use existing data where possible, rather than requiring the need for further data collection. The objective of data collection during this phase of the project was purely to provide a consistent base line dataset, from which legitimate comparisons could be made. In practice, if the methodology were to be implemented at any location, it would be hoped that as many as possible of the variables could be represented by existing data, thus providing a cost effective solution to integrated water resource assessment.

Through the selection of these variables, it was planned that data collected would capture the essence of the five components at the household level, and this could be aggregated for all households to generate a community value for the WPI. During this testing process, some difficulties with some of the components were revealed, and it was also observed that there was some inevitable overlap between variables. In the case of the environment component, variables were selected on the basis of being issues that subsistence householders in general

Table 1 Data selected as WPI component variables for community assessment in pilot sites in Sri Lanka, South Africa and Tanzania

WPI component	Variables
Resources (R)	• Assessment of surface water and groundwater availability using hydrological and hydrogeological techniques
	• Quantitative and qualitative evaluation of the variability or reliability of resources
	• Quantitative and qualitative assessment of water quality
Access (A)	• Access to clean water as a percentage of households having a piped water supply
	• Reports of conflict over water use
	• Access to sanitation as a percentage of population
	• % of water carried by women
	• Time spent in water collection, including waiting
	• Access to irrigation coverage adjusted by climate characteristics
Capacity (C)	• Wealth proxied by ownership of durable items
	• Under-five mortality rate
	• Educational level
	• Membership of water users associations
	• % households reporting illness due to water supplies
	• % of households receiving a pension/remittance or wage
Use (U)	• Domestic water consumption rate
	• Agricultural water use, expressed as the proportion of irrigated land to total cultivated land
	• Livestock water use, based on livestock holdings and standard water needs
	• Industrial water use (purposes other than domestic and agricultural)
Environment (E)	• People's use of natural resources
	• Reports of crop loss during last 5 years
	• % households reporting erosion on their land

Note. In the absence of any acceptable figures to represent environmental integrity or environmental water needs, these alternative proxy data were used
Source. Sullivan *et al.* (2002, 2003)

Table 2 WPI component scores for pilot sites in Sri Lanka and South Africa

South Africa	Resources	Access	Capacity	Use	Environment	WPI
Ethembeni	50.0	36.6	59.8	41.5	27.7	43.1
Latha	20.0	17.0	42.1	24.5	28.9	26.5
Wembezi (informal)	50.0	48.8	46.1	18.0	39.1	40.4
Wembezi (formal)	50.0	86.5	78.0	38.1		63.2
Sri Lanka						
Agarauda	20.0	38.3	64.7	74.9	34.2	46.4
Awarakotuwa	10.0	35.2	79.6	21.2	28.1	34.8
Tharawaththa	20.0	26.5	50.6	16.2	42.2	31.1

would be able to relate to, but in practice, it was found that there was a marked difference in the understanding of these issues between urban and rural households, and in some cases, the data was not possible to collect, due to its not being relevant (e.g., for households without land, no erosion could be reported). It is clear that more work needs to be done to identify appropriate variables to represent the environment component, particularly in urban areas.

A comparison of the pilot sites in Sri Lanka and South Africa are provided in Table 2, where the values for each major component are shown.

These figures show that in South Africa, the community with the best water situation is Wembezi Formal, with a WPI value of 63.2. This reflects the fact that in this official government housing scheme, both access and capacity generate relatively high component scores. In the informal squatter settlement nearby, the WPI score is significantly lower, at 40.4, although this itself is a relatively high score compared to the extremely low score in one of the other two communities. These latter two examples reflect the low component scores on access, capacity and the environment, resulting in low overall WPI scores. This suggests that these two villages (and especially Latha) should be given priority in water provision schemes, especially to increase access and improve the exiting level of negative impacts on the environment. For dissemination purposes, these variations both between and within locations are further illustrated through the use of a pentagram diagram along with the final WPI scores. This is illustrated in Figure 1, where higher scores indicate a lower degree of water poverty.

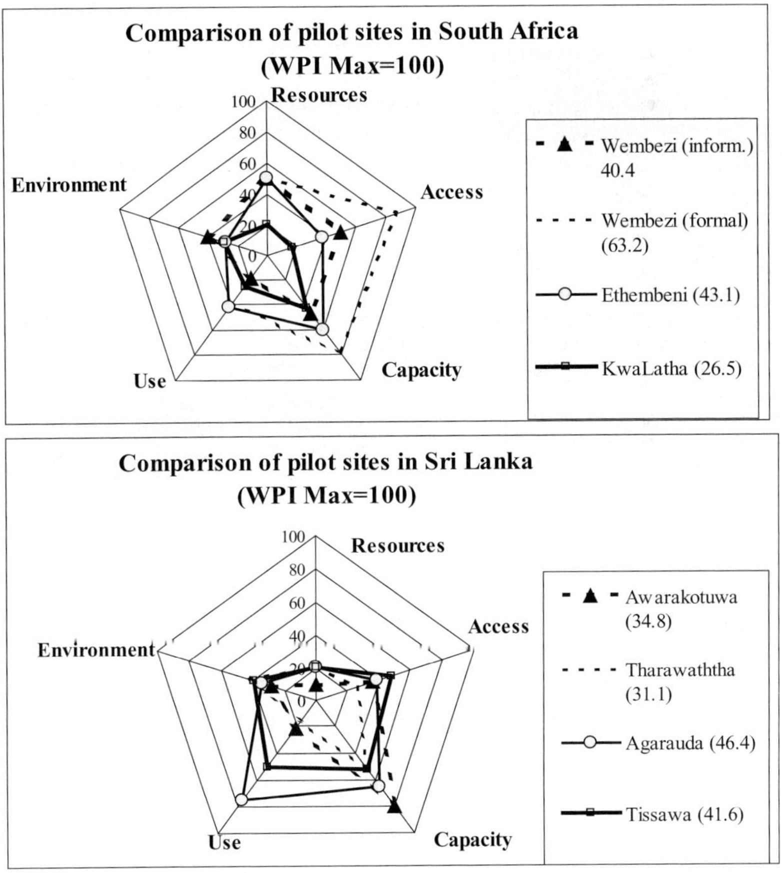

Source: Sullivan et al., 2002

Fig. 1 Using a pentagram figure to illustrate the comparison of WPI scores for four communities in South Africa and Sri Lanka, 2001

Interpretation of the WPI pentagram

The pentagram diagrams show the component scores which indicate the strengths and weaknesses in each location. For example, in Agarauda, Sri Lanka, it seems evident that much more productive use is made of water, as shown by the higher value on the *use* scale. This is in part due to the impact of small-scale brick making, common in that community, as illustrated in Figure 2. At the same time, the environmental impact of that water use in the assessed locations seems to be quite severe, illustrated by the low scores on the environmental axis in Figure 1. The environmental scores are similar to the South African cases, although there is considerable variation within each country. From these examples, we can see that real alleviation of poverty could be achieved in the South African cases, for example, if the development of community gardens (Batchelor *et al.*, 1996; see Figure 2), which can also be successful in peri-urban areas, were facilitated through investment in more effective and equitable water management. Other development options would also be possible in all of these locations, but both consultation with the community, and cost-effectiveness analysis would be needed, to make the most appropriate selection for each particular location.

The productive use of water can generate economic benefits for local communities. The WPI can capture the impact of such benefits in the overall assessment it provides.

Fig. 2 Small scale brick making and community gardens can contribute significantly to poverty reduction in both rural and peri-urban communities

Application of the Water Poverty Index at the district level scale, based on available national statistics

In an attempt to examine how existing national data sources can be used to calculate the WPI, the methodology has been applied to National Census and other data from the Statistics South Africa datasets (Cullis and O'Regan, 2004). Table 3 shows how the WPI components are represented here by a single specific variable already available in national data, and how the resolution of that data is not always of the same scale. In this application of the WPI, the selection of the variables was limited by what was available nationally, and the number of variables were reduced to examine the potential for a rapid application using existing data.

The importance of scale is well illustrated from this example, as it can clearly be seen in Figure 3 that when the data is collected at the sub-catchment scale, one area is identified as the most water poor area, while if enumerator area data is used (i.e. finer resolution data),

Table 3 How WPI variables have been represented by data from national statistics

Component	Variable	Source	Lowest level of aggregation
Resources	Mean annual accumulated stream flows per capita, weighted for the average distance of an enumerator area from the nearest stream	Schulze and Dlamini (2002) using the ACRU model under baseline land-cover conditions	Subcatchment
Access	Percentage of Households with access to a private water source (both water supplied to the house and to private yard taps)	Census 1996	Enumerator Area
Capacity	Percentage of Households below a certain level of annual income (annual household < Rand 2400)	Census 1996	Enumerator Area
Use	Estimation of domestic use based on rural or urban classification of enumerator area	Usage taken from national estimates of average rural and urban usage (DWAF, 2002) and EA type from Census 1996	Enumerator Area
Environment	Combined Land Degradation Index	Hoffman *et al.* (2001)	Municipal District

Source. Cullis and O'Regan, 2004

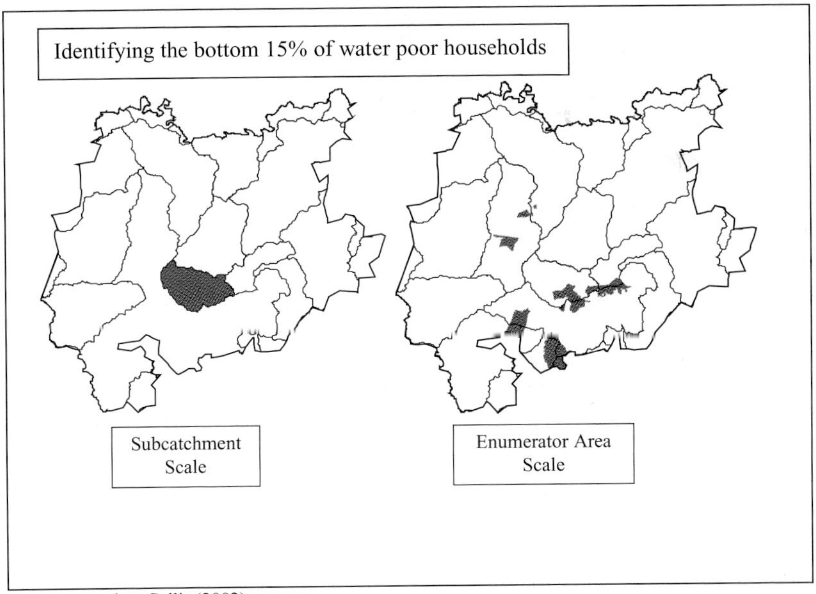

Source: Based on Cullis (2002)

Fig. 3 WPI values generated using census data for the Escourt District, KwaZulu Natal, South Africa a much clearer picture emerges of the location of the most water poor areas. This example not only demonstrates the possibility of using existing national census data, but also the importance of using the finest resolution possible to produce the most accurate results.

Application of the Water Poverty Index at the watershed scale

The WPI methodology has also been applied at the watershed scale in sample catchments in Nepal, Pakistan and India (Merz, 2003). This has been done using the PARDYP[1] datasets, which have been generated from a hydro-meteorological research network in the Hindu Kush Himalaya. Three catchments in Nepal, India and Pakistan were assessed from existing data, with appropriate variables being selected to represent each of the five key WPI components. The results are shown in Figure 4.

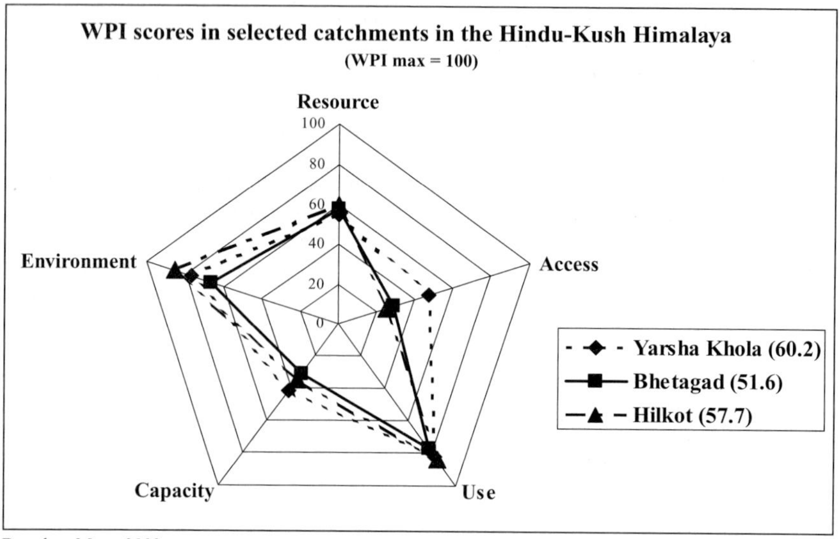

Based on Merz, 2003

Fig. 4 Application of the WPI at the basin scale

This figure shows that taking the WPI scores overall, the situation is worst in the Bhetagad catchment, in India, where the WPI score is 51.6, followed by the Hilkot catchment in Pakistan, with a score of 57.7; and the highest score has been found in the Yarsha Khola catchment in Nepal. By looking at each component, we can see there is some similarity between the catchments in terms of *Resource* availability, but in terms of *Access* and *Capacity,* the Yarsha Khola catchment scores better than the others. In terms of the *Environment*, Hilkot catchment in Pakistan gains the highest score. This suggests that although there are improvements to be made there, the most urgent attention should be given to communities in the Bhetagad catchment, where low scores are found on all the components. This demonstrates

[1] The PARDYP data have been generated from an integrated multi-donor funded watershed project in the Hindu Kush Himalaya. Its aim is to contribute to balanced, sustainable, and equitable development of mountain communities and families in the region.

how the tool can be used to prioritise policy responses at the watershed scale,[2] and the advantage that it offers over previous approaches is that it provides an holistic assessment of the issues underlying any water management decision.

Application of the Water Poverty Index at the national level

As an exercise to demonstrate the application of the method at different scales, the WPI methodology was applied to publicly available national level data for 147 countries (Lawrence *et al.*, 2002). Sources of data used for selected components in this calculation are shown in Table 4. Making comparisons in this way can be of use, but it is important to recognize that variations in the data (produced by the use of different methodologies of data collection in different countries) will weaken the power of comparisons. Data may be both defined and recorded in different ways in different places, and there is much need for standardization of approaches before definitive comparisons can be made. In spite of this, making comparisons between countries on the basis of this national scale can still be of use, particularly for donor agencies and international organizations.

Table 4 Sources of data for national level WPI components

Resources	World Resources Institute (2000) Table FW.1, and Gleick (2000). Shiklomanov (1997) has compiled a comparison of water resources data for a selected, but large, range of countries from different sources, including the WRI, Gleick and the Russian State Hydrological Institute. The original WRI data have been adjusted to take account of the variation in estimates of water resources by taking the modal estimate. The most striking discrepancy was in the case of Peru, which WRI says has 1746 billion cubic metres of internal freshwater flows (69,000 per capita), while all other estimates have at 40 billion cubic metres (1,600 per capita). The World Bank's Development Indicators also quote the former number and the WRI as the source, although earlier years of the WRI's data have the latter estimate. Population data (to obtain per capita estimates) from World Resources Institute (2000) Tables HD.1 and SCI.1 and UNDP (2001).
Access	World Resources Institute (2000) Table HD.3, and UNDP (1999). Irrigation: World Resources Institute (2000) Table AF.2. and Gleick (2000), with cropland areas from World Resources Institute (2000) Table SCI.1.
Capacity	GDP from UNDP (2001); Under-5 mortality rates from World Resources Institute (2000) Tables HD.2 and SCI.1 (7); Education data from UNDP (2001).
Use	Gleick (2000), World Resources Institute (2000) and World Bank (2001).
Environment	World Economic Forum, Yale Center for Environmental Law and Policy, and Center for International Earth Science Information Network, Columbia University (2001)

Source. Sullivan *et al.* (2002)

In this example, by using the WPI structure, international comparisons between countries can be made, as shown in Figure 5. The advantage of this approach is that, since the assessment is based on public data from sources such as the World Resources Institute and the World Bank, comparisons between different locations (although by no means perfect) can be considered to be reasonably robust.

[2] These figures cannot be compared directly to those shown for South Africa or Sri Lanka in Figure 1, as different data was used to generate them.

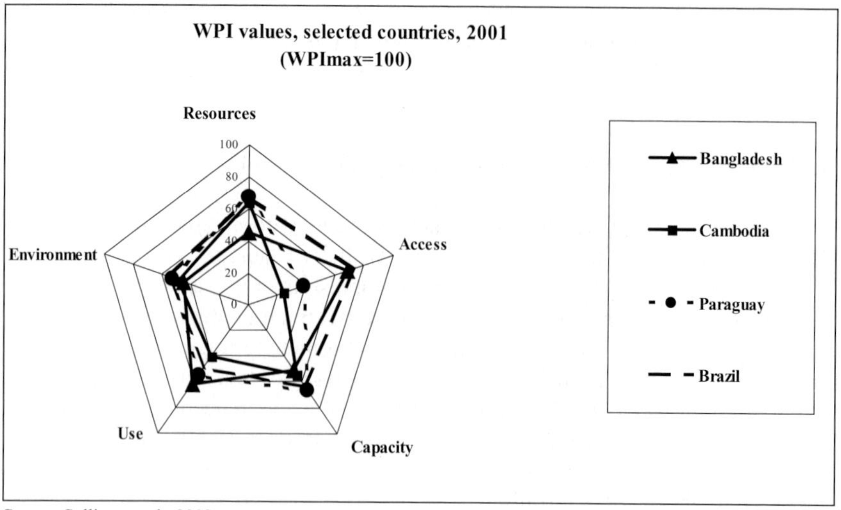

Source: Sullivan et al., 2002

Fig. 5 Comparing countries at the national level

From this figure, we can see how these different countries vary on the different WPI components. It is clear that in Cambodia, the most significant problem is lack of access, while for Bangladesh, the problem is of a low score on resources. In a country which has high precipitation and major rivers flowing into it, this may be counter-intuitive, but in reality this low score on the *Resource* component is a result of the very large population relative to the resource, and it also reflects both the high degree of variability in runoff rates, and the serious problems of both human and naturally-induced contamination of both ground and surface waters. For Paraguay, the most significant problems are concerned with water access issues, and the low productive value associated with water use. Once again this provides support for the premise that use of the WPI in this way can help to target attention to those issues and places where need is most acutely felt. This is achieved by combining the relevant data into a coherent framework in such a way as to make the resulting information more relevant for the purpose of policy making and public awareness raising.

How national level assessments can hide reality at the local scale

National data representing any issue always provides only a broad spectrum analysis, which almost never represents the position of any local individual or group. This is true of data such as per capita GNP, average weight, average number of telephones, etc. In the case of the WPI, the example provided by Zimbabwe in Figure 6 effectively illustrates this point, and its relevance to policy. It has to be noted however, that while the conceptualization of the five main components remains constant in any application of the WPI, the practical calculation of the values at the different scales shown in this figure, inevitably means that direct comparison of the scores is not valid. In this figure, for efficiency of space, the national values are shown here purely for illustrative purposes, not for comparison with the three villages, which may legitimately be compared to each other. It is however possible to observe from this figure however that the overall national score for Zimbabwe imply a higher capacity (reflecting

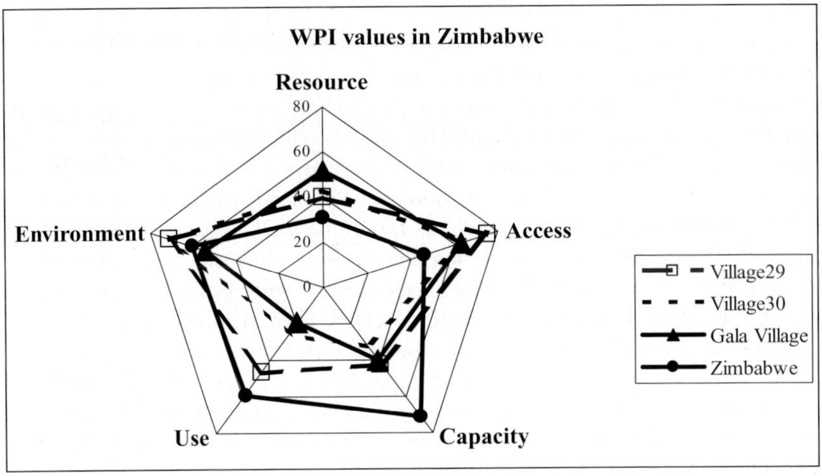

Source: Based on data collected by Dube (2003)

Fig. 6 Comparing villages in Bulilima district, Zimbabwe, with national data

the relatively high level of education, good health and access to finance in the country as a whole), than that found in the local communities.[3] While the use score is high at the national level, it is much lower in the community scores. From these community scores, it seems clear that in terms of poverty reduction, better mobilization of resources, and an increase in capacity, will be more fruitful in these particular rural areas, than would be achieved by direct improvements in access or the environment. Again, final selection of the appropriate measure to be taken in response to this would require economic analysis of the various available options. Clearly, however, from the national level figures, it appears that there is an overall need to improve the resource endowment, (more infrastructure/storage), as well as access to it for the population as a whole, not just selected sectors such as agriculture.

Another example of national figures misrepresenting local issues is provided by the example of Guyana, which features amongst the top 10 countries in the world on WPI scores. In that country, there are many water resources (the name *Guyana* is taken from the Amerindian word meaning '*Land of many waters*'), and the WPI *Resource* score is 90.7. It has a small population of well educated and generally healthy people (*Capacity* = 69.7), and has an environment which in large areas, still remains relatively intact (*Environment* = 54.4). These conditions give rise to the generation of an overall WPI score of 75.8. The reality is however, that the vast majority of these people live along the coastal plain, and so conditions in the urban and peri-urban zones are very different. In those locations, there are high numbers of people with inadequate or unreliable provision, so there is a clear need for investment in improvements in access, storage and distribution systems. This illustrates an important point: in order to develop effective policy guidance, it is essential that any assessment tool be applied at the appropriate scale.

[3] It is important to note that the situation has changed significantly in Zimbabwe since 2003.

Discussion

It has been suggested earlier that ideally, the data for the WPI components should come from existing sources. This is important, as much data exists, often in diverse institutions. It is important however to realize that currently this data may be inconsistent, unreliable or even invalid for what it claims to represent, so results from any assessment or modelling process should be treated with caution. Ideally, nations which consider using this tool as part of the suite of resource management tools they use, would be well advised to create a consistent and clearly structured database from which the variable values can be derived. This would not only facilitate more effective water management strategies, but it would also provide the data required to implement a system which would be able to incorporate the externalities (environmental costs and benefits) of water use into the national accounting system, known by some as 'green accounting' (Ekins and Simon, 1999). Another way to contribute to this process would be to incorporate key questions on water use into the national census structure, to facilitate detailed socioeconomic data at the local to national scale (Sullivan and Meigh, 2003).

There is no doubt that there is much room for improvement and refinement of the WPI structure. This can only be achieved through an iterative process, and the examples shown here are parts of that process. Other work is ongoing to take these issues further, eventually to the point where the tool can be considered fully robust, suitable to be incorporated into macroeconomic planning structures, in the same way as other indicator tools such as the *Human Development Index* (HDI) and the *Retail Price Index* (RPI) currently are. If this can be achieved, and the WPI incorporated with other water management tools, it will contribute to the achievement of Integrated Water Resource Management (IWRM), a stated policy goal for most nations today.

Conclusions

In this paper we have demonstrated the interest in, and relevance of, the use of an integrated indicator approach for more effective water policy making. We have shown how the Water Poverty Index goes some way towards meeting the needs for such a tool, and how the determination of the structure of this index in a participatory manner has given rise to widespread interest in both the scientific and popular literature. We have shown that through the integration of knowledge from both the biophysical and social sciences, more holistic tools for policy making can be developed, and through an iterative process, these can be adapted for use at a variety of scales. The great advantage of the WPI is that it provides a reasonably simple process to combine biophysical, social, economic and environmental data to produce a single index value, with its associated pentagram, enabling more comprehensive understanding of the meaning of the results. The generation of a tool to capture this integrated approach has not been achieved before.

On this basis, we suggest that effective implementation of the WPI as a tool for water management could contribute to poverty reduction. In order to achieve this, it would be necessary to standardize data collection definitions and procedures, incorporate key questions into census surveys, and refine the structure of the WPI to meet local needs. While it is clear that this will take time to achieve, we believe that it would not be a wasted effort. If done effectively, it would result in the situation that by 2015, (when the Millennium Development Goals are to be assessed), the country would be in situation where a more consistent, transparent and equitable process of water management could be achieved. Furthermore, the degree to which the MDGs themselves had been reached could be demonstrated through an ex-post evaluation of the data collated in the process.

Acknowledgements This paper has been generated from work funded by both DFID and The UK Natural Environment Research Council. The author is indebted to all of those who have been involved in the original research on the Water Poverty Index, and to all of the householders, practitioners and policy makers who have contributed to its development so far.

References

Acreman M, King J (2002) Defining the ecosystem component of a Water Poverty Index. In: Sullivan CA, Meigh JR, Fediw T (eds.) Developing and testing the Water Poverty Index: Phase 1 Final Report. Report to DFID, CEH Wallingford, UK

Ahmad QK (2003) Towards poverty alleviation: the water sector perspectives. Int J Water Resour Dev 19(2):263–277

Alcamo J, Döll P, Kaspar F, Siebert S (1997) Global change and global scenarios of water use and availability: An application of WaterGAP1.0. CESR, University of Kassel

Alcamo J, Doll P, Henrichs T, Kaspar F, Lehner B, Rosch T, Siebert S (2003) Global estimates of water withdrawals and availability under current and future "business-as-usual" conditions. Hydrol Sci J–Journal Des Sciences Hydrologiques 48(3):339–348

Arnell NW, King R (1998) Implications of climate change for global water resources. Report to Department of the Environment, Transport and the Regions by Department of Geography, University of Southampton, UK

Baker JL, Grosch ME (1994) Measuring the Effects of Geographic Targeting on Poverty Reduction Living Standards Measurement Study Working Paper 99, World Bank

Batchelor C, Lovell C, Murata M (1996) Simple microirrigation techniques for improving irrigation efficiency on vegetable gardens. Agricultural Water Management 32(1):37–48(12)

Bradbury RH (1996). Are indicators yesterday's news? Institute of Environmental Studies, University of New South Wales, Sydney, Australia

Campbell BM, Jeffrey S, Kozanayi W, Luckert M, Mutamba M, Zindi C (2002) Household Livelihoods in Semi-Arid Regions: Options and Constraints. CIFOR, Bogor

Carney D (ed.) (1998) Sustainable Rural Livelihoods: What Contribution Can We Make? Department for International Development, London

Clarke MJ (1997) Restoration Ecology and Sustainable Development. Edited by Urbanska, Webb, and Edwards. Cambridge University Press

Crenshaw EM, Christenson M, Oakey DR (2000) Demographic transition in ecological focus. Am Sociol Rev 65(3):371–391

Cullis J (2002) Targeting the Water-poor through Water Poverty Mapping. A Case Study using Census Data and the Water Poverty Index for the Estcourt Municipal District, South Africa. MSc. Dissertation Oxford University, Environmental Change Institute

Cullis J, O'Regan DP (2004) Targeting the water-poor through water poverty mapping. Water Policy 6:397–411

Department of Water Affairs and Forestry (DWAF) (2002) Proposed First Edition of the National Water Resource Strategy DWAF, Pretoria

Dube O (2003) The Water Poverty Index in arid regions: The Study of the Nata River basin (the Makgadikgadi salt pans catchment area) in Bulilima district: Zimbabwe-Botswana. M.Sc Dissertation, University of Leicester

Ekins P, Simon S (1999) The sustainability gap: a practical indicator of sustainability in the framework of the national accounts. Int J Sustainable Dev 2:24–58

Escamilla M, Kurtycz A, van der Heim R (2003) Water participation for poverty alleviation – the case of Meseta Purepecha, Mexico. Water Sci Technol 47(6):145–148

Falkenmark M (1989) The massive water scarcity now threatening Africa – why isn't it being addressed. Ambio 18:112–118

Feitelson E, Chenoweth J (2002) Water poverty: towards a meaningful indicator. Water Policy 4:263–281

Gill R (1996) An integrated social fabric matrix/system dynamics approach to policy analysis. Syst Dyn Rev 12(3):167–181

Gleick P (2000) The World's Water 2000–2001: The Biennial Report on Freshwater Resources, Island Press, Oakland, California

Greiner R (2004) Systems framework for regional-scale integrated modelling and assessment. Math Comput Simulat 64(1):41–51

Gurnell AM, Montgomery DR (eds.) (1999) Hydrological Applications of GIS. Wiley, London

Hammond A, Adriaanse A, Rodenburg E, Bryant D, Woodward R (1995) Environmental Indicators: A Systematic Approach to Measuring and Reporting on Environmental Policy Performance in the Context of Sustainable Development World Resources Institute, Baltimore

Harris G (2002) Integrated assessment and modelling: an essential way of doing science. Environ Model Softw 17(3):201–207

Hoffman and Hoffman Ltd (2002) Water Poverty Index media usage report. world water council, Marseilles

Hoon P, Singh N, Wanmali SS (1997) Sustainable livelihoods: Concepts, principles and approaches to indicator development. UNDP, New York

Huby M (2001) The sustainable use of resources on a global scale. Social Policy Adm 35(5):521–537

IISD (International Institute for Sustainable Development) (1999). Beyond delusion: a science and policy dialogue on designing effective indicators for sustainable development

IPCC (Intergovernmental Panel on Climate Change) (2001). Climate Change 2001: the Scientific Basis. Cambridge University Press, Cambridge and New York

Jackson N, Giacomello AM, Sullivan CA, O'Regan D (2004) Managing watersheds for a better future. Research report to DFID, CEH, Wallingford

Jewitt G (2002) Can integrated water resources management sustain the provision of ecosystem goods and services? Phys Chem Earth 27(11–22):887–895

Kammerbauer J, Cordoba B, Escolan R, Flores S, Ramirez V, Zeledon J (2001) Identification of development indicators in tropical mountainous regions and some implications for natural resource policy designs: an integrated community case study. Ecol Econ 36(1):45–60

Kuznets S (1973) Modern Economic Growth: findings and Reflections. Am Econ Rev 63:247–258

Lambin EF (1997) Modelling and monitoring land-cover change processes in tropical regions. Prog Phys Geogr 21(3):375–393

Lawrence P, Meigh JR, Sullivan CA (2002) The Water Poverty Index: an international comparison. In: Sullivan CA, Meigh JR, Fediw T (eds.) Developing and testing the Water Poverty Index: Phase 1 Final Report. Report to DFID, CEH, UK

Mallawaarachchi T, Walker PA, Young MD, Smyth RE, Lynch HS, Dudgeon G (1996) GIS-based integrated modelling systems for natural resource management. Agricult Syst 50(2):169–189

Meigh JR, McKenzie AA, Sene KJ (1999) A grid-based approach to water scarcity estimates for eastern and southern Africa. Water Resour Manage 13:85–115

Merz J (2003). Water balances, floods and sediment transport in the Hindu Kush-Himalayas – Data analyses, modeling and comparison of selected mesoscale catchments. Unpublished PhD Thesis. Bern: Department of Geography, University of Bern

Meybeck M (2003) Earth Systems and Global Water Systems: Linkages and feedbacks. Philosophical transactions. Royal Academy, London

Midgley DC, Pitman WV, Middleton BJ (1994) Surface Water Resources of South Africa 1990. Reports 298/1/94 to 298/6/94, Water Research Commission, Pretoria, South Africa

Molle F, Mollinga P (2003) Water poverty indicators. Conceptual problems and policy issues. Water Policy 5:529–544

Naiman RJ, Décamps H (1997) The ecology of interfaces – riparian zones. Annu Rev Ecol Syst 28:621–658

Naiman RJ, Bunn SE, Nilsson C, Petts GE, Pinay G, Thompson LC (2002) Legitimizing fluvial systems as users of water: an overview. Environ Manage 30:455–467

Nur Y, Fazi S, Wirjoatmodjo N, Han Q (2001) Towards wise coastal management practice in a tropical megacity. Jakarta Ocean Coastal Manage 44(5–6):335–353

Pahl-Wostl C (2002) Towards Sustainability in the Water Sector: The importance of human actors and processes of social learning. Aquat Sci 64:394–411

Postel S, Richter B (2003) Rivers for Life: Managing Water for People and Nature. Island Press, Washington DC, 253 pp

Rees W (1996) Revisiting carrying capacity: area based indicators of sustainability. Popul Environ: A J Interdiscip Stud 34(10):1667–1679

Reynolds CS (1998) The state of freshwater ecology. Freshwater Biol 39:741–753

Ricciardi A, Rasmussen JB (1999) Extinction rates of North American freshwater fauna. Conservation Biol 13:220–222

Richter BD, Baumgartner JV, Wigington R, Braun DP (1997) How much water does a river need. Freshwater Biol 37:231–249

Ruben R, Moll H, Kuyvenhoven A (1998) Integrating agricultural research and policy analysis: Analytical framework and policy applications for bio-economic modelling. Agric Syst 58(3):331–349

Salameh E (2000) Redefining the Water Poverty Index. Water Int 25:469–473
Saleth RM, Samad M, Molden D, Hussain I (2003) Water, poverty and gender: an overview of issues and policies. Water Policy 5:5
Samal PK, Palni LMS, Agrawal DK (2003) Ecology, ecological poverty and sustainable development in Central Himalayan region of India. Int J Sustainable Dev World Ecol 10(2):157–168
Satterthwaite D (2003) The links between poverty and the environment in urban areas of Africa, Asia, and Latin America. Ann Am Acad Pol Soc Sci 590:73–92
Schulze RE, Dlamini DJM (2002) Mesoscale Indicators of Water Poverty in the Thukela Catchment, South Africa, under baseline land cover conditions Appendix 9.14 to
Schulze RE (1999) Transcending scales of space and time in impact studies of climate and climate change on agrohydrological responses. Paper given at the GCTE Focus 3 Conference on Food and Forestry – Global change and global challenges. Reading, Sept. 1999
Scoones I (1998) Sustainable Rural Livelihoods: a Framework for Analysis. IDS Working Paper No. 72, Institute of Development Studies, University of Sussex, UK
Seckler D, Amarasinghe U, Molden D, De Silva R, Barker R (1998). World water demand and supply, 1990 to 2025: Scenarios and issues. Colombo, Sri Lanka: IIMI. vi, 40p
Sen AK (1985) Commodities and Capabilities. North Holland, Amsterdam
Sen AK (1999) Development as Freedom. Clarendon Press, Oxford
Sengupta RR, Bennett DA (2003) Agent-based modelling environment for spatial decision support. Int J Geogr Inf Sci 17(2):157–180
Soussan J (2004) Water and Poverty. Fighting Poverty through Water Management. ADB. Manila
Srinivasan TN (1994) Human Development: a new paradigm or reinvention of the wheel? American Economic Review, May
Stern DI, Common M, Barbier E (1996) Economic growth and environmental degradation: The environmental Kuznets curve and sustainable development. World Dev 24:1151–1160
Streeten PP (1994) Human development: means and ends. Am Econ Rev 84:232–237
Streeten PP (1996) Human Development – the Debate about the Index. Int J Soc Sci
Sullivan CA, Meigh JR (2003) The Water Poverty Index: its role in the context of poverty alleviation. Water Policy, 5:5. Oct 2003
Sullivan CA (2001) The potential for calculating a meaningful Water Poverty Index. Water Int 26:471–480
Sullivan CA (2002) Calculating a Water Poverty Index. World Dev 30:1195–1210
Sullivan CA, Meigh JR and Fediw T (2002) Developing and testing the Water Poverty Index: Phase 1 Final Report. Report to DFID, CEH, UK
Sullivan CA, Meigh JR, Giacomello AM, Fediw T, Lawrence P, Samad M, Mlote S, Hutton C, Allan JA, Schulze RE, Dlamini DJM, Cosgrove W, Delli Priscoli J, Gleick P, Smout I, Cobbing J, Calow R, Hunt C, Hussain A, Acreman MC, King J, Malomo S, Tate EL, O'Regan D, Milner S, Steyl I (2003) The Water Poverty Index: development and application at the community scale. Natural Resources Forum, in press
Sullivan CA, Mutamba M, Kozanayi W (2001) Water Use and Livelihood Security: A Study of Rural Households in Southern Zimbabwe. CEH report to DFID
Tharme RE (2003) A global perspective on environmental flow assessment: emerging trends in the development and application of environmental flow methodologies for rivers. River Res Appl 19:397–441
Turner RK van den Bergh JCJM, Soderqvist T, Barendregt A, van der Straaten J, Maltby E, van Ierland EC (2000) Ecological-economic analysis of wetlands: scientific integration for management and policy. Ecol Econ 35(1):7–23
UNDP (United Nations Development Programme) (1999) Human Development Report 1999. Oxford University Press, New York
UNDP (United Nations Development Programme) (2001) Human Development Report 2003, Oxford University Press, New York
United Nations Educational Scientific and Cultural Organization (UNESCO) (2003) World Water Development Report. UNESCO, Paris
Van Tongeren F, Van Meijl H, Surry Y (2001) Global models applied to agricultural and trade policies: a review and assessment. Agric Econ 26(2):149–172
Vanvoris P, Millard WD, Thomas J, Urban D (1993) Terra–Vision–The Integration of Scientific Analysis Into The Decision-Making Process. Int J Geogr Inf Syst 7(2):143–164
Wallace JS, Acreman MC, Sullivan CA (2003) The sharing of water between society and ecosystems: from advocacy to catchment based co-management. Philos Trans R Soc Lond B Biol Dec 2003, pp 513–528
World Bank (2001) World Bank: 2001, World development indicators 2001, CD-ROM, World Bank, Washington DC

World Commission on Dams (WCD) (2000) Dams and Development: A New Framework for Decision-Making. Report of the World Commission on Dams

World Economic Forum, Yale Center for Environmental Law and Policy, and Center for International Earth Science Information Network, Columbia University (2001) Environmental Sustainability Index (http://www.ciesin.columbia.edu/indicators/ESI), January 2001

World Resources Institute (2000) World Resources Institute: 2000, World Resources 2000–2001 – People and Ecosystems: The Fraying Web of Life, World Resources Institute, Washington

WSSD (2002) World Summit on Sustainable Development, Plan of Implementation. Johannesburg, Sept. 2002

Yoffe SB, Ward BS (1999) Water resources and indicators of conflict: A proposed spatial analysis. Water Int 24(4):377–384

Water Resour Manage (2007) 21:129–148
DOI 10.1007/s11269-006-9045-z

ORIGINAL ARTICLE

Capturing the complexity of water uses and water users within a multi-agent framework

Thomas Berger · Regina Birner · José Díaz · Nancy McCarthy · Heidi Wittmer

Received: 20 April 2006

Abstract Due to the hydrological and socio-economic complexity of water use within river basins and even sub-basins, it is a considerable challenge to manage water resources in an efficient, equitable and sustainable way. This paper shows that multi-agent simulation (MAS) is a promising approach to better understand the complexity of water uses and water users within sub-basins. This approach is especially suitable to take the collective action into account when simulating the outcome of technical innovation and policy change. A case study from Chile is used as an example to demonstrate the potential of the MAS framework. Chile has played a pioneering role in water policy reform by privatizing water rights and promoting trade in such rights, devolving irrigation management authority to user groups, and privatizing the provision of irrigation infrastructure. The paper describes the different components of a MAS model developed for four micro-watersheds in the Maule river basin. Preliminary results of simulation experiments are presented, which show the impacts of technical change and of informal rental markets on household income and water use efficiency. The paper also discusses how the collective action problems in water markets and in small-scale and large-scale infrastructure provision can be captured by the MAS model. To promote the use of the MAS approach for planning purposes, a collaborative research and learning framework has been established, with a recently created multi-stakeholder platform at the regional level (Comisión Regional de Recursos Hidricos) as the major partner. Finally, the paper discusses the potentials of using MAS models for water resources management, such as increasing

T. Berger (✉)
University of Hohenheim, D-70593 Stuttgart, Germany
e-mail: 490e@uni-hohenheim.de

R. Birner · Nancy McCarthy
International Food Policy Research Institute

J. Díaz
University of Talca

H. Wittmer
Center for Environmental Research Leipzig-Halle

transparency as an aspect of good governance. The challenges, for example the need to build trust in the model, are discussed as well.

Keywords Multi-agent systems · Simulation models · Integrated water resources management · Collective action · Trade of water rights · Innovations · Chile

1. Introduction

The management of water resources in an integrated and sustainable way is usually complex. This is due to the fact that the scale of management implied by hydrological characteristics often comprises many layers of social, political, and economic institutions. Even at the sub-basin level there may be manifold users such as local small-scale farmers, large-scale commercial farmers, hydro-electric power companies, other industrial users, municipal water users, and those using water resources for leisure and/or tourism. The complexity of different water users is associated with the complexity caused by different uses to which water resources are allocated, including irrigation, potable water, power generation, industrial production, environmental amenities, and recreation. The associated use and management rights are exercised through a variety of institutions that function at different scales. The governance structures for water resources management are often characterized by overlapping national, regional and local regulatory frameworks and authorities. Moreover, irrigation systems are typically common-pool resources and their management involves classical problems of collective action (see, e.g., Ostrom, 1990). There is a variety of mechanisms for allocating water resources, including administrative allocation, water markets and user-based allocation (Dinar *et al.*, 1997). Local water user associations play an important role in water resource management, and in many countries their role has been strengthened in recent years by decentralization and devolution (Meinzen-Dick *et al.*, 1997; Meinzen-Dick *et al.*, 2001). Moreover, negotiation approaches have gained increasing importance in water resource management (Bruns and Meinzen-Dick, 2000).

The last decades have seen increasing efforts to model the complexity of water resource management with the aim to use the model results for improved management (see, e.g., Jakeman and Letcher, 2003). In this paper, we highlight the development and use of a modeling approach that is particularly well suited to capture the complexities of water resource management described above: multi-agent simulation (MAS). MAS makes it possible to couple sub-models for water run-off, crop growth, economic decisions and network interaction within an integrated modeling framework. This is not a unique feature of MAS since there are a number integrated river basin models that have proven their ability to capture the technical, biophysical and economic complexities of water resource management. The specific strength of MAS, however, is that this modeling approach can also represent social and institutional relations among water users, enabling us to more fully capture social phenomena such as *collective action.*

To illustrate this modeling approach, we take a sub-basin of the Maule river in central Chile as an example (compare Figure 1 below). Chile is very well suited for a case study to show the potential of applying a multi-agent modeling framework for planning decision support. The country is internationally recognized for its innovative approach to water resource management. Chile has introduced a system of tradable private water rights and devolved the management of irrigation facilities to water user groups and their umbrella organizations. At the same time, public sector institutions have retained an important role in providing technical information and collecting data as well as in coordination,

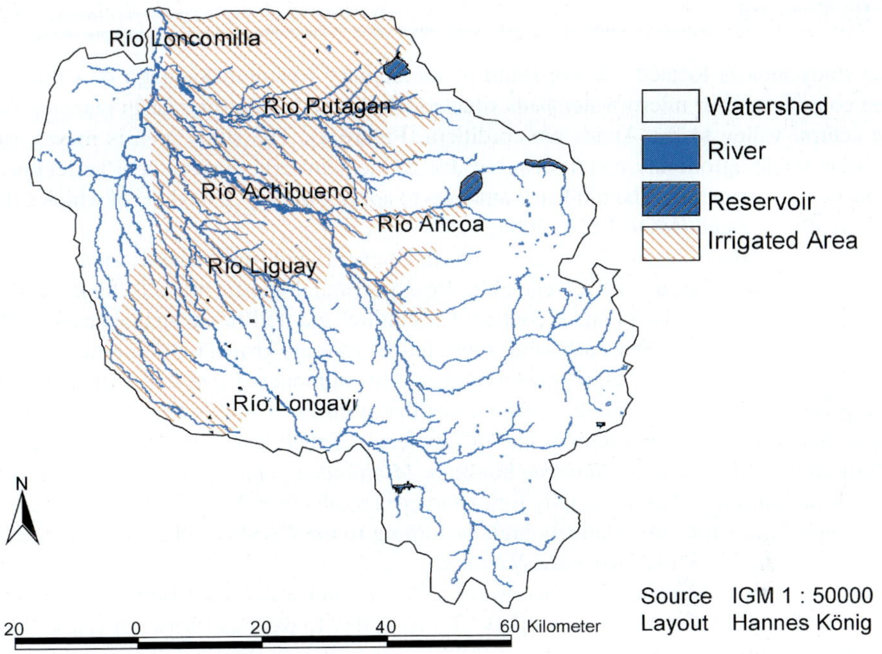

Fig. 1 Study area within the Maule River basin

oversight and conflict resolution. A new trend in water resource management is the privatization of large-scale infrastructure development in the form of concessions. The experience with the Chilean model of water resource management has, however, been mixed. For example, (Bauer, 1998) found that the market incentives to conserve water have been ineffective and that water right trading has been less active than expected. It is also an open question as to how marginalized water user groups in rural areas with low levels of political, bargaining and purchasing power will be affected by the next round of market-oriented management instruments. As we will argue in this paper, multi-agent simulation can provide an important planning tool which helps to better evaluate ex ante different policy scenarios and understand their implications for different groups of water users, including marginalized groups.

This paper reflects an early stage of research and focuses on the conceptual and methodological approach. It is the product of an ongoing research project within the "Challenge Program on Water and Food" of the Consultative Group on International Agricultural Research (CGIAR). The project aims at researching the practical use of a multi-agent simulation model in a multi-stakeholder context. Therefore the model is being developed and applied in a collaborative research and learning framework, which includes local water user organizations, state agencies, and local as well as international research organizations.

The paper proceeds as follows: Section 2 describes the study region. The problems for water resource management to be analyzed in this region are identified in Section 3. In Section 4, we describe the multi-agent simulation approach and its application to water resource management. Section 5 shows how the modeling approach will be applied in the study region. Section 6 concludes.

2. Background

The study area is located 300 km south of the country's capital, Santiago de Chile. The area consists of four micro-watersheds situated within the Maule river basin reaching from the central valley to the Andes pre-cordillera (Figure 1). The study area is mostly rural with favorable agronomic conditions, but due to the semi-arid environment agriculture is dependent on irrigation. The total area amounts to approximately 3,500 km^2 of which a third is currently irrigated (MOP-DGA, 2004).

The area produces high-value export crops in vineyards and fruit orchards, such as apples, grapes and raspberries, as well as vegetables for export and domestic markets. Sugar beet and tomato are mainly produced under contract agriculture, and traditional annual crops such as wheat, maize, beans, potatoes and other subsistence crops are grown for domestic consumption. The farm structure is heterogeneous and consists of modern large-scale export oriented enterprises, traditional hacienda type holdings that produce mainly wheat, sugar beet, and cattle, *campesino*[1] farm households that mainly received their holdings through the agrarian reform in the 1960s, and *minifundista* holdings. *Minifundista* and landless households often provide agricultural labor, especially for export crop production. Poverty levels compared to other regions in Chile are relatively high, according to the CASEN 2002 survey (*Encuesta de Caracterización Socioeconómica Nacional*).

Irrigation relies mainly on surface water resources and water availability varies considerably within years (extreme cases up to 20% from day to day) and between years (due to the phenomenon El Niño La Niña, which has lead to increasing rainfall variability). The irrigation infrastructure dates back to the 19th century; it has been erratically amended and the system is characterized by relatively low water-use efficiency. This is due to traditional irrigation methods, high levels of uncontrolled return flows, losses through infiltration, and lack of storage facilities for buffering overnight water flows. Beginning in 1981 with the establishment of a Water Code, Chile implemented a market-oriented approach to water resource management and assigned water rights to individual users. Water user rights are defined as shares of the total available water volume and tradable if registered. The poor infrastructure, however, makes monitoring of individual user's water uptake and water trade difficult.

2.1. Institutional set-up

Within the private water sector, water user groups are organized as follows: At the lowest local level users are organized in *Comunidades de Agua* (Water Communities); these groups are responsible for water distribution to users' plots, maintenance of tertiary and secondary irrigation channels, and the collection of fees for higher level organizations. The *Asociación de Canalistas* (Canal Associations) is the next highest level and comprises all those *Comunidades de Agua* that receive their water from one irrigation channel. They are responsible for the maintenance of this canal and distributing the water to the secondary canals that provide the water to the *Comunidades de Agua*. *Juntas de Vigilancia* (Watch Committees) are in charge of water distribution from a stream or reservoir to the primary channels. If smaller

[1] The category *campesino* comprises farm households that have land to employ the family members. They include beneficiaries from the Agrarian Reform but also farms in colonization areas and indigenous groups. *Minifundistas* hold too little land to employ the labor of their family and necessarily require off-farm income.

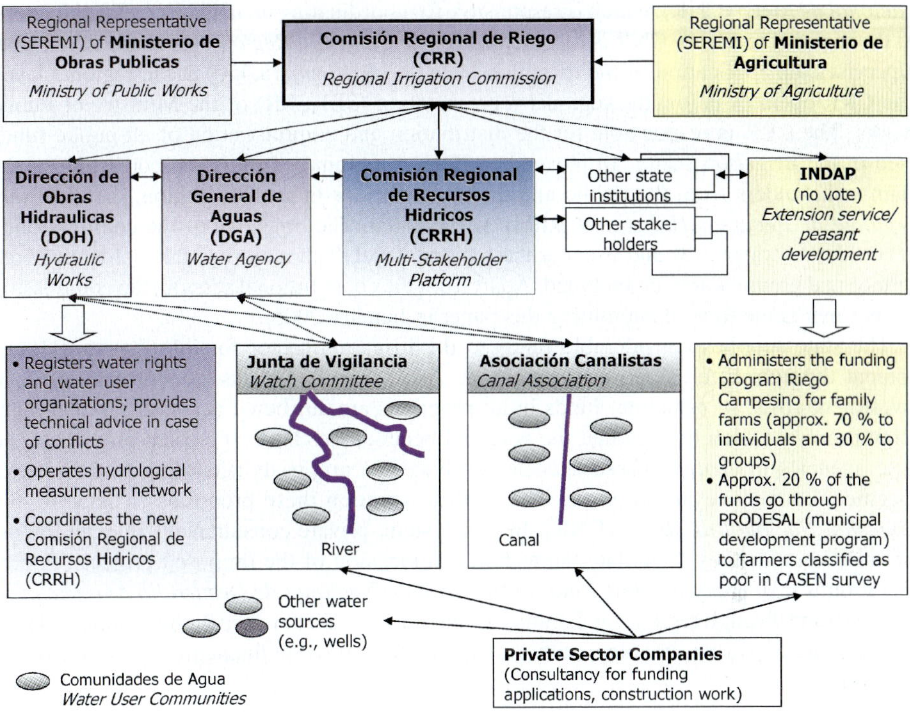

Fig. 2 Institutional set-up of the water sector at regional level

canals receive their water directly from a river or stream there is no intermediate level of an *Asociación de Canalistas*.

The most important surface water use besides agriculture in the 7th Region of Chile – the administrative entity pertaining to the study area – is hydropower generation. Currently there is no hydropower generation facility in the four micro-basins falling within the study area; however, concerns about the security of supply of natural gas from Argentina have increased the pressure to consider new hydropower projects. Conflicts between hydropower generation and irrigation mainly concern the timing of water release. Recreational use of water reservoirs currently occurs to a minor extent, but this may increase, especially in response to increased regional tourism. Another issue of increasing importance is water quality. The main concerns are poorly purified sewage water and contamination with agro-chemicals. Finally, the implementation of 'good agricultural practices' (EU-determined quality standards for high-value export crops) also has implications for both farming practices and water use and management. Certification to these standards has been a prerequisite for export to EU and US markets since 2002.

Within the public sector (Figure 2), there are two units belonging to the Ministry of Public Works (*Ministerio de Obras Públicas*) responsible for water management: The *Direccion General de Aguas* (DGA) and the *Dirección de Obras Hidraulicas* (DOH). DGA is in charge of registration of water rights and water user organizations and in charge of operating a measurement system for water flow. DOH is the technical counterpart in charge of the development of hydraulic infrastructure and of the maintenance of large-scale infrastructure such as reservoirs and bridges. Both entities, DGA and DOH, form part of the *Comisión*

Nacional de Riego (CNR), which is responsible for coordinating all irrigation related matters. At the regional level, the CNR is represented by the *Comisión Regional de Riego* (CRR) that supervises the application of the irrigation law (*Ley de Riego* 18.450) at the regional level. The CRR operates below the regional representation (SEREMI) of the Ministry of Public Works. The CRR is responsible for the distribution and administration of all public funds used in the irrigation sector. In November 2003, a regional commission consisting of the main stakeholders from the public and the private sector in the 7th Region, the *Comisión Regional de Recursos Hídricos* (CRRH), was created. The objective of the commission is to coordinate water uses and users at the regional level. The majority stakeholders of both surface and groundwater are included. Apart from the constitutional session, the commission has not met at the time of submitting this paper in January 2005.

The state supplies considerable funds to the irrigation sector for infrastructure development through different funding programs. As a general rule, users contribute 30% of investment costs. In principle, funds in all programs are attributed according to a competition process, which implies that the potential beneficiaries have to present projects to the state agencies in charge. These agencies evaluate the proposals according to the criteria specified for the relevant program. The final decision on these proposals is made by the *Comisión Nacional de Riego* (CNR). In this system, private consultancies play an important role because they formulate the technical dimension of the proposed projects. Poorer user groups may apply for state funds of the *Instituto Nacional de Desarrollo Agropecuario* (INDAP) and through regional and municipal institutions. However, in highly politicized procedures, municipalities and other authorities sometimes provide financing before or close to elections.

2.2. Market-oriented policy instruments

Although laid out in legal documents much earlier, it is only in the last four years that Chile has actually implemented market-oriented policies in infrastructure development; private concessions to the provision of large-scale infrastructure, and the transfer of existing infrastructure to water user groups. The first experience with concessions to infrastructure development is the Illapel reservoir (*El Bato*) in the 4th Region of Chile, which started in 2001 and was expected to be functioning in September 2004. The company that won the concession to build this reservoir received a state subsidy of about 75% of the investment costs to lower the price that water users will finally have to pay per cubic meter (Chileriego, 2001). Recently, the process of transferring existing infrastructure to water user groups also gained speed. A pre-requisite to these transfers is the legalization of the *Comunidades de Agua*, which is now extensively promoted by CNR and DGA. There is no general rule of how to transfer the infrastructure and how to share the costs. Depending on the socio-economic circumstances and other externalities, infrastructure can be completely transferred without any cost. In other cases, water user groups are asked to refinance the state-built infrastructure, or at least a certain share, which may explain why user groups are often reluctant to make these transfers, especially when rather old and poor infrastructure is to be devolved.

3. Problem statement

The complexity of water uses and users makes it difficult to manage water resources in an efficient and equitable way. Especially the second aspect, equity in access to resources and distribution of benefits, has not been investigated yet for the new market-oriented policy instruments in Chile. In this section, we present three policy research issues that will be analyzed with our multi-agent modeling system; trade of water user rights, development of small-scale infrastructure, and development of large-scale infrastructure.

3.1. Trade of water user rights

Even though trade in water is one of the theoretical advantages of Chile's liberalized water system, only one of the canal associations in the study region (Digua system) has so far developed the capacity to make this trade possible on a significant and regular basis. This leads to a number of policy-relevant research questions:

- What explains the differences across user associations in number of permanent and temporary transactions of water rights? Is trade hampered by poor infrastructure and monitoring facilities? Or are there only small potential gains from trade because of minor differences in water shadow prices of individual users?
- If there are indeed potential gains from trading of water rights, what are the most binding factors that restrict exploiting them more fully? Currently, short-term informal rental contracts are the most frequent form of water trading. What would be the implications of institutional innovations such as secured long-term rental arrangements?
- What is the nature and extent of externalities arising from trade of water rights? In one example, water users (Maitenes channel) report high infiltration losses in a secondary channel and insufficient water flows for irrigation because tail-end water rights were transferred and as a consequence less water flow was diverted into this channel.
- What are the distributional consequences of increased trade in water rights, particularly impacts on smallholders and farm laborers?

3.2. Development of small-scale infrastructure

Investment into and maintenance of small-scale infrastructure is a typical case of *collective action* problems at the lowest level of user organizations (here: *Comunidades de Agua*). Water users have to make individual contributions to jointly-used infrastructure; if a critical mass of users cannot be reached, joint-use facilities are never set up. Additionally, without effective collective action, existing facilities collapse because of lack of maintenance. Examples in the Chilean study region are canal maintenance and improvement, monitoring of individual water uptake and installment of measurement equipment, joint investments such as overnight storage, deep wells, and small-scale hydropower. This again poses a number of policy-relevant research questions:

- What are the incentives for individuals' to participate, given the extent of participation by others?
- What factors affect individual incentives, and what policies/mechanisms could be used to changes these incentives when they inhibit collective action? How are incentives to participate in one activity affected by collective action in other spheres?
- What factors at the water user association level appear to help or hinder collective action in any sphere? (e.g. heterogeneous interests, trust amongst community members, historical experiences)

3.3. Development of large-scale infrastructure

Planning of large-scale infrastructure projects, e.g. the two new multi-purpose dams of Ancoa and Longaví reservoir, involve interactions at higher levels of water user organizations. It is also not clear how the new market-oriented approach of giving out private concessions to build water infrastructure will affect holders of existing water rights versus new holders of additionally assigned water rights. Again, there are several research questions of policy relevance:

- What is the capacity of local user organizations to manage interactions with higher-level organizations and with government agencies? And during the planning process, what is the impact of information asymmetries on the concentration of assets such as land resources?
- How can competition between various water uses (here: hydropower, irrigation, and recreation/tourism) be reconciled in terms of quantity, timing, and quality? What are likely externalities for upstream and downstream water users?
- What are the likely distributional effects of private concessions in terms of access to water and poverty alleviation? Will new infrastructure projects improve the security of water supply for current holders of water rights versus newly assigned water rights?

Addressing these policy issues requires a methodological framework that captures both the complexity of multiple water uses and water users. In the next section we present multi-agent systems as a suitable modeling approach to accomplish this task and to provide policy-relevant information for planning decisions at local and regional level.

4. Multi-agent framework

In recent years much advance has been made in developing computational tools for planning decision support in water resource management. Most research efforts have been multi-disciplinary, for example down-scaling of global circulation models, coupling of meteorology/hydrology models (Kunstmann and Stadler, 2003) and integrated river basin models (Rosegrant *et al.*, 2000; Fischer *et al.*, 2002) to name a few. Integrated river basin models have been instrumental in capturing the complexity of water uses and highlighting the impacts of water reallocation across riparian countries and water-using sectors. These models, however, typically aggregate the categories of water use and do not capture the complexity of multiple water users within the different sectors (Berger and Ringler, 2002). As a result, they provide only broad and general insights for the ex-ante assessment of water management options at lower levels of user organizations. A new class of integrated computational models is therefore needed to explore distributional consequences of market outcomes, simulate corrective policy interventions, and prioritize funding of infrastructure projects.

4.1. Multi-agent systems applied to natural resource management

Here, we utilize a multi-agent system (MAS) model that couples water run-off, crop growth, economic decision and network interaction models at the water user level. MAS applied to natural resource management are generally implemented with fourth generation, object-oriented programming languages. They consist of two components; a cellular model component that links biophysical process models within a grid-cell framework, and an agent-based model component that links socio-economic decision and market models. Their specific characteristic is that all and each real-world actors are represented one-to-one by computational agents. MAS models have been applied to a variety of research questions (for an overview see Parker *et al.*, 2002; Janssen, 2002; Parker *et al.*, 2003), ranging from theorizing about social and spatial dynamics (Gotts *et al.*, 2003; Parker and Meretsky, 2004), simulating diffusion of innovations (Weisbuch, 2000; Berger, 2001; Deffuant *et al.*, 2002), land-use changes (Huigen, 2004) and agricultural policies (Happe *et al.*, 2004), to accompanying role-playing games (Barreteau *et al.*, 2003) and game theory applications (Bousquet *et al.*, 2001).

Of particular importance for this study are *empirical MAS* that are based on empirical data and used for policy analysis (see for example Berger *et al.*, 2006). New statistical approaches are required for the empirical parameterization of these models; common sampling frames that hierarchically link observation units and guide biophysical field measurements and socio-economic surveys (Van de Giesen *et al.*, 2006) along with Monte-Carlo techniques for generating agent populations from sample data (Berger and Schreinemachers, 2006).

4.2. Model specification for case study

The MAS model specification for the case study reported here builds on earlier work of (Balmann, 1997; and Berger, 2001). The model is recursive/dynamic and is used for comparing alternative policy-planning scenarios (simulation horizon 15 to 20 years). It is coded in C++ and runs on Windows and Unix/Linux platforms.

The cellular model component consists of the following sub-models (Figure 3):

1. WaSiM-ETH for the simulation of water flows and balances at micro-watershed level (resolution 100 m, daily time steps). Inputs required are meteorological data, a digital

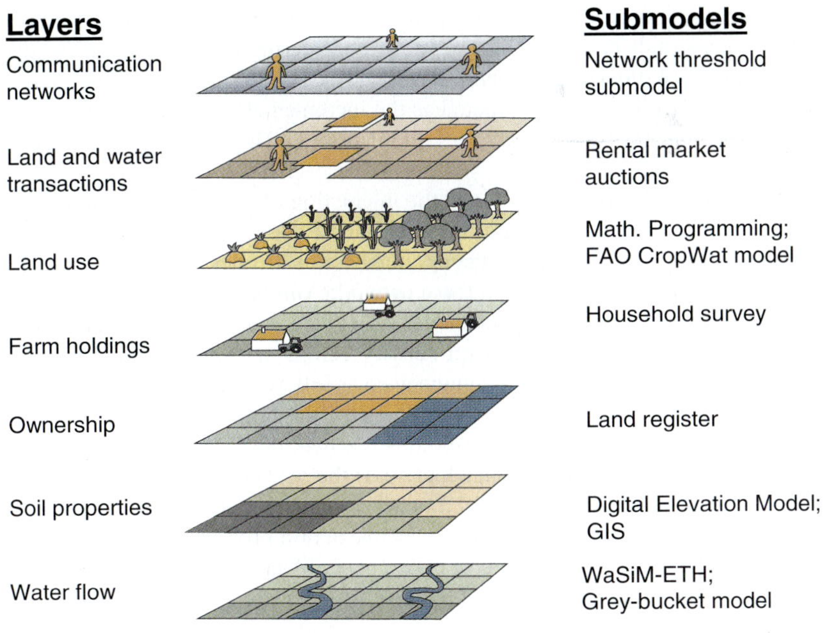

Fig. 3 Components of the Multi-Agent Simulation Model Layout C. Block, ZEF-Bonn. First published in Berger and Ringler (2002)

elevation model and gridded information on soil properties and land use (Jasper *et al.*, 2002).

2. A digital elevation model and GIS coverages such as water infrastructure, location of farmsteads and parcels (taken from land register), soil types (USDA classification), and land cover (Landsat satellite images). Data from field measurements and socio-economic surveys was geo-referenced and incorporated into the database.
3. A grey-bucket model for the diversion of irrigation water within irrigation sections (here: *comunidades de agua*), building on a water-engineering study for the Chilean Ministry of Public Works (see model equations in Berger, 2000). This model will be calibrated with run-off measurements during the peak irrigation season.
4. A crop growth model building on the CropWat model (FAO, 2005), calibrated to regional conditions (about 180 cropping activities including fruit and forestry plantations). Technical coefficients such as labor and fertilizer inputs are being collected from the project's farm holding survey complemented by farm trial data of the University of Talca and INIA (*Instituto Nacional de Investigación Agropecuaria*).

The agent-based model component consists of the following submodels (see detailed description in Berger, 2001; listing of parameters and equations in Berger; 2000):

5. An agent planning and decision model implemented as a sequence of mixed-integer mathematical programming problems (MIPs). The IBM-OSL runtime library is used for solving these problems. The model agents may either represent (a) single water users such as farm households, or (b) the aggregate of all users within one water-using sector. In case (a) recursive MIPs for each and all agents are solved for all years over the simulation horizon, allowing for interactions between them. In case (b) the model specification is equivalent to integrated river-basin models and used for 1 year comparative-static simulation experiments.
6. Transactions of land and water resources are implemented as local auctions submodels. A rental market mediator compiles a list of parcels for renting out and then asks for bids of all potential tenants (bids based on shadow prices, considering the transport costs (distance) between the offered plot and the tenant's farmstead). Provided the highest bid is larger than the landowner's asking price (computed as the incremental return when renting out this parcel), the price is fixed as the average of asking price and highest bid, and the transaction takes place.
7. A network threshold submodel captures the interaction of agents in communication networks. Innovation information, for example about the adoption of water-saving irrigation techniques, is typically exchanged among members of distinct social networks. The position of a particular human actor in these networks can be empirically measured by its network threshold, defined as the percentage of all other peers within its reference group that must previously engage in a novel activity before the actor eventually adopts this behavior. The network submodel is implemented as follows (for details see Berger, 2001): model agents monitor the present adoption level of an innovation and compare it with their individual threshold. If their threshold is reached, they evaluate the innovation and calculate the farm's net benefits from adoption. If the net benefits are positive, they adopt and thereby increase the adoption level monitored by other network members.
8. A critical mass submodel that captures collective action problems of water users within local level user associations, for example maintenance of and water uptake along secondary and tertiary irrigation channels. Model agents form expectations about the other agents' initial dispositions to cooperate and monitor their behavior. Depending on the distribution of incentives over agents the critical mass of cooperating users may not be reached, which in turn would lead to collapse of the maintenance and monitoring systems. Under such

circumstances the MAS model allows to test policy interventions that give sanctions and/or additional incentives to foster cooperative behavior of agents.

Note that comparing the case (a) and (b) specifications of the agent planning and decision submodel (see point 5 in list above) provides useful insights for policy analysis. Case (b) represents the optimal water allocation aggregated over all water users in the absence of information and transaction costs. This optimal solution based on pooling of resources accordingly quantifies the potential gains from a perfectly coordinated allocation of water among multiple uses and users. Disaggregated case (a) scenarios compared against this benchmark indicate the magnitude of information and transaction costs. Simulation experiments would then help to identify the most binding constraints to exploiting these potential gains (see for more details Berger and Ringler, 2002).

4.3. Preliminary simulation results

In the following we present some preliminary simulation results which show the type of quantitative information generated by the MAS. Since the socioeconomic surveys are under way and some of the model components are still under construction – for example WaSiM-ETH and updated GIS coverages –, we used the model specification and data from (Berger, 2001) to analyze the dynamics of water trade, water user income and water use efficiency in the study region. Under the current institutional setting, as mentioned in Section 3, water trade is almost exclusively done informally and on a short-term rental basis. The rental payment is typically due at the beginning of the one-year contract, and the tenant assumes the uncertainties related to actual on-field water availability. Such rental arrangements, therefore, provide only small incentives to the tenant to adopt technical innovations such as high-value permanent crops and new water-saving irrigation techniques. We would therefore expect a very limited effect of water rental trade on water user incomes and water use efficiency. The MAS is used to quantify this effect vis-à-vis the underlying dynamics of technical change in the study region. Again, the simulation results here are preliminary since they are based on data available in 1997 (for a discussion of model validation see Berger, 2001).

Figure 4 depicts the effects of technical change on average household income and on on-field water use efficiency (for details of the technology adoption model see Berger, 2001). Ideal technical change – ideal means here the diffusion of innovations occurs without any adoption costs – could almost double the average household income, compared to the hypothetical situation without innovation. Average water-use efficiency would then considerably improve within a few years and reach a maximum level of about 45%. But under market conditions, the simulation experiments suggest only modest increases of income and water-use efficiency. Household incomes are likely to increase by 11%, compared to the situation without innovation, and water use efficiency may reach a maximum level of about 30%.

The MAS provides further insights into the impacts of current water trade arrangements. We undertook a series of simulation experiments to quantify the likely effects of both technical change and rental markets, which are summarized in Table 1. To isolate the effect of rental markets we either switched the auction module off (first column for each indicator) or on (second column for each indicator). First, note that the scenario 'market solution' in row #2 and column #2 replicates the observed level of rental activities of about 11% (see Berger, 2001). Second, we see rental activities decrease the more households are reached by technical change. Third as column #3 and #4 show, rental markets have a relatively large effect on the average household income. The average household income would drop by about 13%, compared to the baseline scenario with rental markets and without technical change. The

Table 1 Simulation experiments on the impacts of informal rental markets

Values in percent rental markets	Rental market activity		Incremental household income		On-field water use efficiency	
	No	Yes	No	Yes	No	Yes
Ideal technical change	0.0	2.8	78.4	80.4	43.0	42.7
Market solution	0.0	11.5	−13.5	11.0	26.3	28.4
Without technical change	0.0	13.1	−13.5	0.0	26.3	25.3

Notes. Preliminary simulation experiments based on data of 1997. Rental market activity is measured as the percentage of transferred water/land resources per year. Incremental household income refers to the discounted increase of average household income ("Net Benefit Increase"), compared to the baseline scenario (rental markets – without technical change)

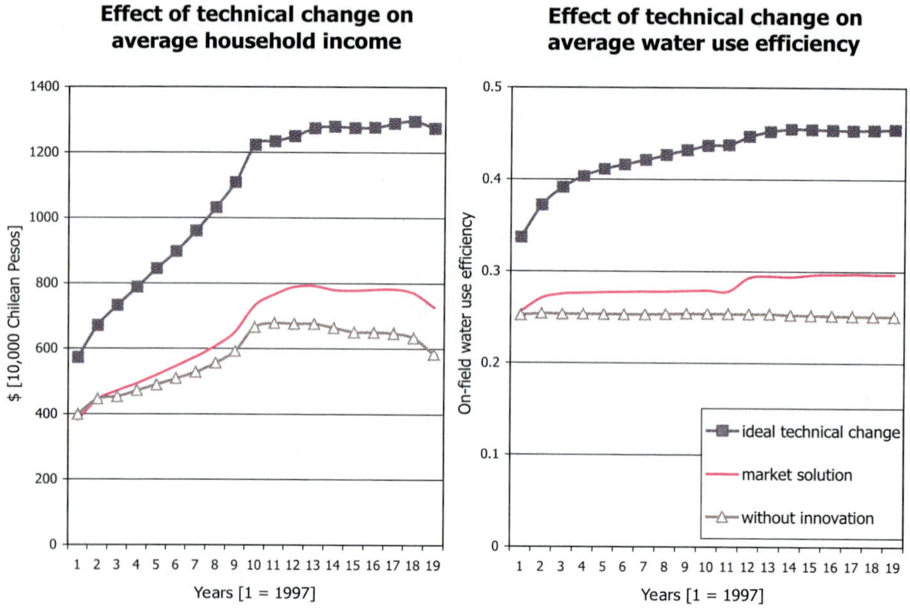

Fig. 4 Simulation experiments on the impacts of technical change

reason is that farm households that are not reached by technical change engage in off-farm activities and even opt for out-migration, which in turn increases supply on the rental markets. However, under the current water trade arrangements, this re-allocation of water/land resources has had rather limited effects on on-field water-use efficiency, as columns #5 and #6 show.

In summary, the simulation experiments highlight the large potential from technical innovation but the limited potential from water trade under current rental arrangements.

5. Outline of simulation approach

In Section 3, we identified three policy relevant research questions; trading water rights and the development of small-scale and large-scale infrastructure. In this section we outline how MAS can be used to address these research questions. This section is divided in five sub-sections. The first sub-section characterizes the collective action problems that are inherent in water resource management and that can be analyzed using MAS. In the second to fourth sub-sections, we specifically discuss collective action problems that may arise in water trading and in small-scale and large-scale infrastructure provision, respectively. The final sub-section describes the collaborative research and learning framework that is being established for the use of MAS.

5.1. Collective action problems in water resource management

As indicated above, the specific strength of MAS is that the model can represent social and institutional relations among water users, enabling us to more fully capture social phenomena such as collective action. And, the nature of the model lends itself to continuous process two-way information flows (user-model-user). Both of these characteristics are extremely important in the development of a planning support tool in the complex institutional environment operating in Chile, where irrigation infrastructure is (often) jointly used and managed, and thus must be jointly maintained and monitored. Such jointly used goods require collective action since all users benefit from a well-maintained irrigation system; however, individuals may not fully provide the public goods and services required for optimal investment and maintenance. This situation constitutes a so-called social dilemma. To solve such dilemmas game theory has made important contributions (for specific reference to public goods provision, c.f. Cornes and Sandler, 1986; Sandler, 1992). Nonetheless, analytical representations of game-theoretic models often lead to ambiguous results, particularly when there are many heterogeneous agents, stochastic parameters, etc. The latter means that empirical data to support parameterization of the model is critically important.

In terms of collective action and the functioning of the irrigation system, there are three critical aspects to consider. The first is characterizing the (different) individual incentives to undertake any given collective activity, given the institutional rules and transactions costs for undertaking collective action. Using MAS, we can capture these incentive structures, determine how changes in various external parameters alter these incentives, and thus develop well-defined and targeted policy options to change these incentives and thus promote collective action. Secondly, we can empirically evaluate the transactions costs of collective action at the level of local water-user associations, and also determine how changes in local institutional rules and mechanisms can alter individual incentive structures. Third, we can evaluate links amongst different stakeholders throughout the system, including those between local level water user associations to higher level irrigation agents, between current rights holders and those who may gain from new rights, and between stakeholders representing very different interests in use, e.g. irrigation, hydro-electric power, and recreation.

While the general institutional framework of water use in Chile has been briefly described in Section 2 of this paper, in the next section, we outline more clearly different types of collective action problems that may arise within the context of trading water, development of small-scale (local) water infrastructure, and development of large-scale (supra-local) infrastructure. Though incorporating the institutional framework determining the extent of collective action into the MAS model remains our most important future activity related to data collection, in Section 5.2, we present an illustrative example of the distribution of net gains arising from the discrete decisions to invest in water flow monitoring equipment; space constraints prevent us from doing the same for sections 5.3 and 5.4 here.

5.2. Collective action problems in water trade

We first consider aspects of water trade that imply a potential role for collective action. An important condition for the emergence of trading in water rights is that the purchasers of these rights are assured that they will indeed receive the water in accordance with the rights purchased. In cases where water flows from all primary, secondary and tertiary canals are monitored using sophisticated water flow measurement equipment, the development of water trading rests on the enforcement of water rights using the information collected; the technology provides the information needed for enforcement and there is little need for collective action. Even under this system, however, negative asymmetric externalities may be generated which affect those not party to the trading itself. For instance, if there is a concentration of a significant number of users renting out their rights in some part of the system, this may alter the flow of water throughout the system thereby affecting other users (see Section 3). Avoiding this type of negative externality requires an understanding of the entire irrigation system, and thus a forum where different users can acquire this information and negotiate new rules regarding transfers.

Where information on distribution is not monitored with equipment, collective action and coordination activities become more important. Consider first trading of rights within the same water user organization. If everyone with rights believes that the distribution system functions well enough in the sense that they indeed receive their share of water, then it may be a simple matter to trade water rights. Even in this case, however, community members as a whole must "support" this trade in the sense that, with traded rights, members do not change their behavior compared to the no-trade (or less trade) situation. If the current "no-trade" system functions because near-neighbors monitor each other's use, then a downstream user considering the purchase of an upstream user's water allocation must be assured that the *seller* will still monitor his/her neighbors' use. Yet, the incentives for the seller to provide monitoring, particularly if payment is made at the beginning of the season, are reduced vis-à-vis the situation where monitoring is necessary to ensure that one's own allocation is received. This monitoring problem can be expected to be reduced to the extent that a seller wants to have the option to rent out rights in the future as well. However, the larger the distance between the purchaser and the seller, the greater are the chances that others not involved in the contract become tempted to take a little bit more, since more water is flowing through. To offset this, someone – most likely the purchaser – will have to increase monitoring.

The monitoring and enforcement problems become even more severe when trading occurs across different communities of water users. Though representatives of community water user associations can attend meetings at the next higher level of aggregation (*Asociación de Canalistas* or *Junta de Vigilancia*), there is often no formal framework for negotiating trades.

A simple example may help to illustrate the information on incentives that will be generated by the model. Consider a simple 6-person game, where the players choose whether or not to invest in water flow monitoring equipment. Additionally, assume that, without the new equipment no trades are made. Each individual then chooses between two options: either to not contribute or to make 1/Nth of the total contribution required. If all do not contribute their share, contributions are returned at no cost. Players differ only in their productivity of water use, with player 1 having the highest productivity, and player 6 the lowest.

		Players 2:6	
		All Contribute	None Contribute
Player 1	Not Contribute	2804, [411- 2403]	1877, [7- 1551]
	Contribute	1877, [7- 1551]	1877, [7- 1551]

Fig. 5 Payoff matrix, investing in water monitoring equipment, Player 1 and Players 2:6

The payoff matrix shows the gains from either contributing or not contributing (Figure 5). Returns to player 1 are always in the first place in each column; for simplicity, we have summed the returns to players 2–6 into one grand "player", and shown the returns to contributing and not contributing for this group as a whole. Underlined returns show the "best response" of each player to the action of the other; the grey cells indicate the two equilibrium of this game. As we can see, there are two equilibria: either all contribute or none contribute; a standard result for a discrete investment decision. In this particular case, the incentives to cheat – the difference between all contributing vs. all contributing except the player in question – are in fact negative (and equal to negative net gains). Incentives to not be taken advantage of–the difference between no one contributing and the player contributing when no one else does–is equal to zero, since contributions are returned costlessly if not enough funds are raised.

Figure 6 illustrates total net gains to each player if the investment is made, under two different scenarios: the first where the fixed investment cost per member is $100, and the second where the investment cost is $200. Though extremely simplified, the Figure 6 highlights two interesting concepts. The first, well-known to trade theorists, is that gains to trade are the greatest for the most and least efficient under both scenarios. Second, while in both scenarios all players find it in their best interest to contribute, the total amount of water traded and total net benefits decrease. If fixed investment costs rise just a bit more, Player 4 will lose money by making the required contribution; and if a strict per person payment system is enforced, the investment will not be undertaken. In other words, it is the "medium" efficiency group that will gain least from trading. Heterogeneity implies that a simple per person charge is less likely to lead to investment; and intra-group bargaining over the investment payment schedule will then ensue. On the other hand, if each member is allowed to choose how much to contribute, then a sub-group may form to provide the investment in any case; here, we would expect both the least and most efficient water users to pay the greatest amount.

5.3. Collective action problems in small-scale infrastructure provision

The second research question addresses development of small-scale infrastructure, such as the purchase and installation of water distribution monitoring equipment, overnight water storage facilities, and the maintenance of secondary and tertiary canals. As discussed above, investing

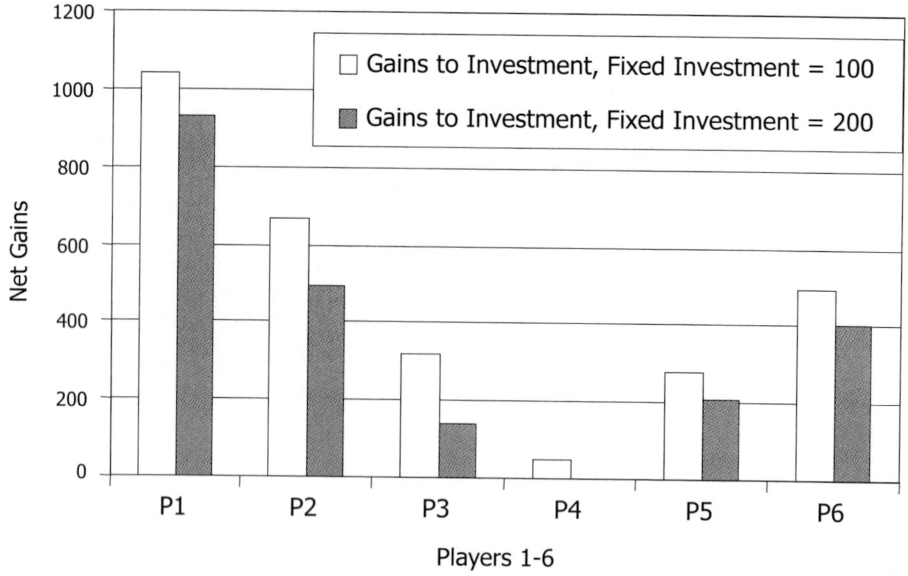

Fig. 6 Net gains from investing in water monitoring equipment and subsequent trade

in water distribution monitoring equipment provides information on actual quantities of water used by each member; such information provides a pure public good even in the absence of water trading. The discrete decision to invest in storage facilities has similar characteristics to the decision to invest in water monitoring equipment.

Additionally, most irrigation investments require periodic maintenance at both the household and community level. Water users must make agreements on maintenance activities, assign responsibilities, and monitor and enforce these agreements. Monitoring maintenance contributions may well be more costly than ensuring everyone contributes their share to a specific investment. In the case of discrete investments described in Section 5.2, as long as net gains to the investment are positive, then there are no incentives to free-ride (in fact, there are "negative incentives"), and no incentives to not be taken advantage of. In the case of maintenance, where each additional unit of labor or money has a marginal impact on the condition of the canal, there are likely to be both incentives to free-ride, and incentives not to be taken advantage of, even when there are net gains to all if all contribute their share. Figure 7 below illustrates a 2 player payoff matrix illustrating this potential outcome.

Fig. 7 Payoff matrix, contributing to canal maintenance, 2 Players

Player 1 \ Player 2	Contribute	Not Contribute
Not Contribute	2000, 2000	1700, 2100
Contribute	2100, 1800	1800, 1800

As with the discrete investment decision, players are indeed better off if all players maintain the canal, but unlike the investment decision, there are now both incentives to free-ride as well as incentives to not be taken advantage of. This result, a "prisoner's dilemma", often occurs with continuous decisions where inputs exhibit diminishing returns and costs expended can not be "returned". In these cases, then, policies or rules need to be constructed that both assure each member that all will contribute (reduce the losses that can occur if you contribute but no one else does), and that reduce your incentives to free-ride (decrease returns to free-riding by selective punishments, for instance).

As with water trading, the MAS model can be used to generate the payoff matrices that characterize the maintenance and investment environment, the distributional consequences of existing investment and maintenance activities, the change in net benefits due to changes in the external environment, and thus the types of policies and/or institutional mechanisms that can be used to alter incentives that currently hinder maintenance and investment.

5.4. Collective action problems in large-scale infrastructure provision

In case of large-scale infrastructure, such as large reservoirs, the investment will affect a much larger group of users than those comprising a water user association and it will often affect users other than farmers, such as hydro-electric power plants, industry, tourists, and municipal authorities. Large-scale infrastructure projects often generate positive externalities, such as a less fluctuation in water flow and new tourism potential, but may also generate negative externalities, such as reduction of biological diversity. Reconciling different water demands will be a much more difficult task, and will require some type of institutional structure within which negotiation and bargaining can take place. The different groups involved may differ in their abilities to organize and articulate their interests, which constitute an inherent problem for bargaining and thus a key role for collective action, especially among resource-poor constituents.

Apart from these local collective action problems, the provision of large-scale infrastructure also involves a number of other problems. For example, the relation between established political decision-making processes and platforms, some created specifically to increase involvement of "all" stakeholders, needs to be clarified in order to capture the distribution of bargaining power and capacity of different stakeholders to shape certain institutional rules. As discussed in Sections 2 and 3, privatization is an option high on the current political agenda. The privatization of large-scale infrastructure provision, based on a concession system, creates additional challenges for decision-making, and for ensuring that public interests are assured.

The MAS can be used to simulate the effects of different types of institutional arrangements for provision and management of large-scale infrastructure. For example, MAS can be used for assessing the effects on all users of giving seasonal priority to hydropower. The MAS can also show the incentives for a private firm to invest in the provision of large scale infrastructure such as reservoirs to increase the amount of water stored. The MAS can also generate different shadow prices for water, thus showing were new water rights might receive the highest prices as well as calculate the distributional effects among the different users. Thus, the effects of different terms of concession contracts can be simulated.

5.5. Collaborative research and learning framework

The main goal of the research project is to develop a modeling system that is of practical use for decision makers faced with the collective action problems identified above. An important prerequisite for being able to provide decision support in practical planning processes is to reach a state in which the different stakeholders have trust in the model and its simulation results. The MAS has a distinctive feature that facilitates trust building. Since the model depicts specific agents, it is easier for model users to identify themselves with these agents and the way in which their behavior is modeled than to understand or trust a set of differential equations.

In the case study under consideration, a collaborative research and learning framework has been set up to provide a platform where the MAS can be developed jointly with those interested in using it. This approach is expected to promote ownership and build trust in the model. The joint development has three objectives: (1) to improve the quality of information used to set up and parameterize the model, (2) to ensure that the relevant questions and criteria can be addressed with the model as well as to identify relevant and feasible policy options and (3) to provide decision makers with access to the model which ideally includes training them to use it.

In order to identify the most appropriate counterpart institution for this joint development, an analysis of the existing governance structures involved in irrigated water use and management was conducted as a first step of the research project. Using information from the governance structure analysis, a stakeholder needs and priorities analysis was undertaken, to ensure that the interests of various stakeholders will be captured by the MAS model. Additional economic and institutional data required to generate parameters to baseline the model are being collected by means of a household survey, a survey at the level of the leaders of the water user associations, focus group interviews within the communities, and key informant and expert interviews at these and higher levels.

6. Discussion and conclusions

The case study presented in this paper shows that multi-agent simulation has a considerable potential for providing planning decision support to policymakers and stakeholders. We identified three areas of water resource management, where MAS can make particularly useful contributions: trade in water resources, small-scale infrastructure and large-scale infrastructure provision. Each of these areas is characterized by specific bio-physical, technical and socio-economic problems. While the bio-physical and technical aspects of multiple water uses are generally well studied and analyzed with existing model approaches, capturing the collective action problems of multiple users and their socio-economic implications in these three areas constitutes a major challenge.

We have argued for each of these three management areas that MAS has a comparative advantage in dealing with this challenge, because it makes it possible to represent and disentangle the diversity of real-world actors and their interrelations. Game theory can be combined with MAS to model the behavior of these actors. Testing different behavioral assumptions and pay-off matrices allows the analyst to gain insights into the underlying distribution of incentives and the dynamics of collective action problems, and to determine how cooperative outcomes may be encouraged by additional policy instruments.

The fact that multiple actors can be included in the model makes it possible to place emphasis on distributional effects of market-oriented policy instruments that may be overlooked otherwise. For example, the implications of infrastructure development for poor and disadvantaged sections of the rural population can be modeled explicitly. Likewise, complex environmental consequences can be better understood and considered in the planning process.

A distinctive advantage of MAS is that it specifies the trade-offs involved in different scenarios, thus allowing policymakers to make better informed decisions. By providing information to the public, the use of MAS may also increase transparency, which is an important aspect of good governance.

The use of MAS also involves challenges which will be briefly discussed here. A major challenge is to ensure the sustainability of using MAS in practical planning situations. Since the model is intended to be used in a multi-stakeholder setting, there are considerable social and management skills required to handle such model-enhanced planning processes (Paassen, 2004). Therefore, there is a need for constant learning and capacity building in the institutions developing, maintaining and using the MAS. As indicated above, another major challenge is to ensure that the stakeholders and the political decision-makers build trust in the model and its simulation results. Since the use of an MAS does not in itself change existing power structures, efforts have to be made to avoid the "capture" of the model by more powerful stakeholders, who may, for example, influence the choice of the scenarios that are – and are not – modeled. Another challenge of this approach is the need for multidisciplinary cooperation between members of various natural science and social science disciplines (Berger *et al.*, 2006). There is also a need to establish an interface for different types of knowledge, including local knowledge and scientific knowledge. This involves communication challenges and requires a learning process that is not trivial. In conclusion, we hope to have shown in this paper that the potential offered by MAS justifies facing these challenges.

Acknowledgements Funding of the CEIAR Challenge Program on Water and Food and of the Robert Bosch Foundation is gratefully acknowledged.

References

Balmann A (1997) Farm-based modelling of regional structural change: a cellular automata approach. Eur Rev Agric Econ 24:85–108

Barreteau O, Le Page C, D'Aquino P (2003) Role-playing games, models and negotiation processes. J Artif Soc Social Simulation 6(2):<http://jasss.soc.surrey.ac.uk/6/2/10.html>

Bauer C (1998) Against the current: privatisation, water markets, and the state in Chile. Kluwer Academic Publishers, Boston, MA

Berger T (2000) Agentenbasierte räumliche simulationsmodelle in der Landwirtschaft. Anwendungsmöglichkeiten zur Bewertung von Diffusionsprozessen, Ressourcennutzung und Politikoptionen. Agrarwirtschaft, Sonderheft 186, Agrimedia

Berger T (2001) Agent-based spatial models applied to agriculture: a simulation tool for technology diffusion, resource use changes and policy analysis. Agric Econ 25(2/3):245–260

Berger T, Schreinemachers P (2006) Creating agents and landscapes of multi-agent systems from random samples. Ecology and Society (in print)

Berger T, Ringler C (2002) Trade-offs, efficiency gains and technical change – modeling water management and land use within a multiple-agent framework. Q J Int Agric 41(1/2):119–144

Berger T, Schreinemachers P, Woelcke J (2006) Multi-agent simulation for development of less-favored areas. In: Van Keulen H (ed) Development strategies for less-favored areas. Agricultural Systems, Special Issue (in print)

Bousquet F, Lifran R, Tidball M, Thoyer S, Antona M (2001) Agent-based modelling, game theory and natural resource management issues. J Artif Soc Social Simulation 4(2):<http://www.soc.surrey.ac.uk/JASSS/4/2/0.html>

Bruns BR, Meinzen-Dick R (Eds.) (2000) Negotiating water rights. Intermediate Technology Publications, London, UK

Chileriego (2001) Hemos creado algo vivo llamado embalse illapel. <http://www.chileriego.cl/revista/rev9/rev9_5.htm>

Cornes R, Sandler T (1986) The theory of externalities, public goods, and club goods. Cambridge University Press, Cambridge, UK

Deffuant G, Huet S, Bousset JP, Henriot J, Amon G, Weisbuch G (2002) Agent-based simulation of organic farming conversion in Allier départment. In: Janssen MA (ed) Complexity and ecosystem management: the theory and practice of multi-agent systems. Edward Elgar Publishers, Cheltenham, U.K., and Northampton, Mass

Dinar A, Rosegrant M, Meinzen-Dick R (1997) Water allocation mechanisms: principles and examples. World Bank Working Paper No. 1779. The World Bank, Washington, DC

Fisher FM, Arlosoroff S, Eckstein Z, Haddadin M, Hamati SG, Huber-Lee A, Jarrar A, Jayyousi A, Shamir U, Wesseling H (2002) Optimal water management and conflict resolution: The Middle East Water Project. Water Resour Res 38(11):1243, doi: 10.1029/2001WR000943

Gotts NM, Polhill JG, Law ANR, (2003) Agent-based simulation in the study of social dilemmas. Artif Intell Rev 19(1):3–92

Happe K (2004) Agricultural policies and farm structures – Agent-based modelling and application to EU-policy reform. Studies on the Agricultural and Food Sector in Central and Eastern Europe, 30: IAMO. <http://www.iamo.de/dok/sr_vol30.pdf

Huigen MGA (2004) First principles of the MameLuke multi-actor modelling framework for land use change, illustrated with a Philippine case study. J Environ Manag 72(1–2):5–21

Jakeman AJ, Letcher RA (2003) Integrated assessment and modelling: features, principles and examples for catchment management. Environ Modell Softw 18:491–501

Jasper K, Gurtz J, Lang H (2002) Advanced flood forecasting in Alpine watersheds by coupling meteorological observations and forecasts with a distributed hydrological model. J Hydrol (in press)

Kunstmann H, Stadler C (2003) Operational high resolution meteorological – and hydrological analysis as decision support system in integrated water resources management. in: IAHS Redbook "Integrated Water Resources Management: Towards sustainable Water Utilization in the 21st century (accepted for publication)

Meinzen-Dick RS, Mendoza MS, Sadoulet L, Abiad-Shields G, Subramanian A (1997) Sustainable water users' associations: lessons from a literature review. in: Subramanian A, Jagannathan NV, RS (eds) User organizations for sustainable water services, World Bank Technical Paper 354. World Bank, Washington, DC, pp 7–87

Meinzen-Dick R, Knox A, DiGregorio M, (eds.) (2001) Collective action, property rights and devolution of natural resource management, exchange of knowledge and implications for policy DSE. Eurasburg, Germany

MOP-DGA (2004) Plan director para la gestión de los recursos hídricos en la cuenca del Río Maule, Diagnóstico

Ostrom E (1990) Governing the commons – the evolution of institutions for collective action. Cambridge University Press, New York

Parker D, Meretsky V (2004) Measuring pattern outcomes in an agent-based model of edge-effect externalities using spatial metrics. Agric Ecosyst Environ 101:233–250

Parker DC, Berger T, Manson S (2002) Agent-based models of land use/land cover change. LUCC report series No. 6. LUCC international project office, Louvain-la-Neuve. <http://www.indiana.edu/%7Eact/focus1/ABM_Report6.pdf>

Rosegrant MW, Ringler C, McKinney DC, Cai C, Keller A, Donoso G (2000) Integrated economic-hydrologic water modeling at the basin scale: the Maipo River basin. Agric Econ (24)1:33–46

Sandler T (1992) Collective action: theory and applications. University of Michigan Press, Ann Arbor, Michigan

van Paassen JM (2004) Bridging the gap: computer model enhanced learning about natural resource management in Burkina Faso. PhD Dissertation, Wageningen University.<http://www.gcw.nl/dissertations/3530/dis3530.pdf>

Van de Giesen N, Berger T, Iskandarani M, Park SJ, Vlek PLG (2006) Integrative water research in the volta basin. In: Ehlers E, Krafft T (eds) Earth system science in the anthropocene: emerging issues and problems. Springer-Wissenschaftsverlag, Berlin:169–186

Weisbuch G (2000) Environment and institutions: a complex dynamical systems approach. Ecol Econ 34:381–391

Water Resour Manage (2007) 21:149–169
DOI 10.1007/s11269-006-9046-y

ORIGINAL ARTICLE

Upscaling field scale hydrology and water quality modelling to catchment scale

Alaa El-Sadek

Received: 20 April 2006

Abstract The aim of the research presented in this manuscript is to model the outflow discharge and nutrient load at the outlet of small scale, mainly agricultural catchments. There to two approaches for the simulation of the transport of water and the transport and transformation of nitrogen in the stream were tested and compared. Both approaches use the DRAINMOD and the DRAINMOD-N models to simulate the hydrology and the nitrogen balance of the land phase at the scale of a field/field block/sub-catchment. Both models are used to generate the drain outflow and the nitrate concentration of the drainage water of the field unit considered. The contribution of the field units to the nutrient load of the river are calculated by multiplying the simulated flow weighted N concentrations with drain outflows. In a first approach, called the lumped approach, the water discharge and the nutrient load of field blocks are routed through the river using an exponential model. In this model the nitrate contribution of an individual field block to the nitrate load in the river outlet is calculated assuming first order nutrient decay/attenuation during the transport of the drainage water from the field outlet to the river outlet. The arrival at the outlet section of the nitrate plumes of the field blocks are phased in time based on the velocity profile in the river. The second approach, herein called the complex approach is using the hydraulic river modeling code MIKE 11. This model is using a complex process ADR (advective-dispersive-reactive) equation to calculate the chemical changes in the river water. The comparative analysis between both routing approaches reveals that the lumped approach is able to predict sufficiently accurate nutrient load at the catchment outlet. The complex approach has the advantage of giving a more accurate estimate of the nutrient load at the catchment outlet, resulting in a more precise modeling of the transport and transformation of the nutrient load in streams.

A. El-Sadek
Researcher, National Water Research Center, Fum Ismailia Canal, P.O. Box 74, Shoubra El-Kheime 13411, Egypt
e-mail: alaa_elsadek@yahoo.com

Keywords Hydrology · Water quality · Field scale · Catchment scale · Rural catchment · Nutrient loadings

1. Introduction

During the last decade, an increasing nutrient load has been registered in shallow and deep aquifers, streams and lakes. Previous is mainly due to an over-abundant discharge of nitrogen and phosphorus from both point (municipal sewage, animal feedlots, etc.) and non-point (intensification of farming practices, agricultural runoff and drainage, etc.) sources (El-Sadek *et al.*, 2000). In this study only the effect of nitrogen load from non-point sources on the river quality of small river basins was examined.

A large number of simple to complex water quality models exist for simulating the physical and chemical processes of contaminant transport in streams. Some of the disadvantages of physically based models are dependency on the availability of site-specific parameters; model parameters require additional effort and expense to measure; and physically-based models impose a theory upon the data, a theory that may or may not be complete (Worrall and Burt, 1999). According to Amatya (2000) mechanistic models can describe the processes of constituent transport and transformation on a finer time and space resolution given the availability and accurate estimates of a large number of input parameters. Therefore such models are less likely to be used in day-to-day planning and evaluation. For those circumstances lumped parameter water quality models are often the choice. The objective of this research was to model the drain outflow and nutrient load of field blocks and to route the drain discharge and nitrate load of those blocks or sub-catchments in the river to the basin outlet using two different approaches, a lumped and a physically-based routing model. The lumped routing concept assumes that the decrease in N loads as water travels from the field edge to the basin outlet is exponentially dependent on time in transit and can be described with a single attenuation coefficient. By predicting the travel times from each field to the outlet on a continuous basis, this approach to nitrogen modeling can be applied on either a distributed (field by field) or aggregated (field block or sub-catchment) basis, in both space and time. In the complex approach, the MIKE 11 model (DHI, 1998) is used to route the nutrient load from the sub-catchments to the concentration of the river water at the basin outlet. For reconstructing the transfer of nitrogen from the soil-crop system of a field/field block/sub-catchment to the river use was made of the quasi two-dimensional mechanistic flow model, DRAINMOD (Skaggs, 1981), in combination with DRAINMOD-N (Brevé *et al.*, 1997).

The DRAINMOD/DRAINMOD-N/GIS/exponential decay model and the DRAINMOD/-DRAINMOD-N/GIS/MIKE 11 approaches were tested using time series of observed nitrate loads at the outlet of the Molenbeek river basin (57.44 km^2). The study was carried out to determine in a rural catchment the contribution from organic and inorganic nitrogen fertilizers to the nitrate load found in the surface water. The DRAINMOD and DRAINMOD-N models allow calculating at the scale of an individual field or field block the daily nitrate leaching for a given soil, crop, climate, geo-hydrological and farming condition. The used water and nitrogen model covers the entire land phase of the hydrological cycle from the source on the soil surface, through the soil profile and the shallow drainage system. The GIS pre-processes the river basin data in field specific data in a format suitable for the simulation models, and summarizes the main simulation results in tables and maps. The time series of the nitrogen load at sub-catchment level is used as input in the lumped (exponential decay-time rating model) and the physically based (MIKE 11) approaches. The model results of both routing approaches were tested versus the NO_3-N load in the river water measured at the basin outlet.

Table 1 The MAP fertilizer standards (as published in 1995)

Crops	Nitrogen (kg ha^{-1})		
	Animal or organic	Chemical or mineral	Total amount
Pasture	150	200	350
Maize	125	150	275
Crops with low nitrogen need	55	70	125
Other crops e.g.: potato, beets, winter wheat	125	150	275

2. Study area

The two routing approaches were implemented and evaluated for the Molenbeek catchment, a tributary of the Dender basin. The river Dender basin is located to the west of Brussels in a region with a rolling landscape. It is a tributary of the river Scheldt and has its springs in the Walloon region. Because of the rolling topography in the source area and the relative small water holding capacity of the soils, the flow in the tributaries of the Molenbeek are characterized by relative large discharge fluctuations. The base flow discharges are small, while the response to rainfall is large. The Molenbeek catchment has a total area of 57.44 km^2 (Willems, 2000). It is a narrow and relatively strong indented catchment. The upstream part is rural, while the downstream part is more urbanized (village of Mere, Erpe and Hofstade). One limnigraphic station (hourly water level data and rating curve) is available at Mere (station 20) (Radwan *et al.*, 2000, 2001). The average monthly rainfall and evapotranspiration in the catchment for 30-year data are 68.7 and 34.8 mm, respectively. A dominant flat to slightly undulating topography and a shallow water table characterizes the catchment. Large areas of the catchment are artificially drained. Since the exact fertilizer package per field for the period of analysis could not be reconstructed the threshold values for N-fertilizers, as specified in the fertilization standards of the Flemish Government were applied. The nitrogen standards according to the Manure Action Plan (MAP, 1995) are listed in Table 1. Those standards were taken as the fertilizer practice for each field block/sub-catchment in which the catchment was divided. The foregoing might have led to an underestimation, respectively to an overestimation, of the nitrate leaching from the sub-catchments. Figure 1 shows the relation between flow and nitrate-nitrogen concentration in the catchment.

For the application of the simulation models, DRAINMOD and DRAINMOD-N, the following information was collected for each of the sub-catchments: climate, land use, soil type, water and nitrogen status of the soil profile at the start of the simulation period, and nitrogen fertilizer practice. The model input was approximated for all fields within each sub-catchment. Soil information was derived from the digitized soil map. Based on the soil series observed in the region, statistical profiles were selected from the soil information system AARDEWERK-BIS (Van Orshoven *et al.*, 1991). The core soil series of the legend of the Belgian soil classification system is built of a three letter combination representing texture of the topsoil, natural drainage class and soil profile development. The series is further specified by qualifiers, which refer to substratum, type of parent material and soil phase (Van Orshoven, 1993). The first letter in the three letter soil code refers to the soil texture, e.g. Z, which stands for a coarse textured cover sand and sand dunes, and is named after the mineral composition of the top 30 cm of the soil profile. Changes in texture with depth due to the presence of a substratum, are given by an alphanumeric symbol in front of the texture symbol in the soil code. The substratum is specified in case it occurs at shallower depth than 80 cm, e.g., sLba. If it occurs between 80–120 cm it is given as a (x)Lba (Van Orshoven,

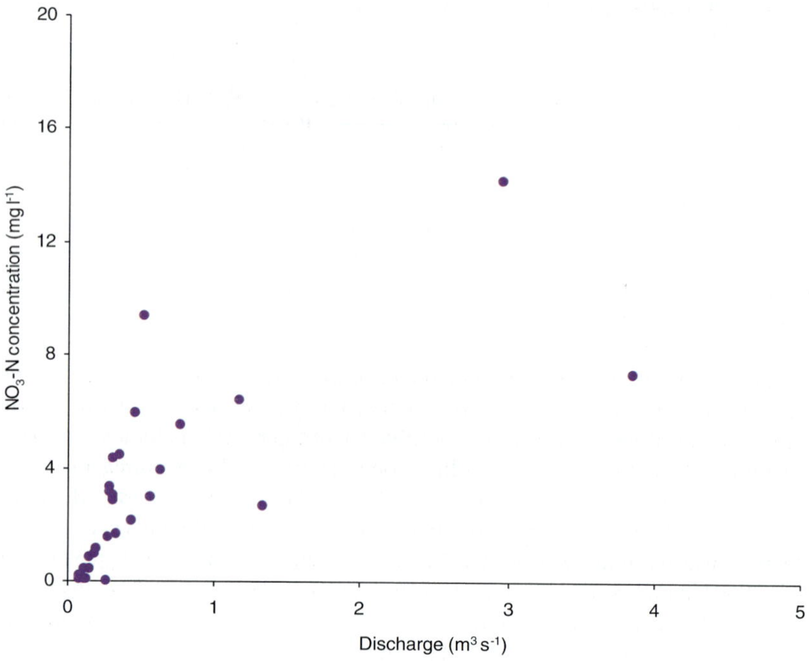

Fig. 1 The relation between flow and nitrate-nitrogen concentration at the catchment outlet

1993; Feyen *et al.*, 2000; Vázquez *et al.*, 2002). The drainage class is determined on the basis of depth to gley phenomena in the soil profile. The third letter in the soil code stands for the profile development. Four different soil types can characterize the soil distribution in the Molenbeek basin. The soil types are primarily silt loam soils (Lca, Ldc, Aba and Ada). The soils have different drainage conditions ranging from dry (a) to medium wet (e), and profile development ranging from poorly (p) to strongly differentiated horizons (g). For the soil profiles the geometry of the profile (boundaries of the pedogenetic horizons), the soil hydraulic parameters and the soil organic carbon content were derived from the soil information system. Land use in the catchment is mainly pasture and arable land. The latter was derived using the TELSAT land use map (1995) of Flanders (OC GIS, Vlaanderen) providing information on land use for grid cells 30 × 30 m. It provides information on the spatial distribution of land use types. Making abstraction of the smaller agricultural land units the main four agricultural uses in the catchment are pasture, maize, fodder beet and potato (Vanongeval *et al.*, 1996). The percentages of the land use areas in the Molenbeek catchment are listed in Table 2. The land use conditions for the simulation periods were assumed constant.

3. Routing models

3.1. Exponential decay based lumped parameter model

The drain discharge rate and the flow-weighted concentration of the drainage water vary from year to year due to change in weather conditions (Amatya, 2000). Their value should

Table 2 The main land use classes in the Molenbeek catchment

Code number	Land use	% of area
2	Discontinuous urban fabric	23.13
3	Industrial or commercial units	0.92
12	Potato	24.96
18	Pasture	10.36
20	Fodder beet	30.14
21	Maize	10.48

be known if one wants to simulate the transport of water and nitrogen and the transformation of the latter in the river. In the absence of measured data they can be simulated using DRAINMOD and DRAINMOD-N. In this study it was assumed that there are no sources or sinks of the constituent other than the natural decay, commonly described by first order kinetics, in contrast to the transformation kinetics in big streams and rivers near the mouth of estuaries. The exponential decay method is based on a gross assumption that dispersion of the constituent concentration along the rivers and ditches is totally negligible due to minimal effects of tidal influence and there are no sources or sinks of the constituent, other than the natural decay (Loucks *et al.*, 1981). The method as developed by Loucks *et al.* (1981) is used to calculate the total nitrate load at the catchment outlet. Mathematically, the cumulative annual load (L) of the nutrient at the catchment outlet can be defined as (Amatya, 2000):

$$L = \sum(L_i) \tag{1}$$

where,

$$L_i = L_{io}^{*}DR \tag{2}$$

$$\text{Delivery Ratio, } DR = \exp(-K^{*}T_i) \tag{3}$$

and L_i is the ultimate annual nutrient load (after attenuation) delivered from a field (i) to the catchment outlet due to an initial field loading L_{io} at that field edge; K is the nutrient decay/attenuation rate during its transport and T_i is the travel time of nutrient for its transport from the field edge (i) in the stream network to the catchment outlet;

$$\text{Initial Field Loading, } L_{io} = \text{Nitrate load}^{*}\text{Area}; \tag{4}$$

The value of the nitrate load is based on a given land management practice (soil, crop/vegetation, and water management). It can be obtained from simulations using DRAINMOD-N for a field of interest. General values on loading can be estimated as a product of annual outflow measured or simulated using DRAINMOD and nitrate concentration, e.g. annual average nutrient concentration, obtained by direct measurements or DRAINMOD-N simulations. The values of nutrient decay or attenuation rate (k) may vary with both the season and location in the river, with generally higher values during low flows and vice versa. The travel time [T] is calculated as:

$$T_i = \text{Distance/Velocity} \tag{5}$$

where distance is the length of the nutrient travel path specified along the canal/stream from the field edge to the outlet [L]. Distance is measured along the specified flow path passing through the nodes of the canal/stream network that is identified and defined during the delineation of the individual fields with its specified outlet in the canal/stream network, and velocity is the average velocity of nutrient movement along the canal [L T^{-1}] and is a function of season or event and location in the catchment.

3.2. MIKE 11 model

The MIKE 11 model consists of different modules of which in this study were used the hydrologic (NAM), hydrodynamic (HD) and water quality (WQ) modules. The MIKE 11 hydrodynamic module uses an implicit, finite difference scheme for the computation of the flow in the rivers. The module can describe sub-critical as well as super-critical flow conditions through a numerical scheme that adapts in time and space according to the local flow conditions. Advanced computational modules are included for the description of flow over hydraulic structures, including possibilities to describe structure operation. The formulations can be applied to looped networks and quasi-two-dimensional flow simulation on flood plains. The water quality module in MIKE 11 was developed by the VKI (Water Quality Institute, Denmark). It describes the basic processes of river water quality in areas influenced by human activities, e.g. oxygen depletion and ammonia levels as a result of organic matter loads. The WQ module solves the system of coupled differential equations describing the physical, chemical and biological interactions in the river.

Concentrations of nitrate are calculated in MIKE 11 by taking into consideration advection, dispersion and the most important biological, chemical and physical processes. The one-dimensional (vertical and lateral variation integrated) equation for the conservation of mass of a substance in solution, i.e., the one-dimensional advection-dispersion equation, reads as follows:

$$\frac{\partial AC}{\partial t} + \frac{\partial QC}{\partial x} - \frac{\partial}{\partial x}\left(AD\frac{\partial C}{\partial x}\right) = -AKC + C_2 \cdot q \tag{6}$$

where C is the concentration (arbitrary unit), D is the dispersion coefficient [$L^2\,T^{-1}$], A is the cross-sectional area [L^2], K is the linear decay coefficient [T^{-1}], C_2 is the source/sink concentration [$M\,L^{-3}$], q is the lateral inflow [$L^2\,T^{-1}$], x is the space co-ordinate [L] and t is the time co-ordinate [T]. The equation reflects two transport mechanisms: advective (or convective) transport with the mean flow and dispersive transport due to concentration gradients. The main assumptions underlying the advection-dispersion equation are that the chemical component is instantaneously mixed over the cross-sections, and the substance is conservative or subject to a first order reaction (linear decay). Fick's diffusion law applies, i.e., the dispersive transport is proportional to the concentration gradient. Nitrification is the process in which nitrogen in the form of ammonia (ammoniac/ammonium) is oxidized to nitrite NO_2^- and further on to nitrate NO_3^-. Denitrification is the process in which nitrogen in oxidized form (nitrite NO_2^- or nitrate NO_3^-) is transformed into free nitrogen (N_2). This free nitrogen is released to the atmosphere.

4. Materials and methods

The 5744 ha large catchment was subdivided into 24 sub-catchments (Figure 2) based on the soil type and cover crop, as shown in Table 3. This table lists the area of each sub-catchment, the distance to the outlet, the cover crop and the main soil type. The main river, with a total

Table 3 Number, area, distance to the river outlet, main cover crop and soil type of the 24 sub-catchments in which the study basin was divided

Sub-catchment	Area (ha)	Distance to the outlet (km)	Cover crop	Soil type
1	179	22.675	Fodder beet	Loam
2	195	22.306	Maize	Loam
3	226	20.647	Maize	Loam
4	234	19.910	Maize	Loam + Silt loam
5	244	19.173	Fodder beet	Loam
6	241	17.883	Maize	Loam
7	245	17.514	Maize + Fodder beet	Loam
8	233	15.671	Maize + Fodder beet	Loam
9	247	15.302	Maize + Fodder beet + Potato	Loam
10	232	14.380	Maize + Fodder beet + Pasture	Loam
11	248	13.643	Maize + Fodder beet + Potato + Pasture	Loam
12	250	12.537	Maize + Fodder beet + Pasture	Loam
13	250	12.168	Maize + Fodder beet + Potato + Pasture	Loam
14	250	10.325	Maize + Fodder beet + Pasture	Loam
15	250	9.219	Maize + Pasture	Loam
16	250	8.482	Fodder beet + Potato	Loam
17	238	7.265	Maize + Fodder beet + Potato	Loam
18	232	5.790	Maize + Fodder beet	Loam
19	250	5.237	Maize + Fodder beet	Loam
20	250	4.776	Potato	Loam
21	250	3.670	Maize + Fodder beet	Loam
22	250	2.011	Maize + Fodder beet	Loam + Silt loam
23	250	1.289	Fodder beet + Potato	Loam + Silt loam
24	250	2.430	Maize	Loam + Silt loam

length of 23043 m, was divided into reaches with 31 nodal points, including the points of lateral inflows from each sub-catchment or block of fields and the hydraulic structures. Input data files were created for each of the sub-catchments (soil hydraulic properties, land use and other parameters). Most of the data on sub-catchment areas and river dimensions could be derived from existing databases. The drain spacing and drain depth were assumed constant at 25 m and 1.25 m below the surface, respectively for the entire region. The measured load at the outlet is equal to 107.5 kg ha^{-1} for the simulation period (four years). Daily flows and nutrient concentrations at sub-catchment level were simulated using the validated DRAINMOD and DRAINMOD-N models for the Belgian agricultural conditions (El-Sadek *et al.*, 2002c, d). The daily flows were multiplied with the nutrient concentration to obtain daily nutrient loading at the outlet of the field blocks. Nutrient load from each of the 24 field blocks was used as input to the MIKE 11 modeling code. The MIKE 11 code was run with the water quality option for a period of four years, 1990–1993. In the first approach, the exponential model was coupled with DRAINMOD, DRAINMOD-N and MIKE 11. In the second approach, MIKE 11, which includes water quality sub-model for nutrient transformation processes, was coupled with DRAINMOD field hydrology and DRAINMOD-N field water quality sub-model to develop a detailed mechanistic hydrology-water quality model.

An experimental field trial with maize, set up by the Belgian Soil Service (Coppens and Vanongeval, 1998) from 1992 to 1995, was used to calibrate and validate DRAINMOD/DRAINMOD-N model. The soil at the farm site, the Hooibeekhoeve in the community of Geel (north-eastern part of Belgium), is sandy and classified as a Haplic

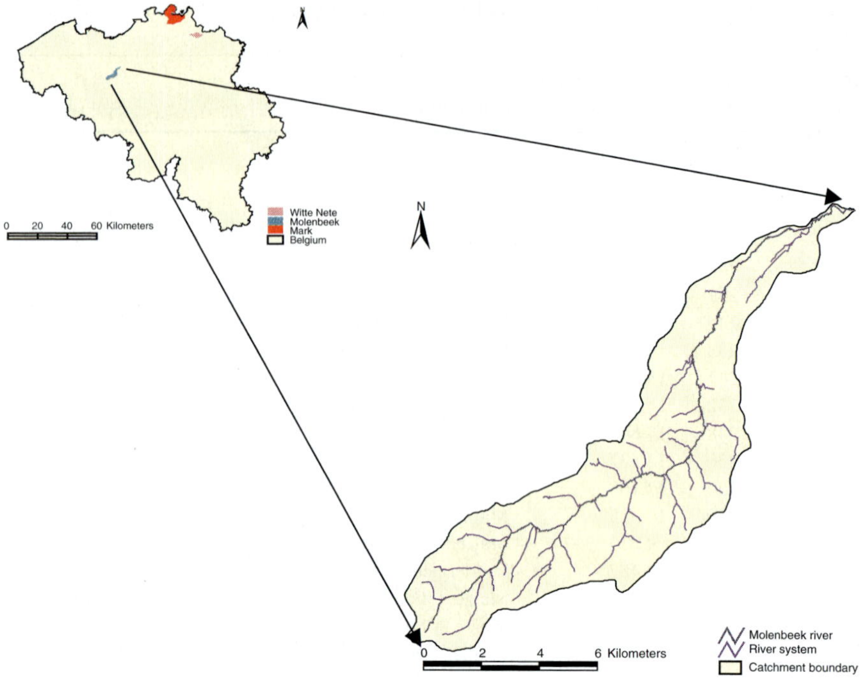

Fig. 2 The Molenbeek river basin with the delineation of the sub-catchments and the main river course

Podzol, mainly sandy soil with a distinct humus and/or iron B-horizon (a Zdg soil according to the Belgian Soil Classification System). The groundwater level fluctuates between 115 and 160 cm below surface. The first two years of the experiment, in 1993 and 1994, maize was sown, whereas in the last season, 1995, the field was fallow. Different pig slurry fertilizer application packages were applied in spring or autumn. NO_3-N in the fertilizer package is added to the soil solution by dissolution of the fertilizer. The maize production practices used in the simulations are characteristic for the sandy region of the Kempen (El-Sadek *et al.*, 2002b). Physical properties of the soil were determined in one plot of the Hooibeekhoeve for each distinguishable soil horizon, using undisturbed soil samples taken with Kopecky rings. van Genuchten-Mualem parameters for describing the hydraulic functions (van Genuchten and Nielsen, 1985) were fitted on both water retention and multi-step outflow data, using the multi-step outflow program (van Dam *et al.*, 1990). Basic water retention and hydraulic conductivity curves were established by averaging individual curves for each soil layer. The field was intensively monitored during the experimental period. Every three weeks, soil samples were taken with an interval of 30 cm to a depth of 120 cm for mineral nitrogen measurements. Mineral nitrogen was measured in groundwater at 200 cm with the same time interval. Organic manure, only as a fertilizer, was applied. Missing data, required to run the model, were either supplementary measured or reconstructed by using the transfer functions of Vereecken (1988), as indicated by Ducheyne and Feyen (1999).

4.1. Statistical analysis

The qualitative judgement of when the model performance is good is a subjective matter (Anderson and Woessner, 1992). Therefore statistical criteria are used for the quantitative judgement (Vàzquez *et al.*, 2002). Statistical based criteria provide a more objective method for evaluation of the performance of the models (Ducheyne, 2000). In this study the following statistical criteria were used to evaluate the model performance:

4.1.1. Mean absolute error (MAE)

$$\mathrm{MAE} = \frac{\sum_{i=1}^{n} |(O_i - P_i)|}{n} \tag{7}$$

where O_i is the observation at time i, P_i is the prediction at time i. The MAE has a minimum value of 0.0.

4.1.2. Relative root mean square error (RRMSE)

$$\mathrm{RRMSE} = \frac{\sqrt{\frac{1}{n}\sum_{i=1}^{n} (P_i - O_i)^2}}{\bar{O}\text{-}} \tag{8}$$

where $\bar{O}$ is the mean of the observed values over the time period (1 to n). The RRMSE has a minimum value of 0.0, with a better agreement close to 0.0.

4.1.3. Model efficiency (EF)

$$\mathrm{EF} = \frac{\sum_{i=1}^{n} (O_i - \bar{O}\text{-})^2 - \sum_{i=1}^{n} (P_i - O_i)^2}{\sum_{i=1}^{n} (O_i - \bar{O}\text{-})^2} \tag{9}$$

EF ranges from minus infinity to 1.0, with higher values indicating better agreement. If EF is negative, the model prediction is worse than the mean observation.

4.1.4. Coefficient of residual mass (CRM)

$$\mathrm{CRM} = \frac{\sum_{i=1}^{n} O_i - \sum_{i=1}^{n} P_i}{\sum_{i=1}^{n} O_i} \tag{10}$$

The CRM has a maximum value is 1.0. If CRM is negative the model overestimates and vice versa.

4.1.5. Coefficient of determination (CD)

$$\mathrm{CD} = \frac{\sum_{i=1}^{n} (O_i - \bar{O}\text{-})^2}{\sum_{i=1}^{n} (P_i - \bar{O}\text{-})^2} \tag{11}$$

The CD describes the ratio of the scatter of the simulated values and the observed values around the average of the observations. A CD value of one indicates to what extent the simulated and observed values match perfectly. It is positive defined without upper limit and with zero as a minimum.

4.1.6. Goodness of fit (R^2)

$$R^2 = \left[\frac{\sum_{i=1}^{n} (O_i - \bar{O}^-)(P_i - \bar{P}^-)}{\sqrt{\sum_{i=1}^{n} (O_i - \bar{O}^-)^2} \sqrt{\sum_{i=1}^{n} (P_i - \bar{P}^-)^2}} \right]^2 \tag{12}$$

where $\bar{P}$ is the mean of the predicted values over the time period (1 to n). R^2 is ranging from 0.0 to 1.0 indicating a better agreement for values close to 1.0 and it is known as the goodness of fit (Shahin *et al.*, 1993; Legates and McCabe, 1999; Vàzquez *et al.*, 2002).

5. Results

5.1. DRAINMOD/DRAINMOD-N calibration and validation

A 3-year time series of data (1992–1995) of the Hooibeekhoeve in the community of Geel was used to calibrate and validate the DRAINMOD/DRAINMOD-N models. The soil, crop and nitrogen parameters were calibrated resulting in a set of representative parameters for the given soil-crop condition. The calibration of the model parameters was carried out by trial and error (Loague and Green, 1991). The calibration of DRAINMOD-N model is based on field data of the fertilizer scenario of 30 ton ha^{-1} pig slurry applied in spring (Figure 3). The calibrated model (DRAINMOD-N) was validated versus data collected on the field fertilizer scenario of 60 ton ha^{-1} pig slurry applied in autumn + 60 ton ha^{-1} pig slurry applied in spring (El-Sadek, 2002a). Finally, the calibrated and validated model was applied to simulate the nitrate transport in the soil profile for the other scenarios (30 ton ha^{-1} pig slurry applied in autumn, 120 ton ha^{-1} pig slurry applied in autumn and 120 ton ha^{-1} pig slurry applied in spring). Summary of inputs and calibrated parameters of the DRAINMOD/DRAINMOD-N model in the Hooibeekhoeve field is shown in Table 4. Validation results are shown in Figure 4. The calibrated and validated DRAINMOD/DRAINMOD-N models were used to assess the nitrate-nitrogen leaching from the 24 sub-catchments in which the river Molenbeek basin was subdivided. In the scenario-analysis, it was assumed that the field blocks were equipped with a subsurface drainage system consisting of parallel, 10 cm diameter, corrugated plastic drains, situated at a depth of 1.25 m below surface and 25 m spaced.

5.2. MIKE 11 calibration

The hydrologic module in MIKE 11 (NAM) was calibrated for two different parameter groups: first the parameter of the transport models (the recession constants or time constants for baseflow, for interflow and for overland flow). Secondly the water balance parameters (maximum water content in the lower zone storage $L_{\max}$, maximum water content in the surface storage $U_{\max}$, and overland flow runoff coefficient). To enable testing if the NAM module is properly simulating baseflow, interflow and runoff, a recursive digital filter was applied (Willems, 2000; Nathan and McMahon, 1990; Chapman, 1991) to separate total flow in three subflows. The working-principle of the filter can be explained physically as the routing of a high frequency signal through a linear reservoir, with the reservoir constant

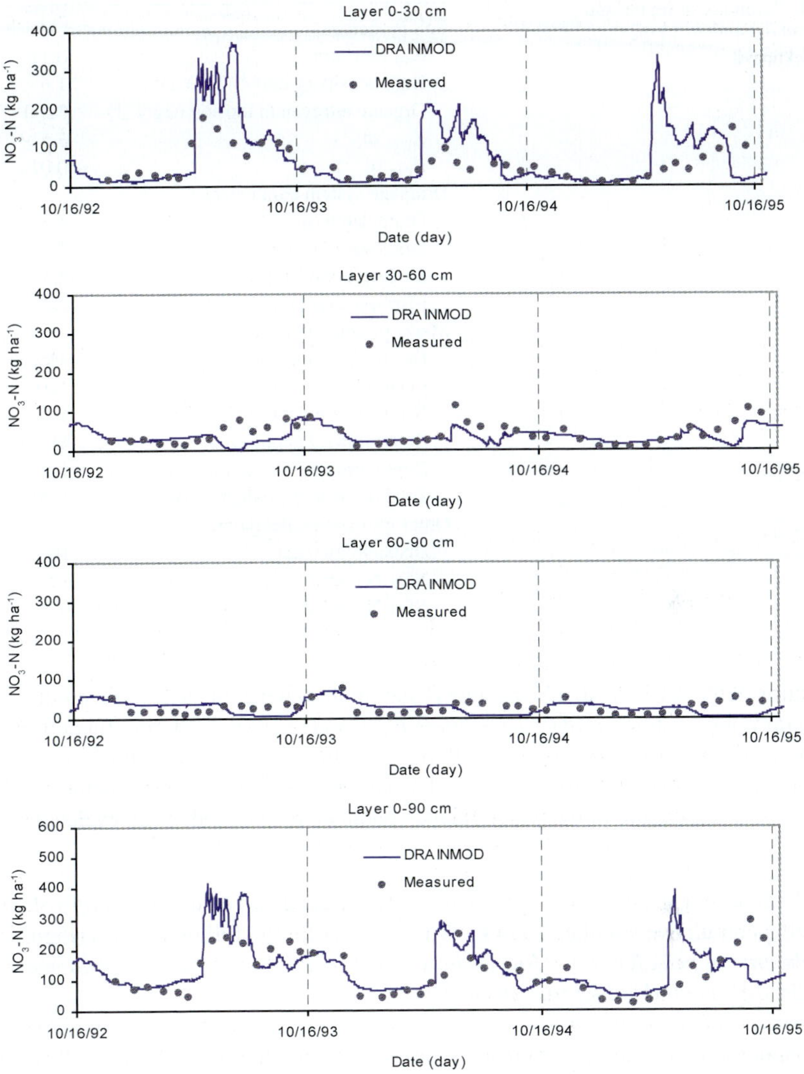

Fig. 3 NO_3-N content in the soil profile at different depths in maize field and sand soil in Geel using fertilizer application of 30 ton ha^{-1} (pig slurry)

equal to the recession constant of the signal that is filtered. In this reservoir routing, the routed signal is considered equal to the filtered signal as it has the same qualitative behavior in recession periods. The subflows with the largest recession constants are separated first. In that way, baseflow is first separated from total rainfall-runoff discharges, secondly interflow is separated from total discharges of surface runoff and interflow. Finally, total flow filtered series is the sum of the three filtered subflows. As time series are often disturbed by random fluctuations that are introduced by the river system (e.g. by regulating structures), the filter is used to get rid of these random fluctuation (Figure 5). The recession constants (time constants) of the subflows are calibrated as the average value of the inverse of the slope of the linear path in the recession periods of a Log (Q) – time graph for the three subflows.

Table 4 Summary of inputs and calibrated parameters in the Hooibeekhoeve

Soil properties	
θ_{wp} (cm^3 cm^{-3})	0.17
Bulk density (g cm^{-3})	1.6
Organic nitrogen in top soil (μg g^{-1})	3200
K_{mnl} (d^{-1})	3.5×10^{-5}
K_{den} (d^{-1})	0.01
Drainage system parameters	
Drain depth (m)	1.25
Drain spacing (m)	25
Surface storage (cm)	2.5
Effective drain radius (cm)	2.5
Maize production parameters	
Desired planting date	May 4
Length of growing season (d)	120
N-fertilizer input (kg N ha^{-1})	160
Date fertilizer application	May 6, May 14
Depth fertilizer incorporated (cm)	10
Total dry matter production (kg ha^{-1})	14500
Other nitrogen model parameters	
Dispersivity (cm)	10
NO_3-N content of plant (per cent)	1.55
NO_3-N concentration of rain (mg l^{-1})	0.8

Calibration of the water balance parameters (maximum water content in surface storage U_{max}, maximum water content in root zone storage L_{max}, and overland flow runoff coefficient) are done by trial and error. The procedure is repeated till the maximization of the agreement between the measured and modeled peak discharges and total volumes is achieved. During the calibration procedure of the water balance parameters, the models are evaluated in three steps:

1. *Evaluation of water balance* (*comparison of simulated and observed runoff volumes*) by plotting cumulative simulated and measured runoff for the full time series (Figure 6).
2. *Evaluation of peak flows and low flows* (*comparison of hydrograph maxima and minima for the different individual rain storms*).
   For such evaluation, the time series is divided into two types of storm periods: first for individual storm events (containing individual hydrograph peaks), and secondly for longer events of shortly successive rain storms (containing a baseflow recession periods of minimum length). The comparison is done by plotting the maximum of each peak and the minimum of each event for both measured and simulated time series as shown in Figure 7 for peak flows, and Figure 8 for low flows. As the plotted points are close to the bisector, the comparison between measured and simulated values is good.
3. *Evaluation of long-term statistics.*
   The observed and simulated discharge values in the full time series are plotted after ranking them in an ascending order (Figure 9). The flattening of the measurement curve is explained by river flooding (which starts at 6 m^3 s^{-1} at station 20). As flooding is described by the hydrodynamic model and not by the hydrological model, these discharges should not be taken into account during the NAM calibration.

The total length of the Molenbeek brook is approximately 23 km. For the first 6 km no detailed data about cross-sections and hydraulic structures are available, but for the next

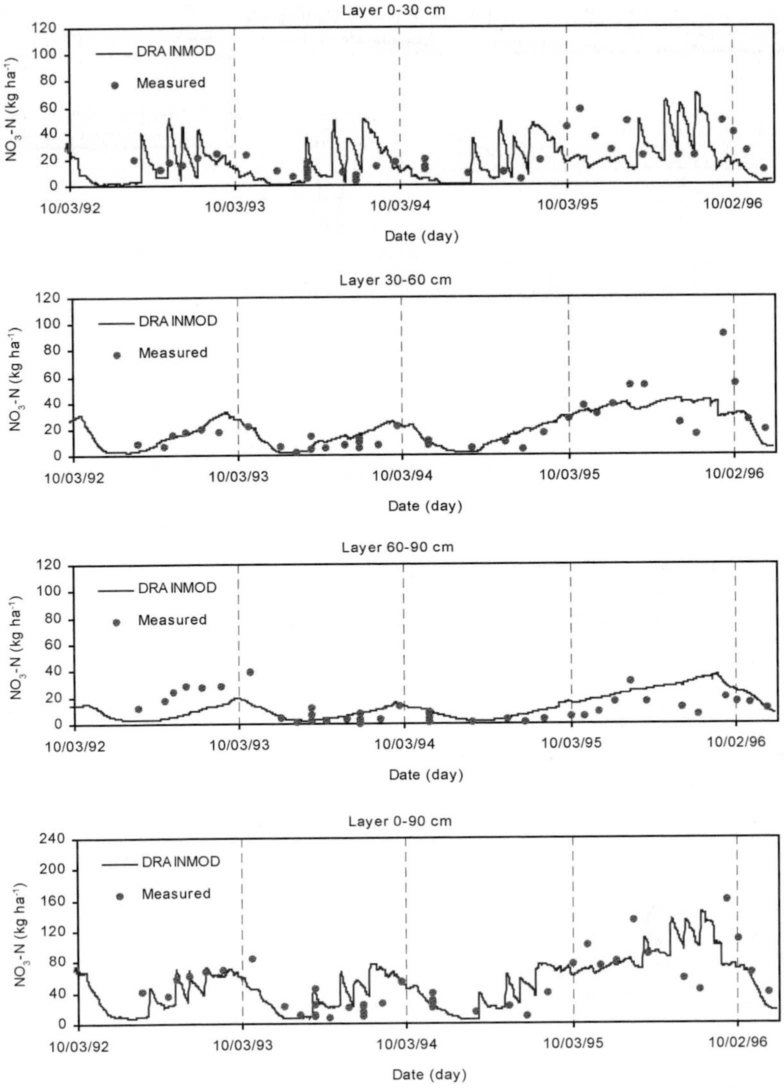

Fig. 4 NO_3-N content in the soil profile at different depths in pasture field and sand soil in Geel using fertilizer application of 20 ton ha^{-1} (cow slurry)

17 km, detailed data about cross-sections and all-important hydraulic structures exist. The latter cross-sections are 440 in total. In this way, the variation in channel geometry along the model branches can be described adequately. For the structures, all 16 bridges, 16 weirs, 6 culverts and 6 control structures over the distance of 17 km are considered. The 6 control structures regulate the water levels at different locations along the brook to prevent areas from flooding, and to limit the flow velocities to avoid erosion of the riverbed. After implementation of the HD model, the simulation result is compared with the measurements at Mere station. More specifically, the simulated Q-H is compared with the measured. The results shown in Figure 10 indicate that the dynamics of the system are presented well.

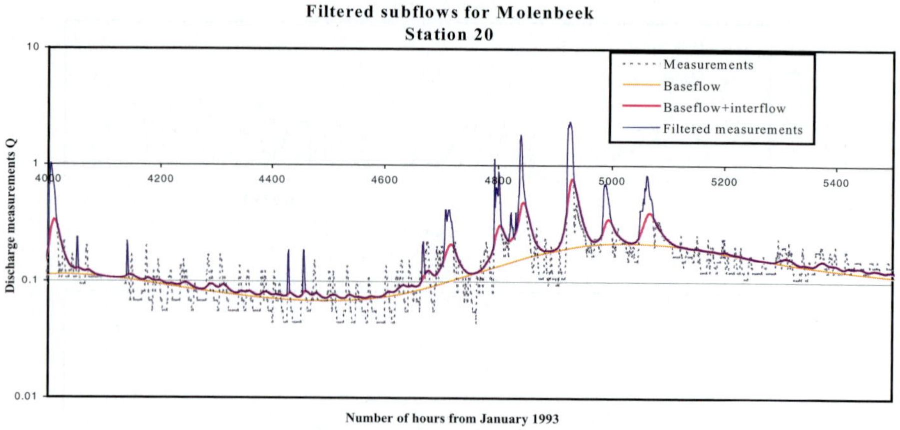

Fig. 5 Splitting of the total discharge of the river Molenbeek in base flow, interflow and runoff

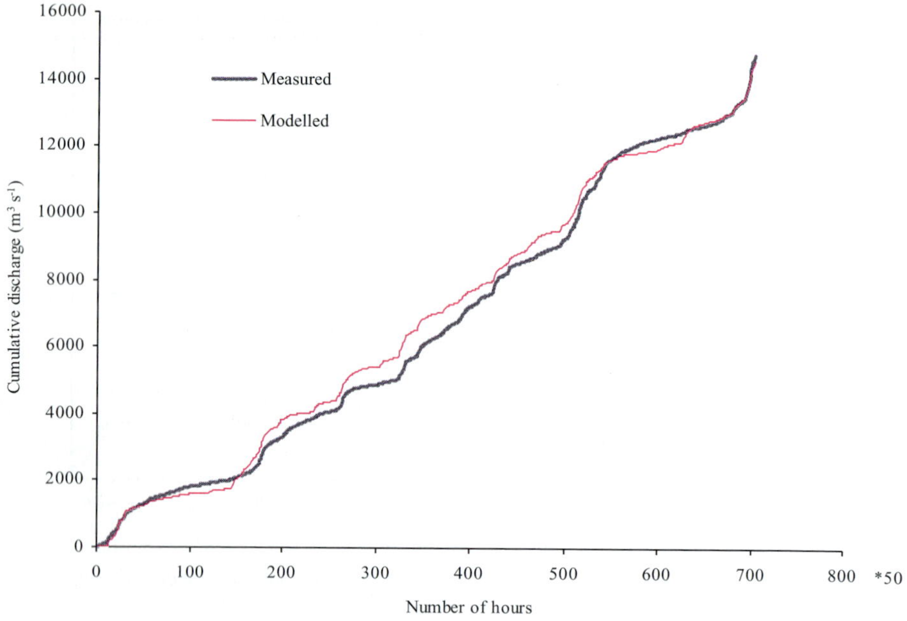

Fig. 6 Comparison between cumulative measured and simulated discharges

The water quality model was calibrated by trial and error to obtain a reasonable fit between the measured and the modeled results. Initial values for the parameters of the process equations are based on standard values found in literature (Brown and Barnwell, 1987; DHI, 1998). Certain parameters are strongly dependent on the location. The calibration was difficult because of the long interval between each of the measurements of the Flemish Environmental Agency (minimum one month). The only parameter calibrated is the nitrification decay coefficient.

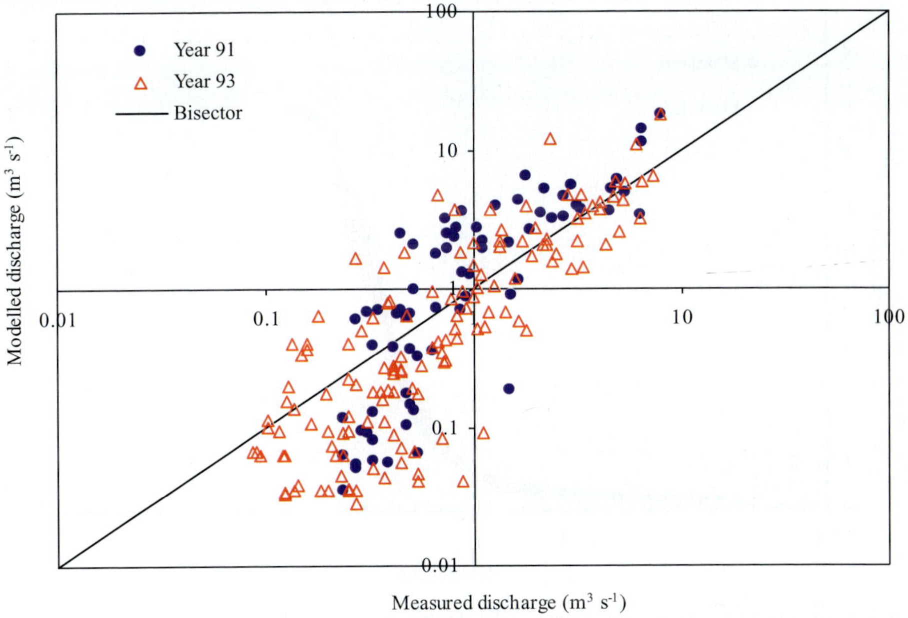

Fig. 7 Comparison between measured and simulated values for maximum of each peak

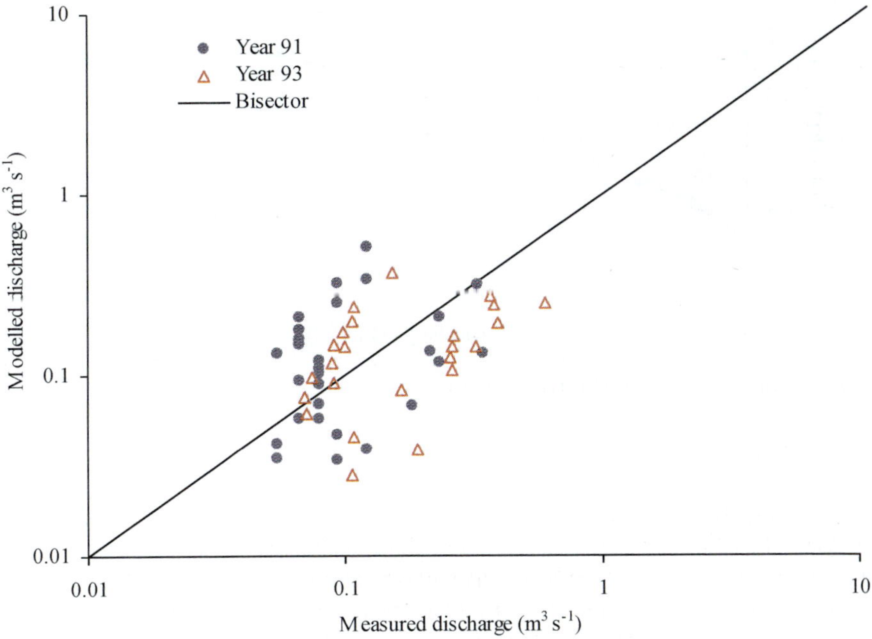

Fig. 8 Comparison between simulated and measured values for minimum of each event

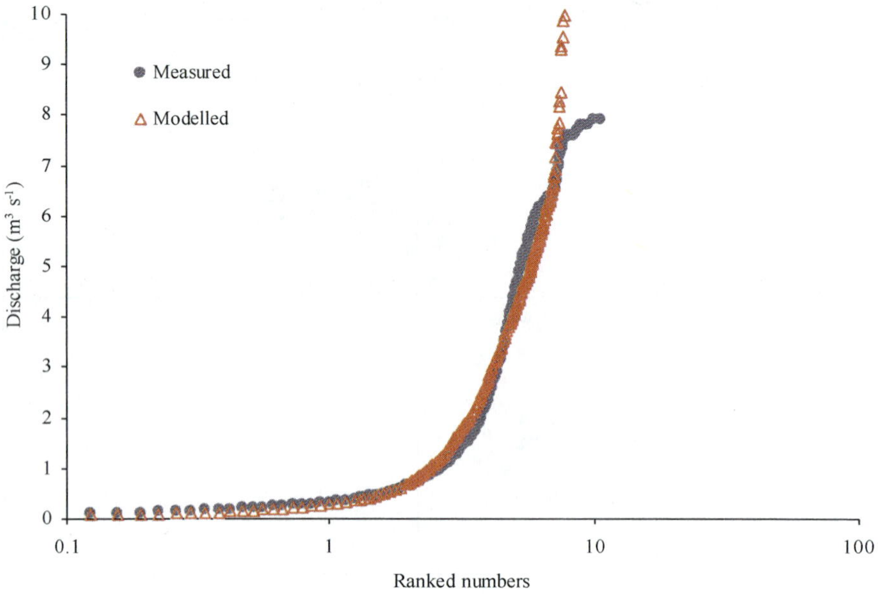

Fig. 9 Comparison between measured and simulated values

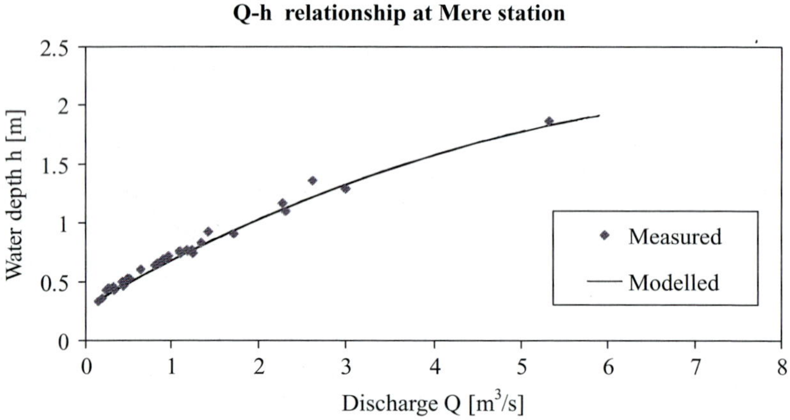

Fig. 10 Simulated and measured Q-H relationship at mere station

6. Discussion

The validated MIKE 11 (Radwan *et al.*, 2000) was run with a 4-year data set to simulate daily flow velocities at all nodes in the river. It was found that average velocity is a function of the season. In general, the wet season had higher velocities. The large velocities occurred infrequently due to larger summer storms. The travel time was estimated as a function of distance from the field edge to the catchment outlet assuming average velocity throughout the catchment. The validated MIKE 11 (using DRAINMOD flow as input) was used to calculate the velocity-input file to the lumped exponential model. MIKE 11 and DRAINMOD outputs of daily discharge rates at Mere station in the Molenbeek catchment were verified using

Table 5 Annual measured and simulated NO_3-N load, in kg ha^{-1}, at the river outlet, for the period 1990–1993

Year	Measured	Simulated with exponential model	Simulated with MIKE 11
1990	2.6	4.0	8.8
1991	40.1	29.8	31.7
1992	17.8	22.4	17.1
1993	47.0	65.6	36.8

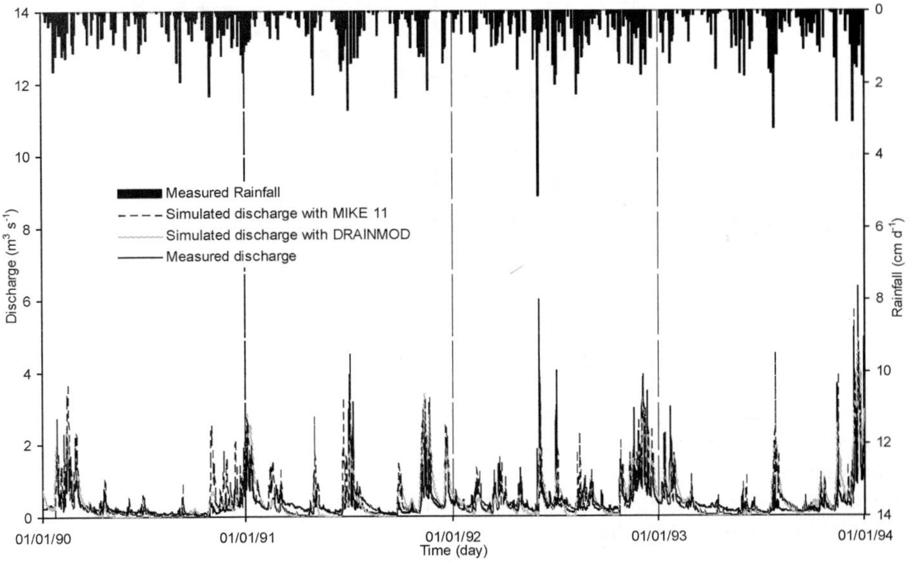

Fig. 11 Daily precipitation, observed and simulated daily discharge at the outlet station for the period 1990–1993

measured data as shown in Figure 11. The nutrient loading or concentration at the edge of a field is simulated using DRAINMOD-N. The predicted discharge outflows are used with a flow-weighted concentration to obtain N loadings at the individual outlets of the fields with different management scenarios. The annual measured and simulated NO_3-N load, in kg ha^{-1}, at the river outlet is given in Table 5 and a comparison of total simulated nitrate load from all fields versus measured data is shown in Figure 12.

Concentration peaks within the period of fertilizer application between March and May can be clearly seen in years 1991, 1992 and 1993. The nitrate load peaks correspond to the rainfall events in these years. This occurs as the upper layers of the soil become saturated allowing the fertilizer to dissolve in the available moisture and to be transported to the river. The low load in 1990 is the result of a dry year. The agreement between the measured and predicted nitrate load in Table 5 and Figure 12 is good. However, it is recognized that daily time series output cannot be compared in detail against monthly point measurements. Plants take approximately 60% of the applied nitrogen with 30% transported via groundwater and surface waters to the river. The time series with predominantly agricultural influences showed the expected winter maximum and summer minimum. The effect of the strong influence of previous year values on the present year can be seen in Table 5 and the time series of the cumulative nitrate load given in Figure 12. The effect of season on the nitrate

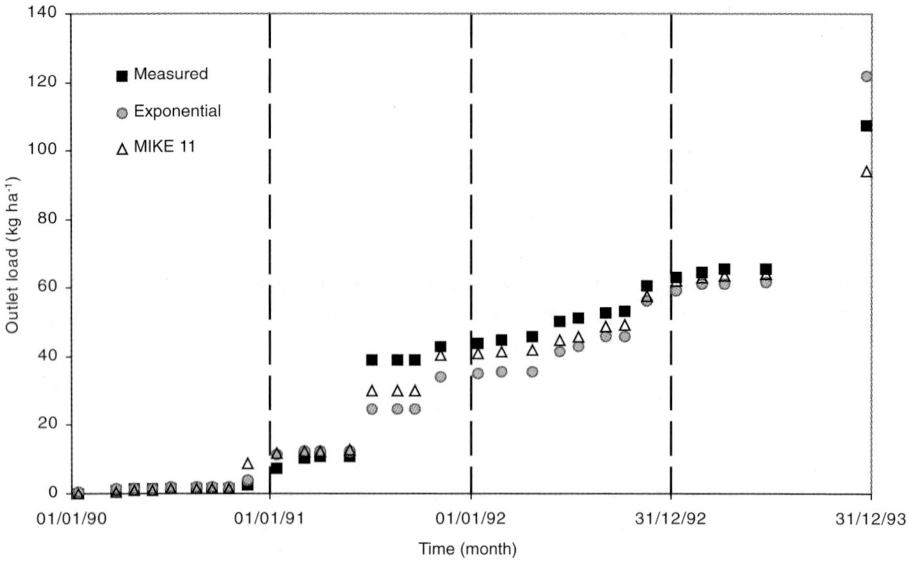

Fig. 12 Predicted and measured cumulative nitrate load at the outlet station for the period 1990–1993

load can be seen in this figure. The residual concentrations in summer tend to have negative effects whereas corresponding concentrations in the winter still have positive effects. This means that the effect of the previous year differs depending upon the rainfall time series. High levels of nitrate in a preceding winter have the effect of lowering concentrations in the following summer. Equally the effects of low values in a preceding summer increase the nitrate concentrations in the following winter. The winter-summer difference in annual nitrogen cycle at the catchment outlet compensates the difference in the following season. For example, in 1992, the difference in nitrate load of the winter season is compensated by the winter-summer difference in annual nitrogen cycle. A wet winter could act to flush out reserves of nitrate leaving little that could be readily removed in the subsequent summer. For example, a dry summer leads to storage of readily mineralizable nitrogen, once the soils wet up again this large reserve of nitrogen stimulates the microbial population to increased activity that enhances nitrate leaching the following winter. The increased microbial activity tends to remove more of the available nitrogen than was preserved from previous summer.

In general, modeling nitrate export purely in terms of the export from present land use tends to overestimate nitrate loads. It is uncertain how much of the nitrate moves directly to the streams in the catchment without going via groundwater. It certainly occurs in the study area, although it seems that the link between soil nitrogen reserves and both river and groundwater is a straightforward. The comparative analysis reveals that both the lumped (exponential model) and the MIKE 11 model are able to simulate with acceptable accuracy the monthly NO_3-N load at the catchment outlet for the 1990–1993 period. The statistical analysis results presented in the Table 6 clearly illustrate that on average DRAINMOD-MIKE 11 model performs better in the prediction of the NO_3-N load at the catchment outlet. The study revealed that for the lumped exponential model the initial field load is the most significant input. Based on published values, a value of (K) (0.10) for nitrate-nitrogen gave reasonable results. A smaller value of (K) may be used for very wet seasons compared to dry seasons. MIKE 11 can estimate travel time using simulated velocity data. The lumped parameter water

Table 6 Statistical performance analysis calculated for the simulation period 1990–1993 for monthly NO_3-N loads

	MAE	RRMSE	CD	EF	CRM	R^2
Measured & Exponential (lumped)	1.941	1.309	0.640	0.713	−0.133	0.831
Measured & DRAINMOD-MIKE 11	1.668	0.963	1.930	0.845	0.123	0.900

MAE: mean absolute error; RRMSE: relative root mean square error; CD: coefficient of determination; EF: model efficiency; CRM: coefficient of residual mass; R^2: goodness of fit

quality model as discussed in this study has to be tested and validated in different locations. Testing of this parameter model is challenging but necessary for agricultural lands with complex land management practices. More field experiments and studies may be required to determine the decay parameter (K) for other nutrients. A large effort should be placed on obtaining its accurate determination and statistical distribution. The lumped (exponential) approach relies on data from readily available databases. The model has relatively few data requirements, and can be easily calibrated, providing a relatively inexpensive, robust means of evaluating the impact of land use and land management on water quality for modeling on an annual basis. The approach reduces the problems inherent in predicting the nitrate load on a daily basis. Another limitation on the use of these models is that the importance of hydrological pathways in determining nutrient delivery to surface waters, and the variations in available transport mechanisms over annual water cycle mean that the models can not predict in real time. Therefore, the approach can be used to predict on a year-by-year basis, the changes in water quality within the catchment.

7. Conclusions

Two approaches, the MIKE 11-DRAINMOD and the lumped, exponential based distributed catchment scale hydrologic and water quality model were tested using four years of measured data of the Molenbeek catchment, Belgium. The modeling approaches were applied in a distributed way and used to model the nutrient load in the river Molenbeek. The statistical analysis indicated that the two approaches are able to reconstruct quite accurately the nitrate load of a primarily agricultural catchment with heterogeneous land management practice. For the analysis the catchment was subdivided in 24, more or less homogeneous, sub-basins with a particular soil-land use management practice. The MIKE 11 prediction of in-stream flow rates (velocities and depths) indicated the potential of the model for being coupled with the DRAINMOD-N model and the Exponential model for predicting cumulative water quality impacts on the Molenbeek catchment. The comparative analysis between both model approaches (lumped and DRAINMOD-MIKE 11) reveals that the lumped model is able to predict sufficiently accurate nutrient load at the catchment outlet. The complex approach (DRAINMOD-MIKE 11) however has the advantage of giving a more accurate estimate of the nutrient load at the catchment outlet, resulting in a more precise modeling of the nutrient load transport and transformation in the land phase and the river of catchments. As such the approach can be used to derive for the study area the fertilizer practice that will result in a NO_3-N load at the river outlet that does not exceed the limits, as specified by environmental considerations. When measured data are not available, calibrated and validated DRAINMOD in Flanders, Belgium could be used to predict water discharges. Efforts should be made to validate DRAINMOD-N for agriculture fields on the poorly drained soils in Belgium. More field studies and experiments are needed in determination nutrient decay parameter for the

Exponential models. Since such models have relatively few data requirements, and can be easily calibrated, they will be the suitable tools to model the total nutrient load at the catchment outlet when Geographic information system (GIS) facilities are not available.

References

Amatya D (2000) Advanced hydrologic and water quality modeling on poorly drained watersheds. Course pack, North Carolina State University, Department of Biological and Agricultural Engineering, North Carolina, USA

Anderson MP, Woessner WW (1992) Applied groundwater modeling simulation of flow and advective transport. University Press, Cambridge, 296 pp

Brevé MA, Skaggs RW, Parsons JE, Gilliam JW (1997) DRAINMOD-N, a nitrogen model for artificially drained soils. Trans ASAE 40(4):1067–1075

Brevé MA, Skaggs RW, Parsons JE, Gilliam JW (1998) Using the DRAINMOD-N model to study effects of drainage system design and management on crop productivity, profitability and NO_3-N losses in drainage water. Agric Water Manage 35:227–243

Brown LC, Barnwell TO (1987) The enhanced stream water quality models Qual2E and Qual2E-uncas: Documentation and user manual, report EPA/600/3-87/007. U.S.EPA, Athens, GA, USA

Chapman TG (1991) Comment on 'Evaluation of Automated Techniques for Base Flow and Recession Analyses' by R. J. Nathan and T. A. McMahon. Water Resour Res 27(7):1783–1784

Coppens G, Vanongeval L (1998) Modellering van de migratie van nutriënten in de bodem.. Activiteitenverslag van de Bodemkundige Dienst van België. Rapport 98/00/3, 40 pp, (in Dutch)

DHI (1998) MIKE 11: a microcomputer based modeling system for rivers and channels. Reference manual of the Danish Hydraulic Institute, Hoersholm, Denmark

Ducheyne S (2000) Derivation of the parameters of the WAVE model using a deterministic and a stochastic approach. PhD thesis No. 434. Faculty of Agriculture and Applied Biological Sciences, K.U. Leuven, Belgium, 123 pp

Ducheyne S, Feyen J (1999) A procedure to reduce model uncertainty by comparison with field data illustrated on a nitrogen simulation model. Proceeding of EurAgEng's IG on Soil and Water Int. Workshop on Modeling of transport processes in soils at various scales in space and time. Leuven, Belgium, 24–26 Nov., pp 457–466

El-Sadek A, de Vos J, Ducheyne S, Feyen J (1999) Analysis of the nitrate leaching to surface waters using a simplified and detailed model approach. International workshop of EurAgEng's field of interest on soil and water, Leuven, Belgium, 24–26 November, Modeling of transport processes in soils at various scales in time and space, pp 569–579

El-Sadek A (2002a) Engineering approach to water quantity and quality modeling at field and catchment scale. Katholieke Universiteit Leuven, Faculty of Engineering, Leuven, Belgium. PhD-dissertation, 256 pp

El-Sadek A, Feyen J, Skaggs W, Berlamont J (2002b) Economics of nitrate losses from drained agricultural land. Environ Eng 128(4):376–383

El-Sadek A, Oorts K, Timmerman A, Radwan M, Feyen J (2002c) Comparative study of two nitrogen models. Irrigation Drainage Eng 129(1):44–52

El-Sadek A, Feyen J, Ragab R (2002d) Simulation of nitrogen balance of maize field under different drainage strategies using the DRAINMOD-N model. Irrigation Drainage 51(1):61–75

El-Sadek A, Willems P, Christiaens K, Casteels C, Ducheyne S, Feyen J (2000) Modeling of the nitrogen balance at catchment scale using a quasi-two dimensional mechanistic model in combination with GIS. Proceeding of the Micro 2000 Congress, Cape Town, South Africa, 22–27 October, CD-ROM

Feyen L, Vázquez R, Christiaens K, Sels O, Feyen J (2000) Application of a distributed physically-based hydrological model to a medium size catchment. Hydr Earth Syst Sci 4(1):47–63

Legates DR, McCabe GJ (1999) Evaluating the use of 'goodness-of-fit' measures in hydrological and hydroclimatic model validation. Water Resour Res 35(1):233–241

Loague K, Green RE (1991) Statistical and graphical methods for evaluating solute transport models: overview and application. In: Wieringa PJ (ed) Validation of low and transport models for the unsaturated zone. J Contam Hydrol 7:51–73

Loucks D, Stedinger J, Haith D (1981) Water resource systems planning and analysis. Prentice-Hall, Inc., Englewood Cliffs, New Jersey, USA

MAP (1995) Decreet van 20 december 1995 (B.S. 30 december 1995) tot wijziging van het decreet van 23 januari 1991 inzake de bescherming van het leefmilieu tegen de verontreiniging door meststoffen (in Dutch)

Nathan R, McMahon T (1990) Evaluation of automated techniques for base flow and recession analyses. Water Resour Res 26(7):1465–1473

Radwan M, El-Sadek A, Willems P, Feyen J, Berlamont J (2001) Modeling of nitrogen in river water using a detailed and a simplified model. The Scientific World 1(S2):200–206

Radwan M, Willems P, El-Sadek A, Berlamont J (2000) Integrated catchment modeling: an application to Molenbeek catchment, Belgium. Fifth International Water Technology conference IWTC-2000, Alexandria, Egypt, pp 83–96

Shahin M, Van Oorschot HJL, De Lange SJ (1993) Statistical analysis in water resources engineering. A. A. Balkema Publishers, The Netherlands, 394 pp

Skaggs RW (1981) Methods for design and evaluation of drainage water management systems for soils with high water tables, DRAINMOD. North Carolina State University, Raleigh, North Carolina, USA

van Dam JC, Stricker JNM, Droogers P (1990) From one-step to multi-step. Determination of soil hydraulic functions by outflow experiments. Rep. 7, De. Of Water Resour., Agric. Univ., Wageningen, The Netherlands

van Genuchten MTh, Nielsen DR (1985) On describing and predicting the hydraulic properties of unsaturated soils. Annales Geophysicae 3(5):615–628

Van Orshoven J (1993) Soil survey data in Belgium and geographical information systems. In 'Assessing hydrodynamic land qualities from soil survey data'. PhD-dissertation No. 237, Faculty of Agricultural and Applied Biological Sciences, K.U. Leuven, Belgium, pp 5–16

Van Orshoven J, Meas J, Vereecken H, Feyen J (1991) A procedure for the statistical characterization of the units of the Belgian soil map. Pedologie XLI-3:193–212

Vanongeval L, Ver Elst P, Boon W, Bries J, Vandendriessche H, Geypens M (1996) De chemische bodemvruchtbaarheid van het Belgische akkerbouw-en weilandareaal (1992–1995). Report No. 96/OO/10, Bodemkundige Dienst van België (in Dutch)

Vázquez RF, Feyen L, Feyen J, Refsgaard JC (2002) Effect of grid- size on effective parameters and model performance of the MIKE SHE code applied to a medium sized catchment. Hydr Proc 16(2):355–372

Vereecken H (1988) Pedotransfer functions for the generation of the hydraulic properties for Belgian soils. PhD-Thesis No. 171, Faculty of Agricultural and Applied Biological Sciences, K.U. Leuven, Belgium, 254 p

Willems P (2000) Probabilistic immission modeling of receiving watercourses. PhD thesis, Faculty of Engineering, Catholic University of Leuven, Belgium

Worrall F, Burt TP (1999) The impact of land-use change on water quality at the catchment scale: the use of the export coefficient and structural models. J of Hydr 221:75–90

Water Resour Manage (2007) 21:171–184
DOI 10.1007/s11269-006-9047-x

ORIGINAL ARTICLE

Linking databases of different sources and scales for groundwater research in the Urema River Basin/Central Mozambique

Franziska Steinbruch · Luis Macario

Received: 20 April 2006

Abstract The Urema River basin is the major sub catchment of the Pungue River basin situated in Central Mozambique. The Urema River basin controls the ecosystem of the Lake Urema, which with its extended flood plains forms a major feature of the Gorongosa National Park. The Urema River influences the flood and low water levels of the Pungue River, which is the main source of potable water of Beira – the second largest city of Mozambique. The area of the Urema river basin is densely populated and intensively used for subsistence agriculture, timber production, widely spread gold mining, tourism and nature conservation.

A university project aiming at developing a model of the groundwater dynamics and assessing groundwater qualities of the Urema River basin was recently initiated. Up to now very few groundwater investigations were conducted in Mozambique. Current database management of groundwater data is poorly developed. Data are kept in tabular formats and serve mainly for record keeping purposes. Confronted with these challenges the presentation demonstrates the project's approaches of linking the various database sources to achieve a large to medium scale concept of the hydrodynamics of the Urema River basin.

Based on current project findings the improvement of the resolution of digital data from current small to larger scales is necessary to enhance information contents. Further geochronological investigations are required to understand the geological evolution of the area and to update the stratigraphic order of geological formations. In future a groundwater-monitoring network needs to be established to capture long-term and baseline data.

Keywords Multi-institutional · Multi-scale · Urema rift · Pungue river basin · Water quality · Groundwater dynamics · Information system · .

F. Steinbruch
Gorongosa Research Center, Beira Office (CIG-UCM, C.P. 821, Rua Marques do Soveral 960)
e-mail: franziskas@carrfoundation.org

L. Macario
Programa de Abastecimento de Aguas Rurais e Saneamento Sofala
e-mail: lmacario@worldbank.org

1. Introduction to groundwater research in Mozambique

Groundwater research neither has roots in Mozambique nor receives major attention. First groundwater studies had as the only objective to assess the economic potential of entire river basins. This has its historical explanation in predominant interests of the former Portuguese colonialism and the slow down in civil activities due to 16 years of war, which culminated in the socio-economic ruin of the country and the exodus of the few higher qualified professionals. Reconstruction of infrastructures, such as the rehabilitation of water wells was for a long time impossible due to the abundance of land mines.

A priority of the current government is poverty alleviation, which implies the facilitation of the supply of clean water in sufficient quantities and in a relatively short distance from the population. The authors believe, that this objective could be more efficiently achieved, if investments would be specifically directed to groundwater research and the improvement of database management.

Recent studies look at groundwater quality and quantification of water to sustain biodiversity as well as to host agro-industries in the Pungue River basin (SAfMA, 2004; SWECO and Associates, 2004). The most comprehensive and conclusive hydro-geological picture of the Urema River area was developed in Owen (2004). However only GEOTECNICA & MINAS, 1972 and the geo-ecology study of Tinley (1977) have actually captured primary data by means of fieldwork and photographical image interpretation. All other studies in the Pungue River basin refer to and re-analyze their findings.

At the level of the Province of Sofala, which is the administrative unit hosting the Urema River basin the necessity of organizing water well data in a database is recognized. It is assumed, that well-organized databases and a policy, which facilitates access to the data will eventually trigger wider use of information in the fields of groundwater exploration and research. As a result of the linkages between geo-databases the success rate of groundwater exploration is likely to improve and thereby decrease the economic risk of investing in well constructions.

The project presented in this paper aims at developing a model of the hydrodynamics and groundwater qualities of the Urema River basin. The paper discusses the linkages of existing databases from the various sources in a multi-scale approach and demonstrates the usefulness in groundwater research. It results in an inventory of existing knowledge, reliability and consistency evaluation, the discovery of contradictions and information gaps, the derivation of a working concept and finally the compilation of a list of further activities.

2. Description of the Urema River basin

The Urema River basin is a major sub-catchment of the Pungue River basin located in Central Mozambique. The Urema River Basin covers an area of 14,000 km^2. The Urema river drains its tributaries along a NE-SW fault structure, the so called Urema rift valley, which belongs to the East African Rift. The distance between the rift margins is about 50 km (Figure 1).

The western tributaries originate in the Sena Plains and the Gorongosa Mountains, the eastern tributaries come from the Cheringoma Plateau. The area West of the Urema rift is made of proterozoic gneisses intruded by the Gorongosa granite and northwards overlaid by cretaceous sandstones. The Cheringoma Plateau consists of tertiary sandstones and dolomites underlaid by cretaceous sandstones. The rift valley is made of recent fluvial and colluvial fanlike depositions. The Gorongosa Mountains reach altitudes of 1,885 m (above sea level) and the highest point on the Cheringoma Plateau is at 352 m (above sea level, Figure 6).

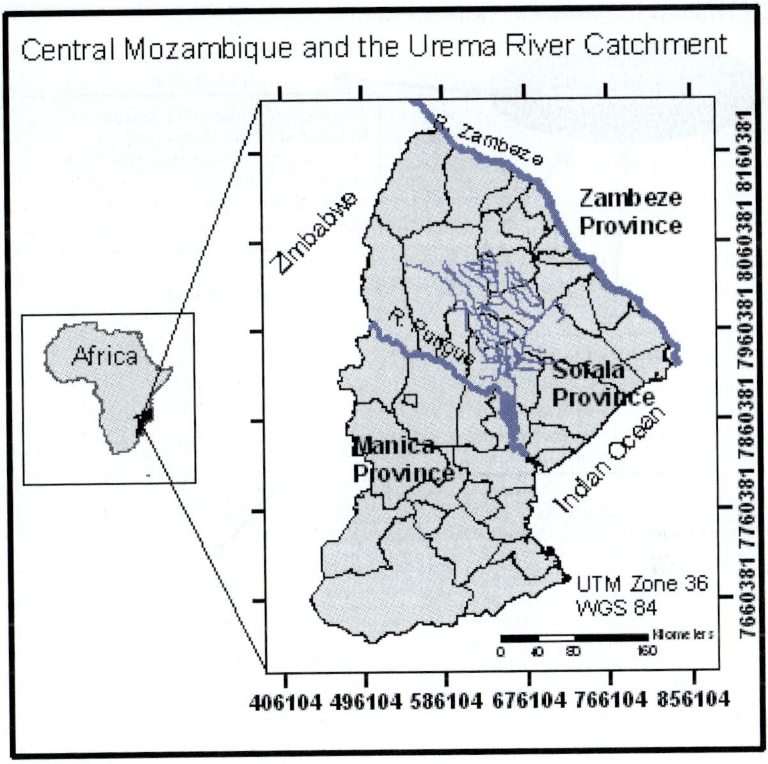

Fig. 1 Map overview of Africa, Mozambique and the Urema River Basin in Central Mozambique (ESRI map, DINAGECA Mozambique)

The Cheringoma Plateau forms the topographic watershed between rivers joining the Urema river and those flowing directly into the Indian Ocean. The elevations in the Urema rift range between 70 and 19 m (above sea level). The deepest parts of the Urema rift host an extended shallow lake, called Urema. The Lake forms a wetland area, which is the central hydrological and ecological feature of the Gorongosa National Park.

The area is characterized by a tropical climate having a dry season from April to October and a wet season from November to March. During normal dry season only Vunduzi and Nhandare River both originating on the Western side of the Urema rift have water.

The Urema River influences the flood and low water levels of the Pungue River. The Pungue River is the main source of potable water for Beira city. The area of the Urema river basin is densely populated and intensively used for subsistence agriculture, timber production, artesian gold mining of placer deposits, tourism and nature conservation.

3. Databases to assess groundwater dynamics and quality

At this stage of the project our approach does not include the setting up of institutional collaborations, but analyses existing inter-institutional relationships and database sharing. Out of existing data a structure for a hydro-geological database and a working concept of the hydrogeology of the area was developed. Following steps were applied:

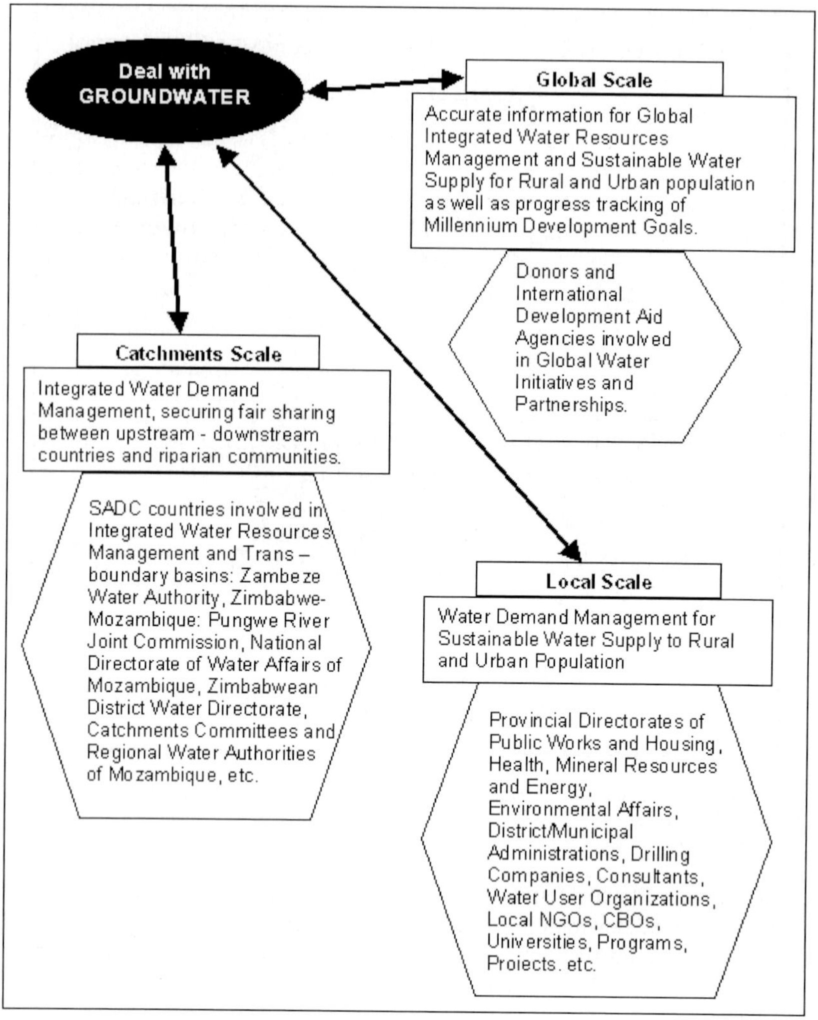

Fig. 2 Multi-scale and multi-institutional approach

(I) Inventory and critical evaluation of required data and existing databases
(II) Identification and evaluation of stakeholders and data levels
(III) Identification of inter-relationships between institutions and databases, re-organization and linkage of data(bases)
(IV) Derivation of concepts and activity plan

3.1. Multi-scale and multi-institutional approach

Groundwater research requires a range of data, which are produced and administered in various Mozambican, regional and international institutions with their respective foci and scales of data representation. The concept of the multi-scale and multi-institutional approach is visualized in Figure 2.

The term multi-scale includes local, regional and global scales of data processing. Various institutions are dedicated to objectives, which are related to one of the scales, e.g. Regional Drought Monitoring Center for Southern Africa (regional/global) or the Water Committee of the community of Nhambita (local).

Multi-institutional is defined by the involvement of various institutions working towards one or more common objectives. In the case of this study the common objective is to facilitate or conduct groundwater research, while each institution maintains its institutional interests. This requires agreements or policies, which define collaborations and data exchange between institutions. Mozambican institutions have discussed the establishment of an information flow and database exchange network. As a result all provinces are compelled to feed the national database in an agreed format, but also in appropriate networks (DNA, 2002).

In Mozambique the state of human and technical resources capacity decreases drastically from the national towards the local scale. Suitable mechanisms must be identified to create connectivity between the levels, which permits the information exchange in the two directions.

3.2. Databases, scales and metadata

Data are facts, which are useless without a context. The organization of data in a database is necessary, as soon as huge data amounts are produced on a regular basis. Generally spatial and non-spatial data are distinguished. Spatial data include geographical locations and spatial relationships. Non-spatial data or attributes are independent of geometric considerations. Both can be brought into relationships in geo-databases, whereby the more complex the matter, the more important becomes the use of a geographical information system as a front program to retrieve and understand the results (Longley *et al.*, 2001).

A database does not need to be in an electronic format. However an electronic database administered by a spatial database management system largely expands functionalities such as simultaneous access of users, multi-institutional data sharing, querying and security (Brail and Klostermann, 2001).

Spatial data are scale-dependent. Many of the data relevant to groundwater investigations are also time-dependent. As a starting point of the current research and the development of a working concept multi-scale data are combined with each other but not merged. This preserves the details of the original input data and allows the down- and up-scaling of information within the predefined limits. A less useful example is DNA (1987), where data of different sources, scales and objectives are merged together. For this reason the few provided details at basin scale are questionable. However, the main objective of DNA (1987) in providing an overall picture of the hydrogeology of Mozambique was met.

Earlier studies with relevance to hydrogeology and hydrochemistry of the Urema River basin follow the traditional approach of data acquisition, which is to conduct large-scale mapping and sampling and then to generalize the information to regional scales (river basin) (GEOTECNICA and MINAS, 1972; Laumanns, *et al.,* 1999; Tinley, 1977).

Recent studies used global and regional scale data (USGS, FAO databases, Landsat 5 and 7 imagery) to draw hydro-geological conclusions on areas at river basin scale (SWECO, 2004; Owen, 2004). It is an efficient desktop approach, which however cannot answer questions related to local water demands like where to search for groundwater.

Knowledge about groundwater in the Urema River Basin can only be enhanced by:

- Considering data from technical reports and borehole records physically stored in archives of drilling companies and partly available from the respective Provincial Directorate.

- Exploring the feasibility to spatially relate water quality reports to the original water well
- Improve the quality of analysis

Data management in Mozambican institutions is still rudimentary and rather called file-keeping. Moreover data in Mozambican institutions are mainly recorded in tabular format, without metadata and lacking spatial references. This applies to any sector, be it water supply and locations of water wells, financial planning in education and locations of schools or planning of vaccinations and locations of health posts. This is, because the usefulness of the data for more comprehensive analysis besides mere institutional record keeping is often not known.

The largest existing groundwater database is at the National Directorate for Water Affairs (DNA) within the Ministry of Public Works and Housing in Maputo, where all country data is stored. That information include detailed geophysical investigation of coastal areas and densely populated settlements, where groundwater is the only supply source. Many provinces have experienced the collection and keeping of groundwater information mainly linked to rural water supply facilities. Till the late 1990s expatriates supported the system and very little capacity building of local personnel took place. When DNA initiated the replacement of the expatriates by national staff, the databases at all levels simply collapsed.

New sector attributions resulting from the approval and implementation of the First National Water Policy (PNA) in 1995 and further sub-sector plans for urban and rural water supply, required effective monitoring tools to assist their implementers. In 1997, the Rural Water Department (DAR-PRONAR), commissioned consultants to conduct a full assessment on rural water infrastructures and establish provincial databases in Zambézia and Sofala using Dbase IV Plus. After such a terrific job, the databases turned out to be inefficient due to weak software maintenance and insufficient users' training. In addition, the lack of a consistent information network through the government structure, the construction of infrastructures by NGO's without authorization from the Provincial Directorates of Public Works and Housing made the existing databases unreliable and not consistent enough for in-depth analyses of groundwater exploitation for the development of rural water supply infrastructures. This situation contributes also to the wastage of financial assets due to repetitive rehabilitations of water infrastructures and in many cases misplacement of new technical facilities, e.g. water wells in unpopulated areas.

By carefully looking at the material stored in archives of Mozambican institutions it becomes obvious, that many data have been lost or have become useless, due to poor data management. Focus on the implementation of suitable database management systems is highly required. Forces need to be directed towards sustainable database management structures, which should serve multi purposes, e.g. supervision and technical documentation of well constructions, groundwater monitoring, water quality control and comprehensive groundwater studies (Figure 3). Data of the Provincial Directorates of Public Works and Housing (DPOPH) should be collected and managed in provincial databases but shared nationwide.

This requires the harmonization of data collection and handling tools, such as the software, database structure and the main field assessment forms. Consistent and consequent codification of water wells from the physical badge throughout all databases is needed. Codification is considered as essential to prevent duplication of data and to facilitate database networking (DNA, 2002).

Metadata for geographic information need to be collected following international standards (ISO19115). An access concept and appropriate quality control measures need to be developed, which assures the administration of different users from the government structures, individuals and other institutions.

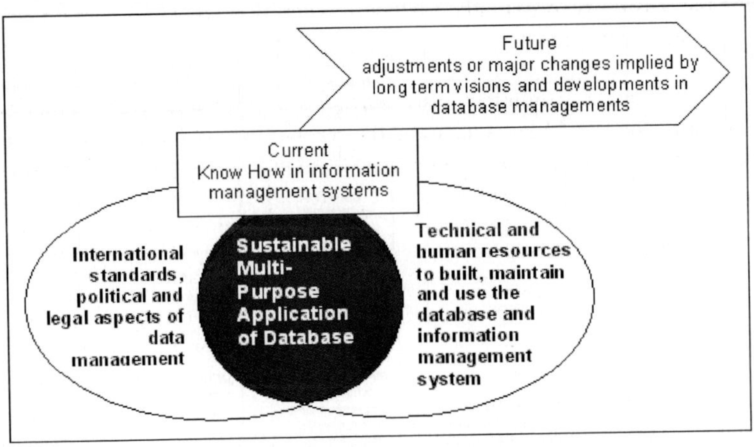

Fig. 3 Factors for sustainable and multi-purpose database management in Mozambique

Groundwater research requires a suit of technical, geological, hydrological, hydrochemical, terrain and physical data. These data are generally obtained from ordinary mapping, in situ and on-site short to long term measurement series, physical and chemical laboratory tests and analysis as well as remote sensing analysis coupled to a geographic information system (GIS). (Mohamed *et al.,* 2003) provide an example of the complexity of a Geographical Information System coupled with a hydro-geological database.

However, geological, aquifer, and water quality studies use volumetric data, which requires an adequate information management system. Current approaches of the Provincial Directorates of Public Works and Housing (DPOPH) in using Access databases are not capable to deal with the nature of the data. The solution for the demand in data and information management systems in a country with weak human and technical capacity such as Mozambique is in the employment of well established and adopted commercial rather than individually developed no-cost products.

The project is using the commercial software Drill & Log, which works on a Visual Fox Pro database coupled with AutoCAD for 2D and 3D geo-referenced visualizations. The database structure is made of related entities including spatial references and attribute data. The program permits multi-user access, data documentation (metadata) and the creation of web archives. Data exchange formats are the common DWG, DXF and DWF. Existing data stored in .dbf or excel format can be transferred directly to the new database. User-defined additional modules can be incorporated later.

3.3. Data hosting institutions, ownerships and responsibilities

Mozambique has due to its political history a highly centralized administration with centralized thinking, responsibilities and data management, which resulted in local incompetence, poorly developed local management skills and the unwillingness to serve central structures.

From the early 1980's to 2000 government-owned drilling companies, GEOMOC and EPARS used to be commissioned at subsidized price for the construction of rural water supply infrastructures by the government. The companies were responsible for data collection, which was presented as technical reports with a copy to the respective provincial directorates. The information from those times is in most cases incomplete, if not disappeared at all.

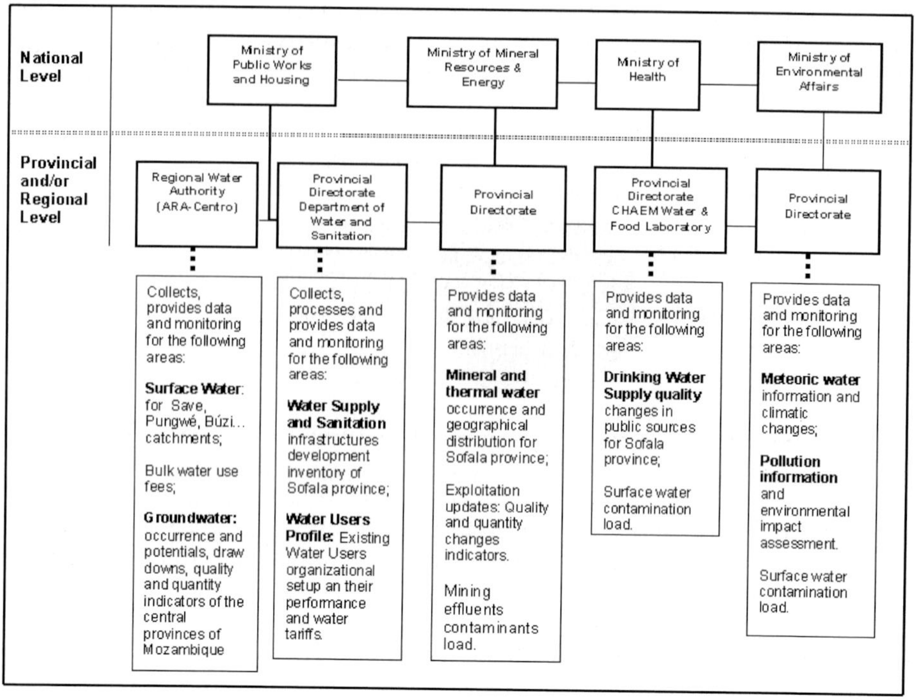

Fig. 4 Mozambican Government institutions in the water sector

Recently reviewed laws empower local decision making and decentralized administrative structures. This political change provides space for multi-institutional approaches, which create a demand for data management concepts and solutions as well as for adequate technical and human resources.

One of the implications of the implementation of the new National Water Policy was to put the state-owned drilling companies onto competition with private companies for water supply infrastructure's construction. This brought changes in the quality of data handling. At the same time the Water Departments (DAS) at Provincial Directorates of Public Works and Housing (DPOPH) took on more responsibility for the data management and information sharing.

Currently there are four ministries and five government departments or authorities involved in the water sector. One of the institutions is acting at catchment level, all others within political administrative boundaries. The current data management constellation in Sofala Province with regard to the water sector is visualized in Figure 4.

There does not exist an established permanent data exchange between the institutions. The current practice for data acquisition is that the government contracts private drilling companies to construct the wells and consulting companies to supervise the works. The drilling companies collect the data. The consulting supervisor has to assure reliability and reports to DPOPH. The DPOPH forwards information to DNA in a digital format or as hard copy reports.

The Ministry of Public Works and Housing through DNA officially announces water supply information to the Mozambican parliament and the public. At the same time the Regional Water Authority (ARA Centro) was established to assume responsibilities of all

DPOPH-DAS (Departments for Water and Sanitation in the provinces) sharing the same river catchment in Central Mozambique (DANN, 1995). However the shift was not conducted and to avoid conflicts ARA Centro limits its responsibilities to the management of surface water at basin level.

Water quality is supervised by the Provincial Directorate of Health (CHAEM), which conducts basic water analysis (main anions/cations, microbiology) in its laboratories. CHAEM issues short commenting laboratory reports. CHAEM keeps an own database, which however lacks spatial references of analyzed water samples. Except of a few cases, it is therefore impossible to relate water quality data to specific boreholes.

Mineral and thermal water is under the responsibility of the Provincial Directorate of Mineral Resources and Energy. Though thermal water provides an important source of information about the origin of groundwater and its exchange and transport mechanism very few specific investigations exist about mineral and thermal water in Mozambique. These are focused to the energy sector (Martinelli *et al.,* 1995).

Although water pollution caused by gold mining has become an obvious problem in the Urema River basin it is not administered by the Ministry of Environmental Affairs. At current there does not exist a database for the registration or monitoring of pollutions. Since pollution problems are affecting public health and security urgent solutions are required, which come at short hand from the sector, that causes the pollution.

4. Linkage between databases with an information system and development of a concept

The combination of information from the different data sources permits the development of a working concept, which serves as the basis for further decisions and focused studies. Figure 5 shows the data management levels made of the database and the database management system, applications and information systems. An example for a soil database is discussed in Meijerink *et al.*, (1994). Databases and database management systems are located at the various institutions. Each institution has its individual applications, which require institutional insider knowledge. The implementation of an information system may provide access to databases of various institutions and facilitates user-friendly and complex analysis of data. In the case of our project it is necessary to structure and link the two more ore less maintained databases of the Provincial Directorate of Public Works and Housing (DPOPH-DAS) and of the Water Laboratory (CHAEM).

These data are analyzed using a GIS. In the light of feasibility, functionality and available local expertise a combination of Visual Fox Pro database system, ArcGIS and AutoCAD is employed. Many of the involved institutions are already familiar with these software products.

For the understanding of the hydro-geology, groundwater dynamics and water qualities or to know, where information is missing it is necessary to develop a concept of the geological evolution of the area.

Summing up all available sources of information about the Urema River basin the following concept and open questions are obtained:

The archaic/early proterozoic basement is made of gneisses and quartz lenses. The gneisses were intruded by proterozoic granites and quartz breccia following old tectonic structures. Gneisses, granites and quartz lenses/veins are outcropping West of the Urema rift in the Southern parts of the Urema basin. Above the basement a strata of 2500 m of createaceous continental so called Sena-sandstones was deposited occurring on both sides of the Urema rift. Sena sandstones crop out–as so called Sena Plains- West of the Urema Rift in the Northern

Table 1 Geological standard profile of the area of the Urema River Basin (compiled from DNA, 1987 and Laumanns *et al.*, 1999)

Stratigraphy	Formation	Facies	Lithology
Q	Quartary	Continental	Alluvial, colluvial
TTs1	Mazamba formation	Continental	Fine textured silty micaceous, purplish violet sandstones, upper parts courser textured
TTi2	Buzi formation (tertiary)	Marine 700 m	Marls, limestones
TTi1	Cheringoma formation (tertiary)	Marine 450 m	Limestones, calcarenite
Ks/Tti	Grudja formation (cretaceous)	Marine 1500 m	Glauconitic sandstones, dolomites, outcropping W-escarpment of cheringoma plateau (60–90masl)
Ks	Sena formation (cretaceous)	Continental 2500 m	Arcose sandstones
G Gamma Qzb Rbeta	Proterozoic	Metamorphic, plutonic	Gneisses Granites Quartz breccia Pegmatites

parts of the Urema river basin. In the rift exists a block with outcropping Sena sandstones (Figure 6, Table 1). This formation is followed by 1500 m of marine glauconitic so called Grudja-sandstones. These crop out only on the Western escarpment of the Cheringoma Plateau. The Grudja formation is followed by 700 m of tertiary marine limestones and dolomites of the Cheringoma Formation. On the East side of the Urema rift – the Cheringoma Plateau – Oligocene features of karstification (caves, collapse structures) along NNW fractures was observed (Laumanns *et al.,* 1999). It is not known how deep the karstification developed and if it also occurred in rocks of the Cheringoma formation in the Urema rift.

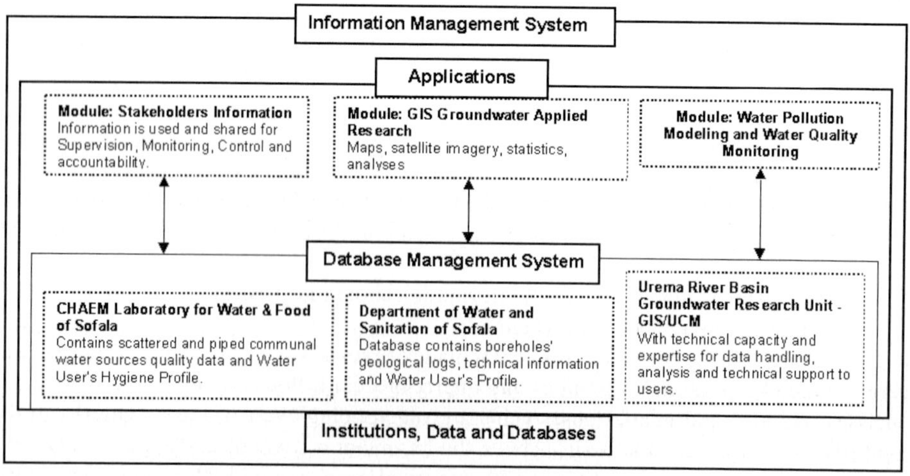

Fig. 5 Linkage between databases of various sources in groundwater research

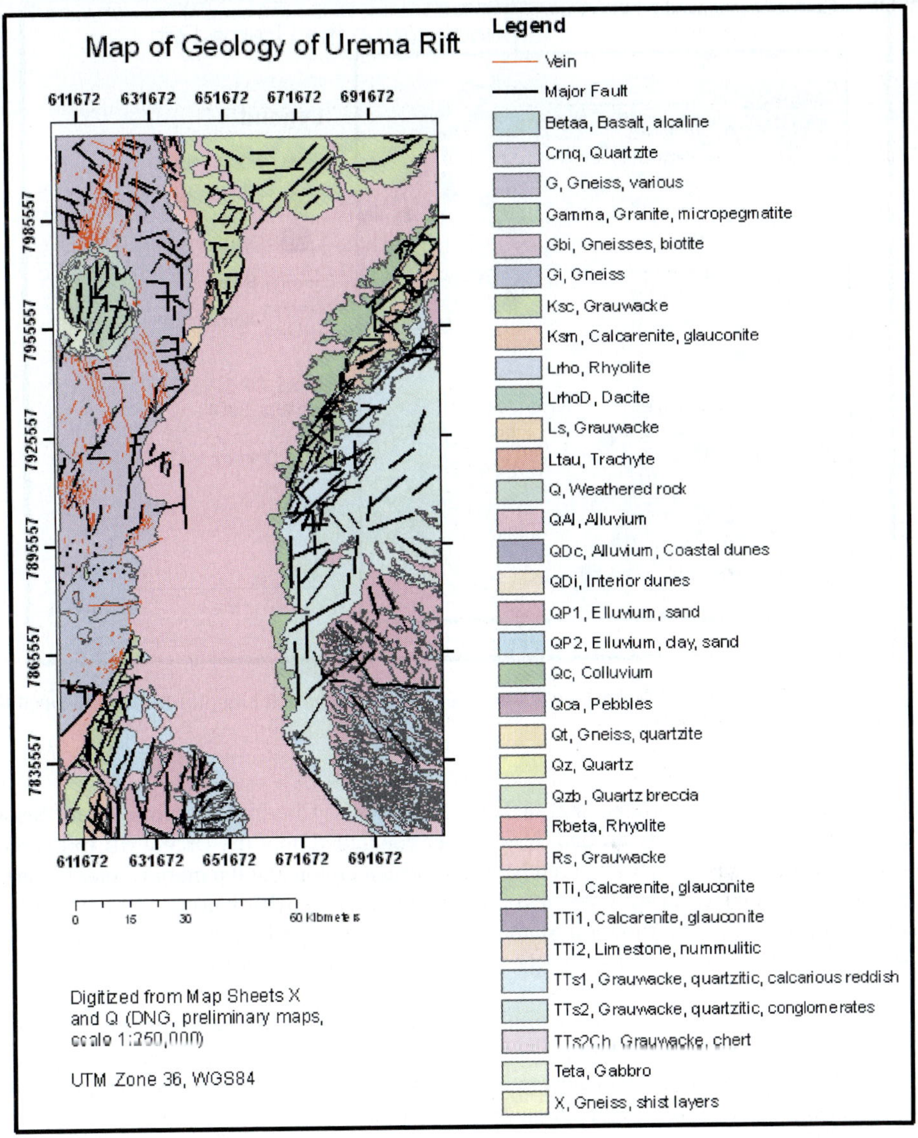

Fig. 6 Map of Geology of the Urema River Basin (digitized from DNG, 1968) including Morphology of the Urema Rift area (Elevations from topographical map of scale 1:50,000, surface factor 30, geological formations from geological map sheets X and Q of scale 1:250,000)

Karst structures were filled with tertiary continental Inhaminga sandstones of the Mazamba formation. Sandstones are characterized by purplish colour and high biotite contents presuming short transport distances and weathered metamorphic basement rocks as the source. There is no information about whether the West side of the Urema rift originally had a depostion of Grudja and Cheringoma Formation, which then was completely eroded due to a faster uplifting process than on the East side of the Urema rift. If the origin of the sediments of the Mazamba formation are gneisses from the basement on the West side of

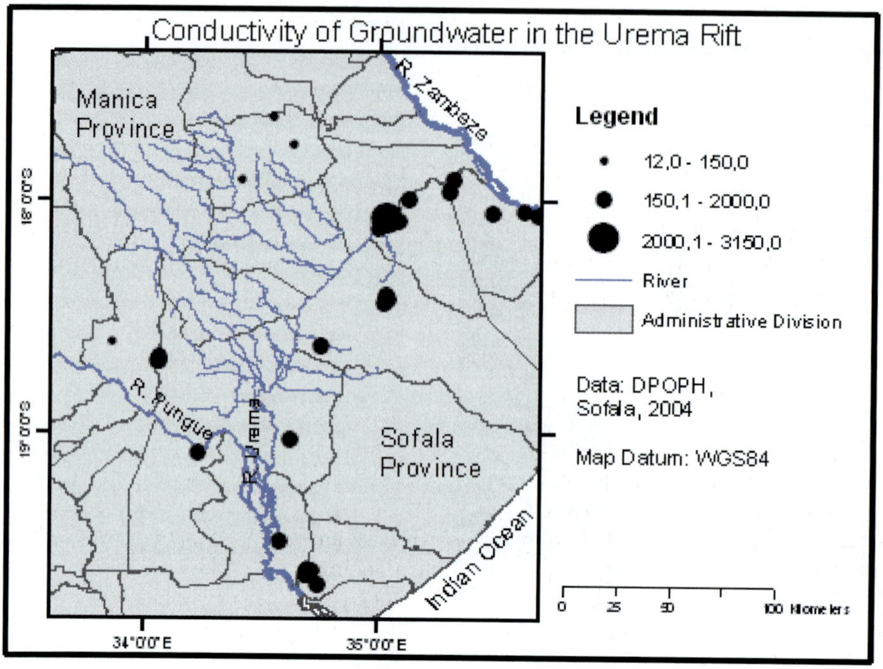

Fig. 7 Map of groundwater conductivities in water wells of the Urema Rift area (in μS/cm, from DPOPH Sofala, DINAGECA Mozambique)

the Urema rift one can assume that at latest in Miocene Cheringoma, Grudja and Sena Formations were completely eroded. It is further concluded, that the Urema rift forms the physical boundary between the old (archaic) Rhodesian craton and the mobile coastal zone. The zone is characterized by transgressions and regressions with marine and continental sediment cycles. It is not known how much of these geological units were deposited and conserved in the Urema rift. During the Quartery erosion and uplifting are the dominant processes.

Today the Urema rift is one of the most active seismic zones of the Southern end of the East African Rift and in Mozambique in general (Chapola, 2000). At least two hot springs ($T > 32°$C) occur along the Western margin of the Urema rift, one in the community of Nhambita and the other one in the community of Tambara.

Rivers draining into the Urema rift are depositing fluvial sediments, whereas the heavier, large particle sizes are deposited on the foot of the rift escarpments and the finer particles are found towards the center of the rift (Owen, 2004). The deposition of suspended particles is controlled by runoff amounts and velocity, which changes with wet and dry seasons. Major amounts of surface water infiltrate the underground along the rift margins. Sediment fans evolving successively from the rift margins towards the center mainly control elevations and thus hydrological dynamics.

Tinley, (1977) assumed, that the Urema Rift served as drainage of the Zambeze River joining the Pungue River near the city of Beira. According to Tinley, (1977) sediment fans have caused the formation of the hydrological watershed between the Zambeze river basin and control the outflow from the Urema rift via the Lake Urema. The extent of hydrogeological recharge-discharge areas (subsurface divide) is not yet investigated. It seems likely, that there still exists a hydro-geological connection between the Zambeze River and the Urema River basin.

Two main aquifer types are expected in the area: poreous sedimentary aquifer and fracture controlled hardrock aquifer in metamorphic and plutonic rock formations. Water tables observed in water wells in the Urema river basin vary between a few meters to up to 82 m below surface. At this stage available information does not permit distinction between aquifers or extents of aquifers. Water wells have a production between 0.9 and 6.33 m^3/h. Available groundwater conductivities are mainly below 2000 μS/cm. A tendency of higher conductivities of up to 3150 μS/cm from deeper groundwater and conductivities up to 149 μS/cm in comparably shallower groundwater is noted (Figure 7).

5. Conclusions and outlook

In general groundwater investigations are virtually non-existing in Mozambique. Database management is poorly developed. Data are kept in tabular formats and serve mainly record keeping. There do not exist inter-institutional relationships regarding database management, because the conditions for multi-institutional database management, i.e. human and technical resources, policies are not in place. As a result overlapping and weakly defined competences and responsibilities are abundant. The data management problem turns into a high economic risk business for groundwater exploration companies, because companies are made responsible for the success of their activities. However the poor data management is also noted in the low performance in meeting water supply plans, i.e. to significantly increase the number of safe water points per population number and administrative unit.

Groundwater research needs to improve the resolution and precision of digital data from current small/medium to larger scales to be able to solve water issues at the sub-catchment level. Further it is necessary to seek to complete the picture of the geological evolution of the entire Urema rift area. This includes geo-chronological investigations to clarify the stratigraphic order of geological features in the Urema River basin.

The creation of a groundwater monitoring network besides the ongoing rehabilitation of meteorological stations is another requirement to capture baseline data on a long-term run. Baseline data provide information on the geological and geochemical background of the area. The knowledge of what is "normal" in the Urema River basin provides an instrument for the detection of abnormalities, such groundwater pollution. Specific ongoing project activities are:

- Implementation of the hydro-geological database,
- Creation of pilot monitoring points in the Urema River basin,
- Analysis of the data using a geographic information system,
- Development of a hydrodynamic model,
- Hydro-chemical characterization of groundwater and
- Production of a detailed hydro-geological map of the Urema River basin

Acknowledgements The access to technical reports was kindly supported by the Provincial Directorate of Public Works and Housing/Department of Water and Sanitations Sofala/Mozambique and the Programme of Rural Water Supply and Sanitation in Beira/Mozambique funded by the Austrian Cooperation. Database management and analysis is promoted by the Center for Geographic Information of the Catholic University of Mozambique.

References

Brail RK, Klostermann RE (2001) Planning support systems-integrating geographic information systems, models, and visualization tools. Esri Press, Redlands

Chapola LS (2000) Country report in Malawi – seismicity and tectonics, geological survey of Malawi. Lilongwe

DNA (1987) Hydrogeological Map of Mozambique, scale 1:1,000,000 and explanatory text, Maputo

DNA (1995) National Water Policy, Boletim da Republica, Resolution Nr. 7/95, Maputo (Portuguese) – www.dna.gov.mz

DNA (2002) Primeira reunião nacional sobre o banco de dados do sub-sector de AAR, September 2002, Msika-Manica

DNG (1968) Geological Map, Sheet Q and X of Central Mozambique, Scale 1:250,000, Maputo

GEOTECNICA & MINAS (1972) Inventariação dos recursos hidraulicos da bacia do Rio Pungoe, Previsão dos recursos de aguas subterraneas, Rel. 205 e205a, Maputo

Laumanns M, Ruggieri R, Germano A, Galletti I (1999) Report on the European speleological project: Cheringoma 1998, July 1999, Maputo

Longley PA, Goodchild MF, Maguire DJ, Rhind DW (2001) Geographic information system and sciences. John Wiley & Sons, Ltd. Chichester

Martinelli G, Dongarra G, Jones MQW, Rodriguez A (1995) Geothermal features of Mozambique-Country update. In: Proceedings of the World Geothermal Congress

Meijerink AMJ, Brouwer HAM, Mannaerts CM, Valenzuela CR (1994) Introduction to the use of geographic information systems for practical hydrology. ITC publication number 23, Enschede

Mohamed B, Frederic H, Pierre N, Ingrid R (2003) Coupling GIS with a Database for Hydrogeological Mapping, 4th Congress on Regional Geoscientific Cartography and Information Systems, 18th June 2003, Bologna

Owen R (2004) The Millennium ecosystem assessment gorongosa area hydrogeology condition and trend report. Ministry of Tourism of Mozambique, Maputo

Tinley KL (1977) Framework of the Gorongosa Ecosystem, Diss., Pretoria

SAfMA (2004) The millennium ecosystem assessment gorongosa area-general report. Ministry of Tourism of Mozambique, Maputo

SWECO & Associates (2004) Development of the Pungwe River Basin Joint Integrated Water Ressources Management Strategy, Monograph Report, Annex I–XII, National Directorate of Water of Mozambique, Maputo

Water Resour Manage (2007) 21:185–198
DOI 10.1007/s11269-006-9048-9

ORIGINAL ARTICLE

Integrated water and food analysis at the global and basin level. An application of WATERSIM

Charlotte de Fraiture

Received: 20 April 2006

Abstract Several recent studies warn that under 'Business-as-Usual' a water crisis is impending, suggesting that appropriate actions need to be taken on the water supply and demand side. While many measures to alleviate water scarcity are within the water sector, it is increasingly recognized that many drivers, policies and institutions outside the water sector have large and real implications on how water is being allocated and used. Important drivers for water use include population and income growth, urbanization, trade and other macroeconomic policies, environmental regulations and climate policy. While some of these processes and trends, especially those at global level, may prove difficult to influence directly, it is important to understand their linkages with water issues to analyze the relative impact of various policies in the agricultural and water sectors on water and food security.

The strong linkages between economic trends, agricultural policies and water use call for an integrated and multidisciplinary modelling approach. The WATERSIM model, developed by the International Water Management Institute (IWMI) is a suitable tool to explore the impacts of water and food related policies on global and regional water demand and supply, food production and the environment. This paper introduces the WATERSIM model and, using some preliminary results, illustrates the importance of global economic trends on food and water outcomes.

Keywords Global water and food model · Integrated modeling approach

1. Introduction

The finite nature of water resources and the increasing pressure on existing sources to meet growing demands, as well as increasing domestic, industrial and environmental requirements have been foremost in the concerns of policymakers and experts for several decades. Studies estimating current future global water diversions and depletion have been carried out over the

C. de Fraiture
International Water Management Institute, Colombo, Sri Lanka
e-mail: c.fraiture@cgiar.org

last decades.[1] Several recent projections warn that under 'Business-as-Usual' a water crisis is impending. To ensure that sufficient water is available, they suggest that appropriate actions need to be taken on both the supply and demand side, including reform of water management, and increasing water use efficiency (Alcamo *et al.*, 1997; FAO, 2003; Rosegrant *et al.*, 1999; Rosegrant *et al.*, 2002; Seckler *et al.*, 1998, 2000; Shiklomanov, 2000; Vörösmarty *et al.*, 2000).

Many of the proposed and implemented measures to alleviate water scarcity fall within realm of the water sector. But many drivers outside the water sector, -such as population and income growth, urbanization, trade and other macroeconomic policies, environmental regulations and climate policy-affect water use. To analyze the relative impact of various policies in the agricultural and water sectors on water and food security, it is essential to understand the linkages between macro level socio-economic drivers and basin level water use.

Traditional water demand and supply projections in the literature tend to insufficiently account for economic processes and feedback mechanisms, or do so only implicitly. The IMPACT-WATER model, developed by the International Food Policy Research Institute (IFPRI), with input from the International Water Management Institute (IWMI) was one of the first global models to integrate an agricultural food supply and demand projections model with a global water supply and demand model to jointly analyze water and food supply and demand into the future under various policy scenarios (Cai and Rosegrant, 2002; Rosegrant *et al.*, 2002). This modelling framework has since been updated and expanded by IWMI and IFPRI into the WATERSIM model to enable a more disaggregated and comprehensive analysis of the future world food and water situations.

This paper lays out the reasons for joint food and water analysis at the global and basin scales. It then introduces WATERSIM as a suitable modelling tool for water and food analysis. The paper presents some preliminary scenario results that underline the importance of socio-economic variables in future food and water demand.

2. Integrated food and water analysis

The strong linkages between economic trends, agricultural policies and water use call for an integrated and multidisciplinary modelling approach. Being an integrated water and food model, WATERSIM is a suitable tool to explore the impacts of water and food related policies on global and regional water demand and supply, food production and the environment.

Water and food demand and supply are intrinsically linked. What people eat and where and how that food is produced, determines to a large extent, how water is being allocated and used. The amount of water required to produce food varies by crop type and region, depending on climate, mode of cultivation (rain fed versus irrigated, high versus low input agriculture), variety and length of growing season, and crop yields. The often quoted rule-of-thumb, which states that every kilogram of wheat depletes 1000 litres of water, masks the enormous variations in crop water productivity.[2] Estimates vary from 400 to 5000 litres, depending on the location (examples in Sakthivadivel *et al.*, 2001; Tuong and Bouman, 2003; Zwart and Bastiaanssen, 2004).

---

[1] Gleick (1999) provides an excellent overview of global water modeling efforts over the past 40 years.

[2] Crop water productivity is defined as the amount of water evaporated by the crop per unit of economic yield, expressed in cubic meter per kilogram of crop.

Renault (2003) and Chapagain and Hoekstra (2003) compare the amount of water required to produce different crops and livestock products. Potato requires the least amount of water per kilogram (90 litres) while, in contrast, one kilogram of beef may take up to 12,000 litres to produce.[3] In terms of water demand it makes a big difference what people eat and where and how this food is produced.

Changes in both the food supply and demand structures have been very rapid in most of the world, and particularly in Asia. Rising incomes throughout much of Asia over the last three decades led not only to increasing consumption of staple cereals, but also to a shift in consumption patterns among cereal crops and away from cereals towards livestock products and high-value crops. Wheat and feed grains increasingly emerged as particularly important cereal crops in a region traditionally dominated by rice consumption. Meat consumption more than tripled, while dairy demand more than doubled from 1967 to 1997. Consumption of high-value crops-such as fruit, sugar and edible oils also increased substantially.

It is clear that both rising incomes and structural changes to consumption patterns will continue to drive trends in food – and hence agricultural water-demand in Asia over the next decades. Rapid urbanization is perhaps the most important ongoing structural shift affecting food consumption, with historical evidence from China indicating that consumption of grains, edible oils and vegetables is higher in rural areas, while consumption of meat, fish and dairy products is higher in urban areas (Huang and Bouis, 1996). Because water requirements to produce high-value crops and meats and oils are generally higher than of cereals water use per kilocalorie consumed will increase over time.

World market prices and distorting subsidies affect -or determine- agricultural production patterns and, hence, play a large role in local water demand.[4] While water scarcity generally does not shape trade flows, trade policies and subsidies have important water implications. By changing incentives to farmers, international trade agreements such as under the WTO and international agricultural policies such as Common Agricultural Policies (CAP) in the EU, affect crop choice and hence water use.

Obviously, these issues call for an integrated water and food analysis. WATERSIM, the model introduced in this paper, is a suitable tool to that purpose.

3. The WATERSIM model

To provide an objective and scientifically sound basis to debates on water for food and environment, the International Water Management Institute (IWMI) and International Food Policy Research Institute (IFPRI) embarked on a joint modelling exercise, resulting in the WATERSIM model. WATERSIM (Water, Agriculture, Technology, Environment and Resources Simulation Model) explores the impact of water and food related policies on water scarcity, food production, and environment. Designed to better understand the key linkages between water, food, and environment; its flexible model structure allows for exploring various scenarios to address key questions and strategic decisions on alternative development paths.

WATERSIM is a global scale model. While the global coverage limits the level of detail and complexity that can be incorporated in the model-due to realities of limited data, computing limitations and limited knowledge- there are compelling reasons to choose a global level

[3] Most of water for meat is needed to produce fodder or feed (see Chapagain and Hoekstra, 2003).

[4] A large body of literature exists on the impact of subsidies and trade distortions on agricultural production. Very few papers deal with the impact on water use. One example is Ioris (2004).

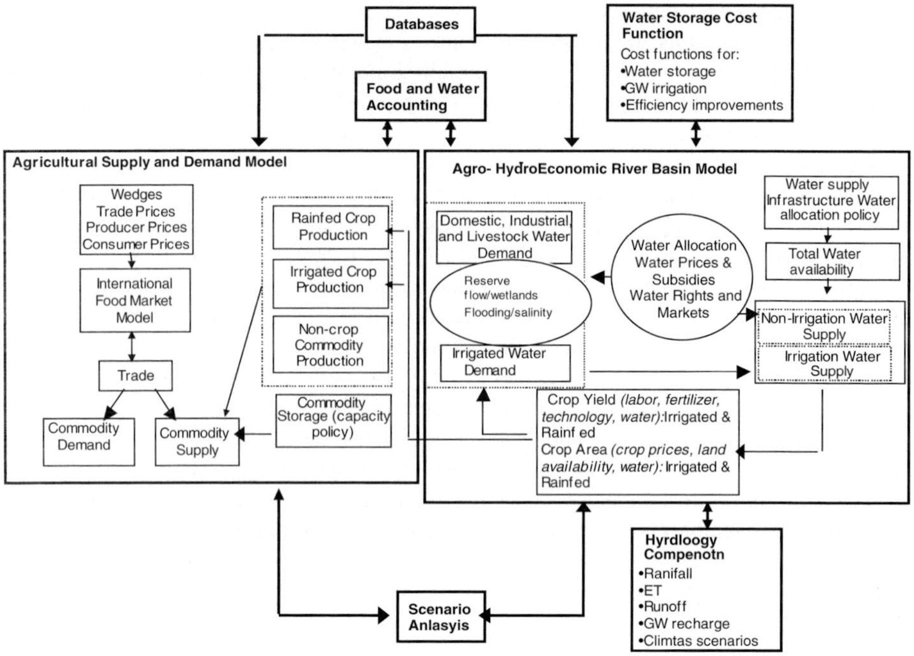

Fig. 1 Schematic diagram of WATERSIM model

framework. Some of the socio-economic processes that the model deals with are of global nature. Examples include the impact of world market prices and subsidies on crop production, or the impact of international treaties and policies (such as WTO, bilateral agreements, CAP). Further, some physical processes important to questions that the model addresses are global (for example, climate change and changes in global hydrological cycle).

Feedback mechanisms between water and food sectors are an important feature of the WATERSIM model which distinguishes it from other water and food projection models. For example, water shortage may lead to a reduction in food production. But this in turn leads to higher food prices, inducing a higher production in the next season and thus partly offsetting food shortage. Another example: higher food demand lead to higher water demand. But increased water demand may provide an incentive to improve on water use efficiency (if feasible), thus offsetting part of the increased demand.

Broadly-speaking the model consists of two integrated modules (Figure 1): the 'food demand and supply' module, which is adapted from IMPACT developed by IFPRI (Rosegrant *et al.*, 2001); and the 'water supply and demand' module which uses a water balance based on the Water Accounting framework (Molden, 1997) underlying PODIUM (Fraiture *et al.*, 2000 and 2003) combined with elements from the IMPACT-WATER model (Cai and Rosegrant, 2002).

3.1. Spatial and temporal scale

3.1.1. Spatial units

To model hydrology adequately, it makes most sense to choose a river basin as basic spatial unit. When it comes to food policy analysis, administrative boundaries should be used-since trade and policy happen at national level, not at river-basin scale. WATERSIM takes a hybrid approach to its spatial units of analysis. Firstly, the world is divided into 125 major river basins of various sizes with the goal of achieving accuracy with regard to the basins most important to irrigated agriculture. Next, the world is divided into 115 economic regions including mostly single nations with a few regional groupings of nations. The food module uses the economic regions as basic modelling unit, while the water module runs at river basin scale. The interaction between food and water modules is facilitated by intersecting basins and regions into 282 sub-basins or Food Producing Units (FPU's). The hydrological processes are modelled at basin scale by summing up relevant parameters and variables over the FPU's that belong to one basin. Similarly, economic processes are modelled at regional scale by summing up relevant variables over the FPU's belonging to one region.

3.1.2. Temporal scale

Economic processes are modelled at an annual time step, while hydrological and climate variables are modelled at a monthly time-step. Crop related variables are either determined by month (crop evapotranspiration) or by season (crop yield, area). The food supply and demand module runs at region level on a yearly time-step. The water supply and demand module runs at sub-basin level at a monthly time-step. For area and yield computations the relevant parameters and variables are aggregated over the months of the growing season.

The year 2000 is taken as base. Projections are made for the year 2025, but the model allows for shorter or longer projections as well.

3.2. Food module[5]

The food module is a partial-equilibrium agricultural production and trade model, that simulates global food production, consumption and trade levels consistent with observed technologies, input prices and water available for crop production. The model incorporates 32 food commodities – including all cereals, soybeans, roots and tubers, meats (including beef, pig meat, sheep and goat, and poultry), milk, eggs, oils, meals, fruits, sugarcane, sugar beet, cotton, eight fish commodities, fish oil, and fish meal. Despite its primary focus on the agricultural sector, the structure of the food module considers income growth in both the agricultural and non-agricultural sectors. It permits long-term projections of food prices, supply, demand and commodity trade.

The food supply and demand relationships respond to regional economic conditions and world food prices through a system of supply and demand elasticities, that translate price signals into food production and consumption levels. These elasticities are embedded in a system of non-linear equations representing the underlying behavioural and technological characteristics of each region. The supply and demand relationships include regional income and population levels, among other important socio-economic variables – such as agricultural technology. Local prices differ from world market prices by local subsidies and taxes expressed by the Consumer and Producer Subsidy Equivalent. By iterative adjustment of the world prices, the solution algorithm of the food module finds the price levels at which the aggregate agricultural supply and demand levels for the world are equalized – thereby

[5] Largely based on IMPACT work, documented in Rosegrant, Agcaoili-Sombilla and Perez (1995); Rosegrant, Meijer and Cline (2002); and Rosegrant, Cai and Cline (2002).

achieving a global partial equilibrium in food consumption and production. Consequently, food commodity prices are endogenous to the model.

3.2.1. Crop supply functions

Domestic crop production is determined by the area and yield response functions, formulated separately for production under irrigated and rain fed conditions. Harvested area is specified as a response to the crop's own price, the prices of other competing crops, the projected rate of an exogenous (non-price) growth trend, and water. The projected exogenous trend in harvested area captures changes in area resulting from factors other than direct crop price effects, such as expansion through population pressure and contraction from soil degradation or conversion of land to non-agricultural uses. Yield is a function of the commodity price, the prices of labour and capital, water, and a projected non-price exogenous trend factor. The trend factor reflects productivity growth driven by technology improvements, including crop management research, conventional plant breeding, wide-crossing and hybridization breeding, and biotechnology and transgenic breeding. Other sources of growth considered include private sector agricultural research and development, agricultural extension and education, markets, infrastructure, irrigation, and water. Annual production of crop commodity is estimated as the product of its area and yield. If water availability falls below a certain threshold, crop area and yield are reduced. This production response mechanism to water availability provides the essential link between the food and water modules.

3.2.2. Livestock supply functions

Livestock production is modelled similarly to crop production except that livestock yield reflects only the effects of expected developments in technology. Total livestock slaughter is a function of the livestock's own price and the price of competing commodities, the prices of intermediate (feed) inputs, and a trend variable reflecting growth in the livestock slaughtered. Total production is calculated by multiplying the slaughtered number of animals by the yield per head.

3.2.3. Demand functions

Domestic demand for a commodity is the sum of its demand for food, feed, and other uses. Food demand is a function of commodity price, prices of other competing commodities, per capita income, and total population. Per capita income and population increase annually according to region-specific growth rates. Feed demand is a derived demand determined by the changes in livestock production, feed ratios, and own- and cross-price effects. The feed demand equations incorporate a technology parameter reflecting improvements in feeding efficiencies. The demand for other uses is estimated as a proportion of food and feed demand.

3.2.4. Prices

World prices for food are endogenous to the system of equations that represent the underlying food production and consumption relationships.[6] Domestic prices are a function of world prices, and are expressed in terms of the local regional currencies, through exchange rates, and

[6] Based on earlier work by Agcaoili, Oga and Rosegrant (1993); Oga and Gehlar (1993).

are further adjusted by the effect of regional price policy regimes and market characteristics. The effects are expressed in terms of the producer subsidy equivalent (PSE), the consumer subsidy equivalent (CSE), and the marketing margin (MI). The PSE and CSE measure the implicit level of taxation or subsidy borne by producers or consumers relative to world prices and account for the wedge between domestic and world prices. The marketing margin (MI) reflects other factors such as transport and marketing costs or product quality differences. To calculate producer prices, the world price is reduced by the MI value and increased by the PSE value, whereas the consumer prices are obtained by adding the MI value to the world price and reducing it by the CSE value. All the PSE, CSE and MI values are expressed as percentages of the world price.

3.2.5. International linkage through trade

In the global food module, regions are linked through food commodity trade, which is the difference between domestic production and demand for each region. Regions with positive trade balances in a commodity are net exporters of that good, while those with negative balances are net commodity importers. The specification of commodity trade within the food module is non-spatial, i.e. it does not permit a separate identification of importing and exporting regions of a particular commodity, nor the analysis of explicit spatial trade patterns and flows.

At the global aggregate level, the food module reaches an equilibrium in which net trade equals zero. The world price of a commodity is the equilibrating mechanism. When an exogenous shock is introduced in the model, the world price will adjust such that each adjustment is successively passed back to the effective producer and consumer prices via the price transmission equations. Changes in domestic prices subsequently affect commodity supply and demand, necessitating their iterative readjustments until world supply and demand balance, and the world net trade balance, again, equals zero.

3.3. Water module

The methodology adopted in the water demand and supply module is based on water balance at basin level using the concepts of the water accounting framework (Molden, 1997) relating water demand to available water supply. Water demand for human purposes, besides environmental and in-stream purposes, is derived from four sectors-agriculture, domestic & municipal sector, industry and livestock. At sub-basin level, water availability is simulated using a water balance approach, considering internally generated runoff, inflow from other units, groundwater contributions existing infrastructure and management practices. Sub-basins are connected in such a way that outflow from upstream becomes inflow into the lower sub-basin. When supply falls short of demand, the shortages are distributed over months, sectors and crops using an optimization model and allocation rules.

3.3.1. Agricultural water demand

WATERSIM differentiates between depletion and total diversions. Water depletion is defined as a use or removal of water from a basin that renders it unavailable for further use (Molden, 1997). Water is depleted by four processes: evaporation, flows to sinks, pollution and incorporation into a product (for example, water taken up by crops incorporated into plant tissues). Total depletive demand consists of depletion in four sectors: irrigated agriculture, industry, domestic use and livestock.

Depletive water demand in agriculture is a function of the irrigated area, cropping pattern, crop water requirements, effective precipitation and effective efficiency. Crop water requirements are determined at a 0.1 × 0.1 degree global grid using cropping pattern information from AQUASTAT [7] crop coefficients taken from FAO (Allen *et al.*, 1998; Doorenbos and Kassam, 1979) and reference evaporation data from the IWMI-water-and-climate atlas.[8] The effective precipitation, defined as that part of the rainfall that is beneficially used by crops, is computed according to the SCS method (USDA, 1967), using data on total precipitation from the CRU TS 2.0 dataset (Mitchell *et al.*, 2003) determined at a 0.5 × 0.5 degree level of spatial resolution on the global grid.

The Effective Efficiency (EE) – indicating how efficient *depleted* water has been utilized – is computed from the amount of water beneficially used by the intended process divided by the total amount of freshwater depleted during the process of conveying and applying water (Keller and Keller, 1995). The upper limit of EE is 100% but in practice this is never reached due to prohibitively high costs to achieve this. Volumetric water pricing may induce improvements in EE. To facilitate this option in the model, EE is formulated as a function of water price in some scenarios, though in the baseline scenario this option is not used, because of data limitations.

3.3.2. Industrial water demand

Water demand in the 3 industrial sectors in the WATERSIM model – namely manufacturing, energy and agro-industry – is expressed as a function of income, the price of water and a time trend. Industrial water demand is simulated under a variety of growth scenarios with respect to population and per capita income.

3.3.3. Domestic water demand

Water demand in the domestic and municipal sector, including both urban and rural areas, is calculated on the basis of income and population growth rates. The relationship between demand and population and income growth rates was found by regression analysis on time-series data obtained from Shiklomanov (1999) and World Bank (1998).

3.3.4. Monthly water balance at sub-basin level

The total inflow into a sub-basin consists of internally generated runoff, groundwater recharge, inflow from inter-basin transfer and other sources such as desalinization. The total inflow is stored in the basin, or if the inflow is greater than the existing storage capacity, spills to a lower basin or sink. The storage capacity in the sub-basin is simulated by the Basin Equivalent Storage (BES), reflecting the maximum amount of controllable surface and groundwater available for use at one point in time. It is equal to the real storage (surface and groundwater) plus the 'storage' equivalent to the sum of water lifting, gravity diversion, and other forms of water diversion from the water system, discounted for the internal return flows. The BES is a function of investment in infrastructure. Groundwater is function of natural recharge from precipitation and seepage from irrigation fields and canals.

[7] http://www.fao.org/ag/agl/aglw/aquastat/main/index.stm

[8] http://www.iwmi.cgiar.org/WAtlas/atlas.htm

The amount of water available for different uses depends on the basin equivalent storage, water management and the amount of monthly inflow. As long as available storage is small in comparison to inflow, additional storage capacity will increase the amount of available water, up to a certain limit where the amount of inflow becomes the limiting factor. For example, in the Colorado basin where in dry years all potentially utilizable water is depleted or committed to downstream uses, a new dam would merely change the distribution of available water over the basin without augmenting its quantity. Where reservoirs account for big part of the storage, reservoir operational rules impact water availability. For example, if reservoirs are filled at the beginning of the rainy season, inflow from rainstorms cannot be captured and flows out without being made available for later use. The modelling setup allows for reservoir operation rules for other purposes-hydropower and flood retention- but detailed information on these water allocation rules is lacking. Water in the reservoirs is either stored for later use or released. While part of the release is depleted or transferred out of the basin as part of inter basin transfer scheme, the remainder flows out as return flow to a lower sub-basin or sink.

3.3.5. Optimizing water supply according demand

Supply is matched to demand adopting an optimization approach, based on a traditional reservoir operation model (described in Cai and Rosgrant, 2002) with the objective to maximize the ratio of depletive supply over demand. In the absence of detailed information on water allocation rules, an optimization formulation is chosen that –implicitly- assumes a rational water management with perfect information, in which water is allocated in accordance to demand. The optimal allocation is constrained by physical limits, operational rules and environmental concerns which may be different in the various scenarios.

With growing agricultural and industrial development, balancing the needs of aquatic ecosystems and human uses is becoming critical in many water basins. To assess that balance WATERSIM uses in-stream environmental water requirements, expressed as percentage of the total river flow, derived from Smakhtin *et al.* (2004). The percentages are applied at a monthly time step, assuming an equal percentage for each month. While this assumption is debatable, a monthly global dataset on environmental flows is not available. Policies on meeting environmental water requirements can be entered as hard constraints, in which environmental requirements are always met. But they can also be treated as soft constraints to simulate a more realistic situation in which a certain degree of environmental damage is tolerated in extreme water short periods. Priorities between different constraints are determined by an explicit weighting scheme. Weights are assigned by trial and error until allocations to the environment match available empirical evidence.

The result from the optimization procedure is a monthly estimate of the total amount of water actually available for depletion.

3.3.6. Allocation rules to sectors and crops

If water available for depletion falls short of demand, WATERSIM can use different sets of allocation rules to distribute the shortage over the sectors, depending on the scenario and country. For most countries practice shows that the industrial and domestic sectors take preference over agriculture. If the amount of water available for depletion is insufficient to cover industrial and domestic demands, the domestic sector gets priority. But to test different assumptions and scenarios the structure of WATERSIM allows for a variety of allocation rules.

For example, water shortage, if occurring, can be distributed over the sectors proportional to demand or priority can be given to the agricultural sector.

The allocation of irrigation water to crops is based on the profitability of the crop, sensitivity to water stress and net irrigation demand. Higher priority is given to crops with higher profitability, higher drought sensitivity and higher irrigation water requirements.

3.3.7. Yield and area reduction due to water stress

Water shortages, occurring in irrigated or rain fed agriculture, reduce crop yields and harvested areas. When irrigation water is scarce farmers have the choice of reducing the water layer on the field, or reduce the cropped area to increase the water layer on the remaining area. The model simulates the reduction in areas and yields using the formulation developed by FAO (Doorenbos and Kassam, 1979), while considering this trade-off between reduction in area and water layer.

3.4. Integration of water and food modules

The basic assumption in the food module is that each year the world market for agricultural commodities clears, i.e. production equals demand plus change in stocks. The water module is based on a water balance approach, i.e. inflow equals outflow plus change in basin storage. Both modules are connected through two variables: (1) agricultural area, which determines food supply and water demand; (2) crop price which determines food demand and crop profitability which in turn affects water allocation. The food module estimates food production (area and yield) as a function of socio-economic driving forces. Where water limits agricultural production, the model accounts for the effects of water stress through a reduction factor for area and yields, in both irrigated and rain fed agriculture. Updated areas and yields are then fed back into the food module and the market equilibrium recalculated. The model iterates between the water and food modules until market equilibrium and water balance is reached.

3.5. Model implementation

The modelling approach followed here is very data intensive with information derived from a variety of sources. Where possible, GIS techniques are used to format the data into the right spatial resolution. The model is calibrated on data for the base year 2000 and, where available, for the year 1995.[9] Calibration and validation are an essential part of modelling. Yet, for models of global and multidisciplinary scope, such as WATERSIM, it is the most challenging part. Baseline data on water use variables are incomplete and often outdated. Further, the number of model parameters and variables makes calibration a time-consuming and computationally-intensive task. Water use data for calibration of the water module are derived from the best available global dataset, i.e. Aquastat, and where available, complemented with information from national statistics on water use. Data to calibrate the food module mainly come from the FAOstat database. The base line projections are chosen such that growth rates in the coming 25 years are consistent with the trends observed in the past 40 years (1961–2000 data, mainly taken from FAOSTAT[10]). This implies that the baseline

[9] Previous models IMPACT-water (IFPRI) and Podium (IWMI) use 1995 as base year. Direct comparison of modeling results is not always possible due to differences in spatial units and definitions of terms

[10] http://faostat.fao.org/default.jsp

Table 1 Modelling results for the year 2025, China

	2000	BAU	OPT	PES
Population in billions	1.28	1.44	1.39	1.48
Income per capita in US$	780	3351	4491	898
Pork demand in millions of tons	42.0	81.0	92.7	58.9
Cereal food demand in millions of tons	232	316	320	283
Cereal feed demand in millions of tons	88	152	181	117
Water depletive demand agriculture in km^3	238	307	336	256
Industrial water demand in km^3	14.1	41.4	51.3	28.1

projection reflects the Business-as-Usual scenario, continuing past trends. Validation of the baseline scenario is done by visual inspection by plotting time series data and projections of food production and demand in one graph.

While the model is solved in GAMS, input and output files are in Excel format to facilitate easy analysis of modelling results. It takes about 10 hours to solve one scenario on a high-end PC (3.4 GHz CPU).

4. An example

For the globe, total depletive water demand in the base year at 1830 km^3 of which 1450 km^3 is comprised of agricultural demand.[11] The importance of economic trends in water use is illustrated by a closer look at the case of China, which is one of the most important model regions and highly spatially heterogeneous in terms of the cropping and water use patterns in its sub-basins. In the base year 2000 China's per capita income is 778 US$, with a population of 1.28 billion. The food demand for cereals in China is 196 million tons (mainly rice and wheat), while pork consumption amounts to 42 million tons (pork is the most eaten meat in China). Cereal feed demand is 81 million tons (mostly maize). Total water depletive demand amounts to 311 km^3 of which agriculture accounts for 252 km^3 or 81%.

In order to examine the impact of various macro economic variables on model results, three income and population scenarios are borrowed from the Millennium Ecosystem Assessment.[12] The three scenarios on population and income growth are "business-as-usual" (BAU), a pessimistic scenario ('PES') and an optimistic scenarios ('OPT').[13] Modelling results for China are presented in Table 1.

The optimistic scenario foresees a more than fivefold increase in income over the coming 25 years, while in the pessimistic scenario income hardly grows at all. Population growth is slightly higher in the pessimistic scenario as compared to the optimistic scenario. Higher income growth levels leads to a change in diets towards meat consumption (mainly pork) which, in turn, entail more demand for feed grains. This leads to 31% more water depletion in the optimistic scenario as compared to the pessimistic income scenario. Interestingly, the additional water depletion to produce this feed is more than the predicted increase in industrial water as a result of increase of income.

[11] Watersim computations calibrated to IWMI 2000 and Aquastat 2005.

[12] Website:

[13] These correspond with Millennium Ecosystem Assessment story lines 'Techno Garden' for BAU; 'Econo Opt' for optimistic and 'Fortress' for pessimistic scenario.

While the modelling results are consistent with observed trends, the absolute numbers need to be interpreted with care. The results in Table 1 assume an equal area and yield growths in all three scenarios. In reality, part of the increase in water demand may be off-set by improvements in water productivity as a result of income growth. Further, the increase in total crop water demand does not necessarily represent an increase in irrigation water demand – as these increases could also be met from rainfed sources.

Keeping these caveats in mind, this small illustration serves to highlight the significance and potential impact of an increase in income growth levels, especially when combined with moderate population growth levels, as under the optimistic scenario. The rapid rise in China's agricultural water use under increased meat consumption is interesting and deserves further study.

5. Conclusions and discussion

Economic and agricultural policies and trends within and outside the direct realm of water resources management have substantial impacts on global and basin level water demand. Designed as an integrated global water and food model with sufficient spatial and temporal detail, WATERSIM fulfils an important aspect of basin and national studies in providing the global setting for water demand and use studies at basin and country level. The global economic context is linked to local water use at basin scale in many ways. For example, low world market prices for food commodities may render local investments in water development projects less favourable from an economic point of view. This may impact food security in the longer run. Or, developments in crop technology through international research efforts may boost yields and water productivity, reducing water use. Improvements in income and living standards will lead to changes in diets towards more water intensive commodities as meat, oils and sugar and will have consequences for water demand, as the example presented in this paper showed Furthermore, the liberalization of world markets for agricultural commodities, as discussed in the latest WTO rounds, will impact local agricultural economy by creating new opportunities or damaging existing markets. WATERSIM is able to simulate trade scenarios by adjusting values of Producer and Consumer Subsidy Equivalents.

The modelling approach chosen by WATERSIM is very data-intensive. With new data retrieval techniques emerging (higher resolution satellite images and better image interpretation techniques) new global datasets are becoming available at a regular interval, while older datasets are being updated. Maintaining and upgrading the database underlying WATERSIM is thus a continuous process, improving the quality of the analysis over time.

Understanding and exploring the linkages between agri-economic policies and water demand enables the analysis of the relative impact of various policies in the agricultural and water sector on food and water security. In a globalizing world, global change processes such as changes in international trade, WTO agreements, world markets, climate change, economic growth or stagnation, will play an increasingly important role in food and water demand and supply. The WATERSIM modelling framework chooses a global approach to complement national and river basin level studies. But being a global model, the level of detail is limited in comparison to small scale basin level models. Limits on data availability and computing power make further spatial disaggregation difficult at this stage. Exogenously (or 'soft') linking detailed basin models to WATERSIM runs is an appropriate approach to capture both the global context and detailed basin level processes.

Acknowledgements The author would like to thank Claudia Ringler and Siwa Msangi from the International Food Policy Research Institute for their valuable input and contributions. This work was made possible through the financial support from the Comprehensive Assessment of Water Management in Agriculture.

References

Alcamo J, Döll P, Kaspar F, Siebert S (1997) Global change and global scenarios of water use and availability: an application of water GAP 1.0. Center for Environmental System Research (CESR), University of Kassel, Kassel, Germany

Allen RG, Pereira LS, Raes D, Smith M (1998) Crop evapotranspiration, guidelines for computing crop water requirements. FAO Irrigation and Drainage Paper No. 56

Agcaoili MC, Oga K, Rosegrant MW (1993) Structure and Operation of the International Food Policy and Trade Simulation (IFPTSIM) Model. Paper presented at the second workshop of the research project on projections and policy implications of medium- and long-term rice supply and demand. International Rice Research Institute, Los Baños, Philippines

Cai X, Rosegrant M (2002) Global water demand and supply projections. Part 1: A modeling approach. Water Int 27(3):159–169

Chapagain AK, Hoekstra AY (2003) Virtual water flows between nations in relation to trade in livestock and livestock products. Value of Water Research Report Series no. 13, UNESCO-IHE, Delft, The Netherlands

Doorenbos J, Kassam AH (1979) Crop yield versus water. FAO Irrigation and Drainage Paper No. 33

FAO (Food and Agriculture Organization of the United Nations) (2003) Unlocking the water potential of agriculture. ftp://ftp.fao.org/agl/aglw/docs/kyotofactsheet_e.pdf

Fraiture C de, Cai X, Rosegrant M, Molden D, Amarasinghe U (2003) Addressing the unanswered questions in global water policy: a methodology framework. Irrigation Drainage 52(1):21–30

Fraiture C de, Molden D, Amarasinghe U, Makin I (2001) Podium: Projecting global water supply and demand for food production in 2025. Physics and Chemistry of the Earth (B. Hydrology Oceans and Atmosphere) 26(11/12):869–876

Huang J, Bouis H (1996) Structural Changes in the Demand for Food in Asia. Food, Agriculture, and the Environment Discussion Paper 11. International Food Policy Research Institute, Washington, DC

Keller A, Keller J (1995) Effective efficiency: A water use concept for allocating freshwater resources. Water Resources and Irrigation Division Discussion Paper 22. Winrock International, Arlington, Virginia, USA

Mitchell TD, Carter TR, Jones PD, Hulme M, New M (2003) A comprehensive set of high-resolution grids of monthly climate for Europe and the globe: the observed record (1901–2000) and 16 scenarios (2001–2100). J Clim: submitted

Molden D (1997) Accounting for water use and productivity. System-Wide Initiative on Water Management (SWIM) Paper No.1, International Water Management Institute, Colombo, Sri Lanka

Oga K, Gehlar C (1993) The International Food Policy Simulation (IFPSIM) Model: Documentation and Applications. Mimeo. International Food Policy Research Institute, Washington, DC

Renault D (2003) Value of virtual water in food: principles and virtues. In: Hoekstra AY, Hung PQ (eds) Proceedings of the international expert meeting on virtual water trade, IHE, Delft, the Netherlands

Rosegrant MW, Agcaoili-Sombilla M, Perez ND (1995) Global Food Projections to 2020: Implications for Investment. 2020 Vision for Food, Agriculture and the Environment Discussion Paper No.5. International Food Policy Research Institute, Washington, DC

Rosegrant MW, Paisner MS, Meijer S, Witcover J (2001) Global Food Projections to 2020: Emerging Trends and Alternative Futures. 2020 Vision Food Policy Report. International Food Policy Research Institute, Washington, DC

Rosegrant MW, Meijer S, Cline SA (2002) International Model for Policy Analysis of Agricultural Commodities and Trade (IMPACT): Model Description. International Food Policy Research Institute, Washington, D.C. <http://www.ifpri.org/themes/impact/impactmodel.pdf>

Rosegrant MW, Cai X, Cline SA (2002) World Water and Food to 2025: Dealing with Scarcity. International Food Policy Research Institute, Washington, D.C.

Rosegrant MW, Ringler C, Gerpacio RV (1999) Water and land resources and global food supply. In Food security, diversification and resource management: Refocusing the role of agriculture? In: Peters GH, von Braun J (eds) Proceedings of the 23rd international conference of agricultural economics, held at Sacramento, California, 10–16 August, 1997, Ashgate, Oxford

Sakthivadivel R, de Fraiture C, Molden D, Perry C, Kloezen W (1999) Indicators of land and water productivity in irrigated agriculture. Int J Water Resour Dev 15(1/2):161–179

Seckler D, Amarasinghe U, Molden D, de Silva R, Barker R (1998) World water and demand and supply, 1990–2025: scenarios and issues. Research report no. 19. International Water Management Institute, Colombo, Sri Lanka

Seckler D, Molden D, Amarasinghe U, de Fraiture C (2000) Water Issues for 2025: A research perspective. IWMI's contribution to the 2nd World Water Forum. International Water Management Institute, Colombo, Sri Lanka

Shiklomanov I (2000) Appraisal and assessment of world water resources. Water Int 25(1):11–32

Shiklomanov IA (1999) Electronic data. Provided to the Scenario Development Panel, World Commission on Water for the 21st Century. Mimeo

Smakhtin V, Revenga C, Döll P (2004) Taking into account environmental water requirements in global-scale water resources assessments. Comprehensive Assessment Research Report no. 2. International Water Management Institute, Colombo, Sri Lanka

Tuong TP, Bouman BAM (2003) Rice production in water-scarce environments. In: Kijne JW, Barker R, Molden D (eds) Water productivity in agriculture: Limits and opportunities for improvement. CAB International Publishing

USDA (United States Department of Agriculture) (1967) Irrigation water requirements. Technical release No. 21. Engineering Division, Soil Conservation Service (SCS)

Vörösmarty C, Green P, Salisbury J, Lammers B (2000) Global water resources: vulnerability from climate change and population growth. Science 289(5477):284–288

World Bank (1998) World Development Indicators 1998. Washington, DC

Water Resour Manage (2007) 21:199–210
DOI 10.1007/s11269-006-9049-8

ORIGINAL ARTICLE

The adaptive integrated data information system (AIDIS) for global water research

Wolfgang-Albert Flügel

Received: 20 April 2006

Abstract Global research programs related to river basin water resources have at least two things in common: (1) they assess and model hydrological process dynamics on a macro scale and (2) research partners jointly working on such research issues are internationally distributed in different institutions. These prerequisites require a sophisticated and scale bridging data assessment and information management comprising geo-referenced distributed data components, measured or simulated time series, and socio-economic information. Networking such international research structures by means of the internet pose new challenges to Geoinformatics in respect to the design of a Web based distributed database system, metadata and GIS-information management, geo-referenced data query and visualization. Such data management must include powerful and efficient data exchanging software tools and information sharing policies to ensure that decision making can jointly be done on the base of the best information available. Geoinformation includes raster and vector GIS coverages, measured process time series data and associated metadata. Furthermore there are needs to integrate multidisciplinary information and research knowledge related to IWRM comprising information obtained by remote sensing, GIS analysis, modeling, and socio-economic assessments for vulnerability and mitigation. Addressing these challenges and to cope with such data organization and management tasks the Adaptive Integrated Data Information System (AIDIS) has been developed by the DGHM at the FSU-Jena. It is based on open source software (OSS) and a multi tier class hierarchy structure. AIDIS has implemented the full ISO 19115 metadata model, and enhances its structure if required e.g. for time series or documents. A first prototype was developed for the Challenge Program "Water and Food" (CPWF) of the CGIAR and has been improved and refined for the Tisza River basin within the "Tisza River" EU-project comprising at present about one hundred GIS maps and more than 5000 measured and simulated time series.

Keywords Integrated water resources management (IWRM) · Adaptive data information system (AIDIS) · Object-relational data model · Geo-spatial data management ·

W.-A. Flügel (✉)
Department of Geoinformatics, Hydrology and Modeling (DGHM), Friedrich-Schiller University (FSU-Jena), Jena, Germany
e-mail: c5wafl@uni-jena.de

Distributed client-server network · River basin information system · Holistic systems approach · Hydrological response units · Regional modelling · ISO 19115 · Metadata

1. Introduction

Integrated Water Resources Management (IWRM) as defined by the GWP-TAC (2000) has widely been accepted as the proper strategy to handle river basins water resources and numerous literature has been published in this regard (Heathcote, 1998; Jain and Singh, 2003). IWRM, however, it is not always easy to implement and one of the reasons is that the majority of semi-arid river basins in this world are ungauged basins as defined by the IAHS-PUB initiative (Young and Romanowicz, 2004). On a global scale such basins at present show ongoing population growth, and already suffer from land degradation, floods and droughts causing severe socio-economic damages, decreasing of livelihood, increasing poverty, and even loss of human lives. Furthermore they are highly vulnerable to impacts from climate change (Fischer *et al.*, 2002; IPCC, 1997a, 2001a,b), and scenario based simulations (Alcamo and Henrichs, 2002; IPCC, 1997b) indicate that their scarce water resources will likely come under even more stress in the forthcoming decades (IPCC, 1997c; Querner, 2002).

2. Integrated water resources management (IWRM)

IWRM in semi-arid regions is dominated by a generic problem scenario assembly: (1) A growing population demands increasing food production but simultaneously rural areas and industrialized cities compete for increasing water supply. (2) Expensive water diversion are required to sustain and even extend irrigation agriculture to alleviate poverty and improve livelihoods in rural areas. (3) Fast growing cities need increasingly more water for their boosting population and industrialized economic development. (4) The worldwide recognized necessity to reserve water for the environment is becoming a further priority IWRM factor, that enhances the present water allocation dilemma.

Pahl-Wostl (2002) states that such a controversial situation applies to the problem of "common pool resources" and concludes that social learning has to be promoted in addition to decision making based only on factual knowledge. Both, however, require comprehensive information readily accessible for the appreciation of the interaction between the natural, biological and socio-economic components related to the river basins hydrological dynamics which has to steer the IWRM decision making processes, but in reality is often lacking. Remote sensing in this regard produces extremely valuable information (Stolz and Mauser, 1996) about land use and land cover (LULC) and topography (Schmullius *et al.*, 2000). The design of Clearinghouse Networks (FDGC, 1997) is another step towards building a global available data information environment to support IWRM on a basin scale.

In conclusion sustainable and adaptive IWRM is depending on the availability and accessibility of information for system analysis and modelling, social learning, and sustainable decision making. These different aspects are related to a comprehensive basin assessment that requires an efficient data and information management such as is realised by AIDIS comprising the following considerations:

(i) A holistic systems approach must be applied that differentiate between generic components and their regional variation.

(ii) A methodical concept must be implemented that accounts for the systems core components within its natural environment (NE) and human dimension (HD)
(iii) The system assessment must account for basins heterogeneity in terms of spatial distribution and determining underlying generic rules.
(iv) The process must interface multidisciplinary assessment techniques and analysis methods in relation to scale issues in an interdisciplinary manner.

3. Holistic system approach

IWRM must be seen as a decision making process that is based on comprehensive, spatially distributed information (GWP-TAC, 2000; Staudenrausch and Flügel, 2001), and adopting a holistic systems approach (Flügel, 2000) as shown in Figure 1. The latter integrates components of the internal systems dynamics and its exchange across the systems borders (IUCN, 2000) into the evaluation of the river basins water resources balance. By means of innovative techniques such as remote sensing, GIS-analysis, process studies, questionaires and process based regional modelling their interaction will be analysed and used for the design of "what-if?" scenarios in IWRM (Heathcote, 1998) and predictive water balance analysis (Jain and Singh, 2003).

Such a comprehensive systems analysis approach must comprise the different system's abiotic, biotic and socio-economic components, has to address the quality and quantity of their respective resources in order to define the system status and finally must classify heterogeneous system patterns (Turner, 1989). Such an approach has been realized by the regionalisation concept of Response Units (RU) as defined by Flügel (1996) and (1997) that

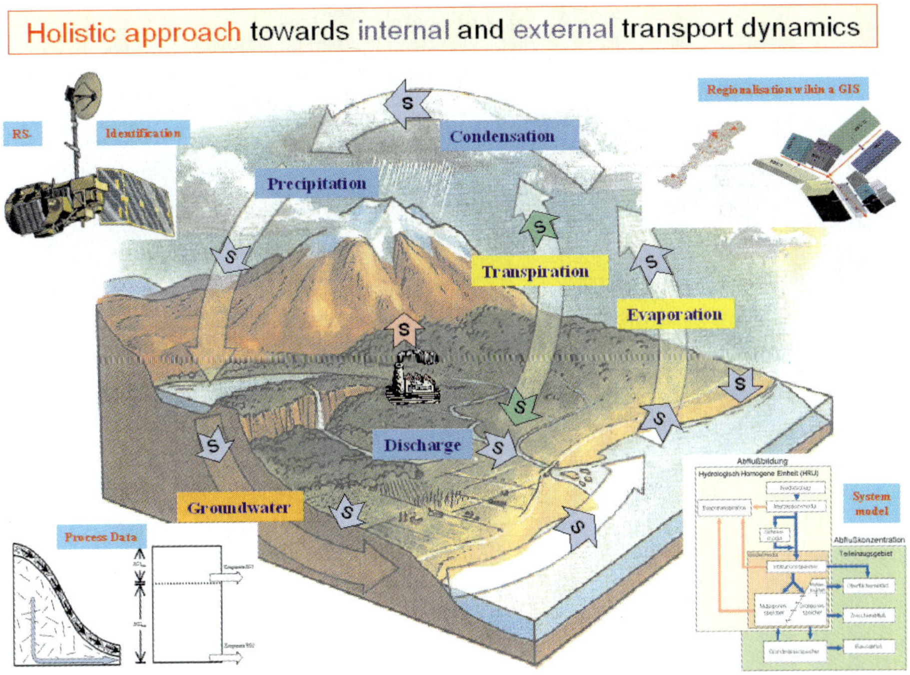

Fig. 1 Holistic systems approach

classifies the basin's heterogeneity in relation to its dominant process dynamics. RU provide the methodical means to complement the Top-Down-Approach applied by (Alcamo and Henrichs, 2002) for global scale modelling with a Bottom-Up-Approach suitable for process oriented modelling on the river basin scale (Flügel, 2000).

4. Multi-layer methodical concept

The methodological challenge is to transfer the concept of a holistic IWRM basin approach into a corresponding information system in such a way that the information demand for IWRM is reflected by a corresponding data model. The latter must feature the generic holistic IWRM components, provide different formats for data input, services for data handling, manipulation and presentation and offer the calculation of IWRM indicators depending on the river basin scale.

Such a methodical information system concept that has been developed based on experience obtained from numerous hydrological river basin systems analyses, and is schematically shown in Figure 2. It is adopting as a multi-layer approach, that integrates both data concepts and data manipulation methods, and can be described as follows:

(1) The data model is build from information related to the three interactively linked abiotic, biotic and socio economic systems components, that comprise the cultural landscape of the respective river system.

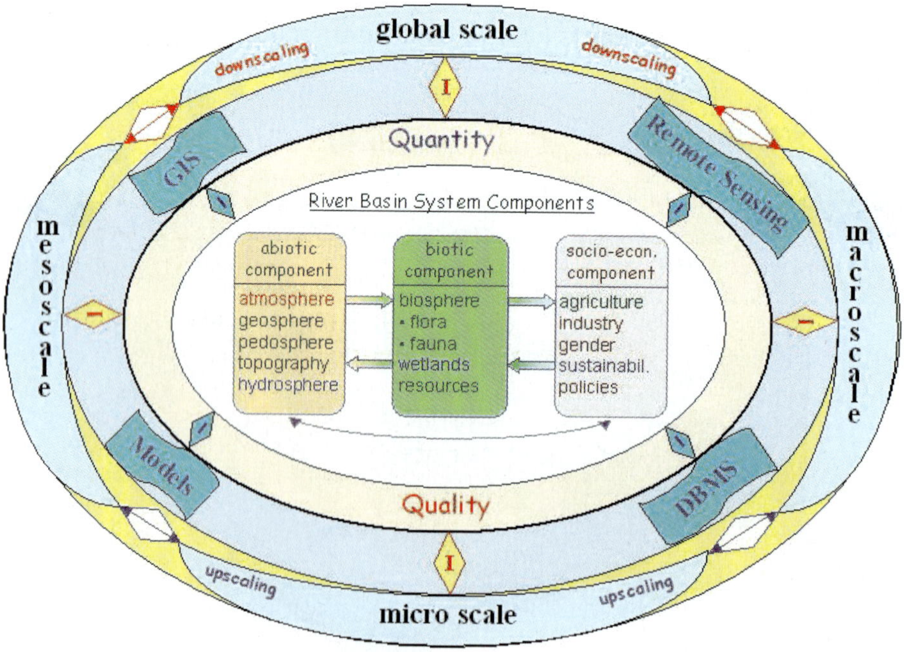

Fig. 2 Conceptual multi-layer methodical ellipsoid for MCL integrated system analysis, evaluation and modelling

(2) These core elements have a specific regeneration dynamics that together constitutes the system's status, which in turn is expressed by data and information that have aspects related to quantity and quality.
(3) Remote sensing, GIS, data base management systems (DBMS) and models compose the toolset to identify, analyse, model and regionalize the generic and regional aspects related to the core components in (1). Furthermore they provide the methodical link between the three core system components, and with the system's climate input dynamics for model based "what-if?" scenarios predictions (Gonzales, 2002).
(4) Techniques provided by the third layer are also used to regionalize the IWRM system's analysis in respect to quantity and quality of the three core components. By means of the well tested concept of Response Units (Flügel, 1996, 1997, 2000; Bende *et al.*, 1997; Bongartz, 2003) the problem of regional aggregation (up-scaling) and disaggregation (down-scaling) can be dealt with (Turner, 1989; Bartel, 2000).
(5) Conceptual interfaces between the four layers are shown as rhombi in Figure 2. They are in fact represented by IWRM indicators (Schulting, 2000; Pykh *et al.*, 2000; EPA, 2004) to be calculated within the information system by algorithms that assess the data stored within the three core system components.

5. Assessing the system status

Water resources within a river basin are regenerated by rainfall, glaciers, snow covered mountains, lakes, reservoirs, rivers, wetlands, groundwater aquifers and marine waters (Costanza *et al.*, 1997). In their regional distribution and process controlled, interactive dynamics they comprise the natural river basin water potential, which in turn is the base for IWRM and economic development. Population growth, food production, rise of incomes, industrialization and urban development are priority criteria for IWRM decision making. In the past years increasing attention has been given to the water demand claimed for the protection of environmental systems, and the services they provide for the people directly or indirectly depending on their existence. The term 'sustainable management' has been introduced to describe the attempt to balance the use of the river basin resources in favor of human development without compromising the basin's environment. There is agreement that this attempt has not always been implemented successfully, and the competition for water frequently becomes emotionally loaded in irrigation agriculture where water is used for crop production and as a priority resource to sustain human livelihoods.

When including the issue of scales and regionalisation it becomes obvious, that our scale dependent knowledge about the interactions between system heterogeneity, process dynamics, and systems response is incomplete. These interrelationships are reasonably well understood on the micro- and meso-scale, but research is required to reveal insight how changes in large scale water balances will impact water allocation on smaller scales (Alcamo and Henrichs, 2002). Furthermore the lag time of system response in relation to scale is only in smaller catchments reasonably well understood, but the response dynamics in large scale river basins is mostly less known.

6. Data and information management in river basin research

The Global Water System Project (GWSP) is dealing with river systems stretching over different scales, and if a holistic approach is adopted comprehensive information is required about the systems heterogeneity and the interdisciplinary knowledge associated with each basin scale. In reference to the requirements for data and information management in river basin research this includes the following aspects:

(i) Georeferenced, distributed information is specifying the spatial heterogeneity within the river basin. It is used for distributed modeling and the analysis of transport systems such as the interaction between upstream and downstream basin regions, i.e. in sediment transport and deposition.
(ii) Time series data can either be connected with a point of observation as it is the case for measurement stations, or can describe the change of a landscape such as land use obtained and classified by means of remote sensing.
(iii) Beside these hard data used to quantify the status of a system, soft information is obtained by socio-economic census. The latter is less quantitative but describes the system component status, aiding decision making from the human perspective. Peoples judgement about the status and quality of water supply are an example for such information.
(iv) Knowledge is another important part of information that assesses and specifies a river basin. The interpretation of the term knowledge might often be different, but it is understood that such information should be available for decision makers to improve the system status. Classifying the erosion potential (Flügel and Märker, 2003) based on measurable landscape criteria is an example for such a knowledge information.
(v) All data and information mentioned so far can be assembled and presented in digital thematic maps which have to be administered together with their metadata within the database. Maps are not only presenting the data but are also appropriate means for spatial data queries.

7. AIDIS objectives, implementation and test

As discussed above IWRM should be based on the principles of integrated systems analysis (Flügel, 2000) and on knowledge obtained form previous research. This requires the collection and managing of diverse data and information comprising beside others (1) station time series related to water quantity and quality, (2) socio-economic census data, and (3) digital maps that presenting the spatial distribution of such information. From this diversity the following *conceptual objectives* for AIDIS have been identified:

(i) The system must have a geospatial extension as hard and soft data both have a geographic reference that can be specified by a Cartesian coordinate system.
(ii) The data and information management must reflect the differentiation of the river basin's real world subsystems in a manner as discussed above. The latter in turn comprises (1) generic components based on the global natural environment, and (b) basin related components which reflect the regional basin conditions.
(iii) The system must be able to cope with the hierarchical structure of the water and solute transport systems within the natural environment and human development (NEHD) in respect to energy, water quantity and quality or economic productivities.
(iv) It must integrate and present the different kinds of data and information in such a way that they are available for the decision making process and for the design of "what-if-scenarios" for prognostic system modeling.

(v) The system's structure must be flexible enough to allow for extensions to be made depending on the progress of the system's understanding within the researchers and decision maker community.
(vi) To ensure compatibility between distributed AIDIS installations the metadata associated with the respective geo-spatial datasets must be encoded according to the ISO 19115 standard (ISO, 2003).

In addition to these conceptual objectives some important *technical objectives* must be met if AIDIS should be successfully applied in present international research programs:

(vii) Arc-View shape files and other formats must be supported for data import, export and visualization.
(viii) Distributed installations must be realized according to a Web-based 'thin client-server' design structure.
(ix) Replication strategies between the distributed server installations must account for specific needs in respect to data security and information distribution policies.
(x) Software for the data base management system (DBMS) and its components should be available on demand.

A literature survey (ANZLIC, 2000; FGDC, 1997; UNRCC-AP, 1997; UNRCC-Americas, 2001; PCGIAP, 1998) revealed that there are numerous attempts to organize and coordinate national, regional and global geo-spatial data. However, integrating diverse data and information as specified above in a structured object-relational information system has not been reported so far in the context of such a program.

7.1. AIDIS software structure

Adaptive IWRM strategies require decision making based on reliable data, river system information, and model simulation (David *et al.,* 1997). They have been organized and are easily accessible in AIDIS. As shown in Figure 3 it consists of different components installed at the AIDIS backbone server on the left side of Figure 3. It comprises the following components that account for the conceptual and technical objectives in the following way:

- AIDIS is based on open source software (OSS) components which are complemented by the developments done by the DGHM from the FSU-Jena, Germany (http://www.geogr.uni-jena.de).
- The PostgreSQL data base (Worsley and Drake, 2002) has been selected as a object-relational data base management system (RDBMS). It is highly extensible, and permits the use of the procedural Standard Query Language (SQL).
- PostGIS (http://postgis.refractions.net/) is implemented as the geo-spatial extension. It provides import and export of Arc-View©shape files and OpenGIS Simple Feature Specification for SQL (http://www.opengis.org/specs/).
- The APACHE (http://httpd.apache.org/) Web-Server is used for the client to server communication, and the visualization of maps is done by means of the Minnesota Map Server (MMS) (http://mapserver.gis.umn.edu/).
- On the right side of Figure 3 the AIDIS Web access for data input and output as well as the AIDIS map publisher are shown, both developed by the DGHM at the FSU-Jena.

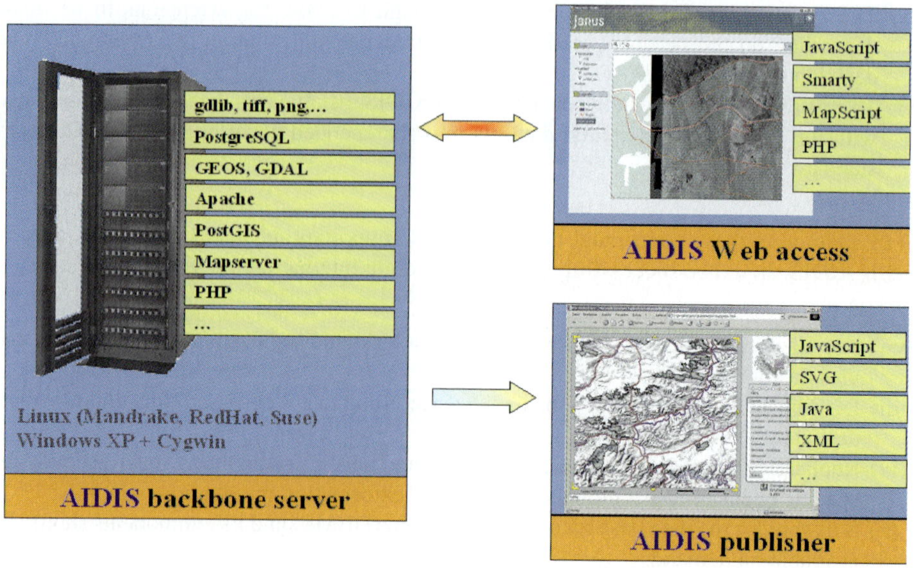

Fig. 3 OSS components of AIDIS

7.2. AIDIS data model components

AIDIS is using a table model that is generic in its design structure and therefore can be adapted to account for data, information, and indicator requirements of river basins in all climates. The object oriented design of AIDIS permits such an adaptation as its multi-tier class layer structure shown in Figure 4 provides a sufficient flexibility to adapt AIDIS by

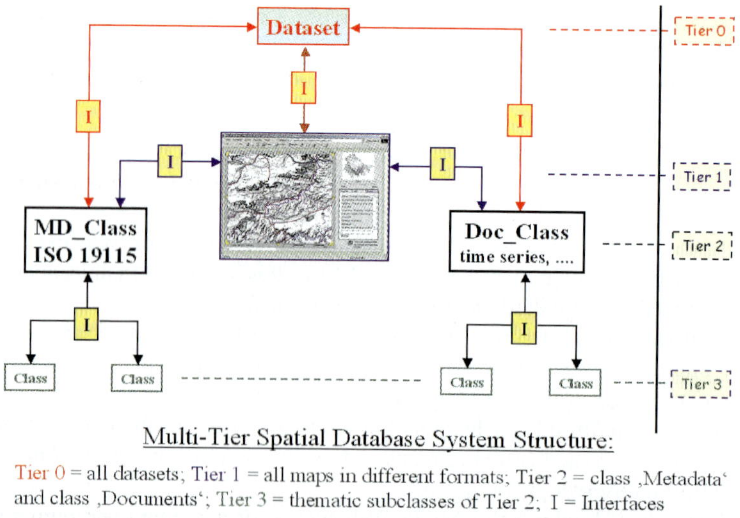

Fig. 4 Multi tier relational information strategy of AIDIS

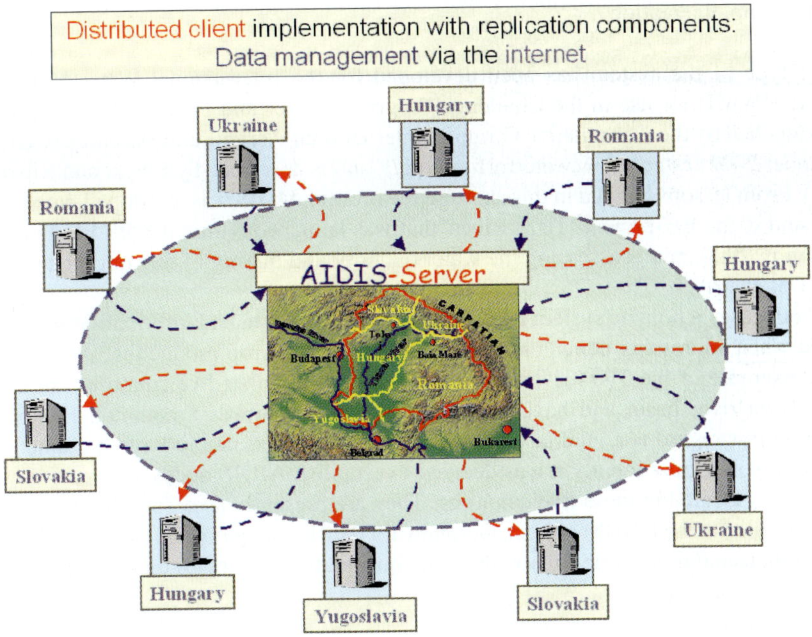

Fig. 5 AIDIS Tisza River Information System (TRIS)

adding additional class hierarchies or modifying existing ones. The structural analysis of AIDIS reveals the following characteristics:

- Designed by means of the Unified Modeling Language (UML) AIDIS has an easy to understand class oriented hierarchical structure.
- Classes are linked by interfaces (I) which store the organization of relationships.
- A central role is given to the 'map-class' on Tier 1, which is used not only for the visualization of results but also for spatial queries that refer to metadata and the document-class hierarchy.
- All data stored in AIDIS are described by means of the ISO 19115 metadata model. Extensions are realized by extended lookup tables (LUT).

As the OOS components of AIDIS although freely available from the internet are not always easy to install, and the use of the MMS requires some training for the design of separate map templates to control the visualization and data queries, the following complement AIDIS components have been developed by the DGHM:

1. A Data Base Networking System (DBNS) that automatically installs all required software components including the Linux operating system from a installation CD on the server. Detailed log files are written during the installation for error checking.
2. The River Basin Information System (RBIS) is based on the MMS and makes use of the GIS functionalities provided by PostGIS. RBIS is also handling the map related data import and export via the Web and thereby allows to generate new GIS maps for dissemination i.e. within a clearinghouse network (CN). At present more GIS functionality is developed for RBIS by means of the OpenGIS "Simple Feature Specification for SQL" (http://www.opengis.org/specs/).

7.3. Test implementation

A prototype of the system has been developed for the International Water Management Institute (IWMI) for use in the Challenge Program "Water and Food (CPWF)", which has been launched by the Consultative Group of International Agricultural Research (CGIAR) in November 2003 (http://www.waterforfood.org/), and is described by Flügel and Rijsberman (2003). From lessons learned in this exercise the present AIDIS was developed, implemented and tested in the EU-project 'Tisza River' that was launched within the 5th EU-Framework program in 2002, and is assessing the water quantity and quality of this important tributary of the Danube river.

In both cases a holistic system analysis and respective data and information strategy was applied when addressing deficits in respect to IWRM and crop production. As can be seen for the example of the Tisza river basin in Figure 5 altogether five countries are tributary to the Tisza river basin and have agreed to contribute towards a common IWRM. Data sharing in this regard is a vital task to ensure that all partners can rely on the same amount of information. Consequently it was decided to establish AIDIS nodes as clients in various partner sites located in these five countries. They are presently assessing the AIDIS central server installed at the DGHM in Jena, Germany for data input, query and export, and use more than fivethousand times series, maps and other information to support their IWRM activities.

8. Conclusion

River basin water resources research and IWRM requires a comprehensive and holistic systems analyses and process understanding, which in turn relies on powerful and effective data management and information sharing strategies to support decision making based on best knowledge about the river basin. Such information comprises geo-spatial data components, resulting from measured and simulated systems dynamics of the natural environment (NE) and human development (ND). Remote sensing and GIS additionally is providing spatial distributed information of the basin's heterogeneity in respect to its abiotic, biotic and socio-economic components.

To address this demand the Adaptive Integrated Data Information System (AIDIS) has been developed and was implemented for an EC-Project in the Tisza river basin. AIDIS accounts for the methodical and conceptual system requirements, and is based on open source software complemented by the RBIS and DBNS extensions provided by the DGHM of the FSU-Jena, Germany. A object-relational data model is used for AIDIS which permits a modular component structure required for the regionalization of the distributed system dynamics.

The modular component structure of AIDIS provides sufficient flexibility for distributed node installations as required by international research programs. It also accounts for the different security requirements of data providers having their own dissemination policies that can be implemented by the AIDIS administrator in each individual BB node. AIDIS is under continuous development by the DGHM which also provides a comprehensive training program. As the data model and the software components are adaptive to various river basin setups AIDIS will continuously be improved and implemented in EU-projects and regional research programs.

References

Alcamo J, Henrichs T (2002) Critical regions: A model-based estimation of world water resources sensitive to global changes. Aquat Sci 64:352–362

ANZLIC, Australia New Zealand Land Information Council (2000) Discussion and background papers. In: Proceedings of the ANZLIC clearinghouse workshop, 3–4 May, Adelaide, Australia, 2000
Bartel A (2000) Analysis of landscape pattern: towards a 'top down' indicator for evaluation of landuse. Ecol Modell 130(1–3):87–94
Bongartz K (2003) Applying different spatial distribution and modeling concepts in three nested meso-scale catchments of Germany. Phys Chem Earth 28:1343–1349
Constanza et al. (1997) The value of the world's ecosystem services and natural capital. Nature 387
David O, Busch C, Küspert K, Flügel W-A (1997) Objektorientierte Datenbanken in der Hydrologie: Anwendungsbeispiel OMS. – In: Geiger W, Jaeschke A, Rentz O, Simon E, Spengler Th, Zillox L, Zundel T. (eds) Umweltinformatik'97, 11. Intern. Sympos. der Ges. für Informatik, GI, Bd. 1, Vol. 1:225–234
EPA, Environment Protection Agency (2004) Index of Watershed Indicators: An Overview. – http://www.epa.gov/iwi/iwi-overview.pdf
FGDC, Federal Geographic Data Committee (1997) Framework, introduction and guide. Book of federal geographic data committee, Washington
Fischer G, Shah M, van Velthuizen H (2002) Climate Change and Vulnerability. IIASA 152 p
Flügel W-A (1996) Hydrological Response Units (HRUs) as modeling entities for hydrological river basin simulation and their methodological potential for modeling complex environmental process systems. DIE ERDE 127:42–62
Flügel W-A (1997) Combining GIS with regional hydrological modelling using Hydrological Response Units (HRUs) – An application from Germany. Math Comput Simulation 43:305–312 1997
Flügel W-A (2000) Systembezogene Entwicklung regionaler hydrologischer Modellsysteme. Wasser and Boden, 52. Jg H 3:14–17
Flügel W-A, Märker M (2003) The response units concept and its application for the assessment of hydrologically related erosion processes in semiarid catchments of Southern Africa, ASTM-STP 1420:163–177
Flügel W-A, Rijsberman F (2003) The challenge program water and food for river basin scale water resources assessment. Proc MODSIM'03 1:434–439 2003
GWP-TAC, Global Water Partnership – Technical Advisory Committee (2000) Integrated Water Resources Management, TAC Background Paper No. 4, 67 p
Heathcote IW (1998) Integrated watershed management. John Wiley and Sons, New York, 414 p
International Organization for Standardization ISO 19115, (2003)(E) Geographic information – metadata. 1st edition, Geneva, Switzerland
IPCC (1997a) The regional impacts of climate change: An assessment of vulnerability, In: Robert T, Watson RT, Zinyowera MC, Richard H, Moss RH, David J, Dokken D-J (eds) Special report of IPCC working Group II, (http://www.ipcc.ch/pub/regional(E).pdf)
IPCC (1997b) An introduction to simple climate models used in the IPCC second assessment report. In: Houghton JT, Gylvan Meira Filho L, Griggs DJ, Maskell K (eds) Technical Paper for IPCC WGI, httpr://www.ipcc.ch/pub/IPCCTP.III(D).pdf
IPCC (2001a) Climate Change (2001) The scientific basis. Contribution of working group I to the third assessment report of the intergovernmental panel on climate change. In: Houghton JT, Ding Y, Griggs DJ, Noguer M, van der Linden PJX Dai Maskell K, Johnson CA (eds) Cambridge University Press, Cambridge, United Kingdom and New York, NY, USA, 881 p (http://www.grida.no/climate/ipcc_tar/wgr1/index.htm)
IPCC (2001b) Climate change 2001: Impacts, adaption and vulnerability. Contribution of working group II to the third assessment report of the intergovernmental panel on climate change. In: McCarthy JJ, Canziani OF, Leary NA, Dokken DJ, White KS (eds) Cambridge University Press, Cambridge, United Kingdom and New York, NY, USA, 881 p (http://www.grida.no/climate/ipcc_tar/wg2/index.htm)
IUCN (2000) Vision for Water and Nature. IUCN – The World Conservation Union, Report 2000
Jain SK, Singh VP (2003) Water resources system planning and management. Elsevier, Amsterdam, 858 p
Krause P (2001) Das hydrologische Modellsystem J2000, Schriften FZ-Jülich, Bd. 29
Leavesley GH, Lichty RW, Troutman BM, Saindon LG (1983) Precipitation Runoff Modelling System: User's manual, Water Resources Investigations 83-4238, USGS, Denver, Colorado
Pahl-Wostl C (2002) Towards sustainability in the water sector – The importance of human actors and processes of social learning. Aquat Sci 64:394–411
PCGIAP, Permanent Committee on GIS Infrastructure for Asia & the Pacific (1998) A spatial data infrastructure for the Asian and Pacific region. PCGIAP Publication, No 1, Canberra
Pykh YA, Kennedy ET, Grant WE An overview of systems analysis methods in delineating environmental quality indices. Ecol Modell 130(1–3):25–38
Querner EP (2002) Analysis of basin response resulting from climate change and mitigation measures. In: Van Lanen HAJ, Demuth S (eds) Regional hydrology: bridging the gap between research and practice. IAHS Publication no. 274:77–84

Schmullius C, Flügel W-A, Frotscher K, Hochschild V, Müschen B (2000) The Shuttle Radar Topography Mission (SRTM) and Applications in Europe, Africa and Siberia, PFG, Jg 5, H 5:361–366

Schultink G (2000) Critical environmental indicators: performance indices and assessment models for sustainable rural development planning. Ecol Modell 130(1–3):47–58

Staudenrausch H, Flügel W-A (2001) Development of an integrated water resources management system in southern african catchments. Phys Chem Earth (B) 26(7–8):561–564

Stolz R, Mauser W (1996) A fuzzy approach for improving land cover classifications by integrating remote sensing and GIS data. In: Parlow E (eds) Progress in environmental remote sensing research and applications, Rotterdam, pp 33–43

Turner MG (1989) Landscape ecology – the effect of pattern on process. Ann Rev Ecol Syst 20:171–197

UNRCC-AP (1997) Resolutions of the 14th United Nations regional cartographic conference for Asia and the Pacific, 3–4 February, Bangkok

UNRCC-Americas (2001) In: Proceedings of the United Nations regional cartographic conference for Americas, January, USA

Worsley JC, Drake D (2002) Practical PostgreSQL, O'Reilly and Associates

Young PC, Romanowicz RJ (2004) PUB and data-based mechanistic modelling: the importance of parsimonious continuous-time models. In: Proceedings of iEMSs – Complexity and Integrated Resources Management, Osnabrück, 14–17. Juni 2004

Water Resour Manage (2007) 21:211–232
DOI 10.1007/s11269-006-9050-2

ORIGINAL ARTICLE

Policy implications of a pan-tropic assessment of the simultaneous hydrological and biodiversity impacts of deforestation

Ellen M. Douglas · Stanley Wood · Kate Sebastian · Charles J. Vörösmarty · Kenneth M. Chomitz · Thomas P. Tomich

Received: 20 April 2006

Abstract Tropical deforestation has many consequences, amongst which alteration of the hydrological cycle and loss of habitat and biodiversity are the focus of much public interest and scientific research. Here we examine the potential biodiversity and hydrological impacts of an extreme deforestation scenario – the loss of all tropical forest areas currently identified by the World Wildlife Fund as being threatened. Existing tropical forest areas are first classified according to two categories of biological distinctiveness – high and low – using indicators developed by the WWF. We apply the tropical deforestation scenario to a macro-scale hydrologic model, keeping track of the share of change in basin runoff that originates from the deforestation of areas of high versus low biological distinctiveness and where that change could impact human populations. Of particular interest are those basins where loss of the most threatened tropical forest areas would give rise to significant biodiversity loss *and* to potentially large hydrological impacts. In such cases it is conceivable that biodiversity conservation could "free-ride" on the concerns of resident populations to maintain the forests for the purpose of minimizing hydrological change. Where such an outcome seems likely, biodiversity conservation efforts might be better targeted elsewhere, perhaps to basins where the loss of forest areas with high biological distinctiveness would have less population impacts, hence requiring an alliance between biological and hydrological interests to gain sufficient social and financial support for conservation.

E. M. Douglas (✉) · C. J. Vörösmarty
Water Systems Analysis Group, Institute for the Study of Earth, Oceans and Space, University of New Hampshire, Durham, New Hampshire 03801, USA
e-mail: ellen.douglas@unh.edu

S. Wood · K. Sebastian
International Food Policy Research Institute, 2033 K St NW, Washington, DC 20006, USA

K. M. Chomitz
Development Research Group, The World Bank, Washington, DC 20006, USA

T. P. Tomich
Alternatives to Slash and Burn Program, World Agroforestry Centre, Nairobi, Kenya

Keywords Deforestation · Hydrology · Biodiversity · Human vulnerability · Conservation

1. Introduction

The world is rapidly losing its tropical forests, placing a significant fraction of terrestrial biodiversity at risk of extinction. Between 1990 and 2000 alone, 73.5 million hectares (Mha) of closed tropical forest were destroyed (FAO, 2001). Of these once forested areas, about 70% have been completely converted to other land cover types (largely pasture) and to agriculture. Only about 23% has some semblance of tree cover remaining, either in the form of fragmented forests (predominantly forest/agricultural mosaics) or timber and tree crops. In general, there are two main driving forces behind tropical deforestation: (1) *unplanned*, generally more gradual, forest degradation due to rural population pressure and its corresponding subsistence and energy needs, and (2) *planned* conversion of forest to other land uses as part of government-driven programmes to stimulate resettlement, cattle ranching and permanent agriculture, as well as commercial plantations (Bruijnzeel *et al.*, 2005). It is now widely appreciated that environmental costs are not taken into account when tropical forests are converted to cropland or pasture. In the face of tremendous social and economic pressures, conservation of these richly biodiverse landscapes is hampered by many operational challenges.

In principle, society can use regulation, taxes, or environmental service payments to harmonize forestholders' incentives with social goals of maintaining environmental service flows from the forest. (Millennium Ecosystem Asssessment, 2005). In practice, however, there has to be social and political support to impose regulations, levy taxes, or raise funds for environmental service payments (World Bank, 2002). Unfortunately, there is often little local support (Tomich *et al.*, 2004) and inadequate global financing for conservation of globally significant biodiversity. Hence, attention has turned to the role of forests in providing hydrological services. Because flood prevention and sediment mitigation are thought to be highly valued local services, conservationists hope to use these forest functions to motivate the conservation of biodiversity-rich forests (Dudley and Stolton, 2003). The efficacy of this strategy, however, rests on a largely untested and contentious assumption: that protection of biodiverse upstream forests would in fact yield benefits palpable enough to attract the interest of a large and comparatively wealthy downstream urban population. Although policymakers and the general public believe that upland deforestation causes downstream flooding in large river basins, contemporary hydrological science casts doubt on this assumption (Chomitz and Kumari, 1998; Calder, 2005; Bruijnzeel, 2004; FAO/CIFOR, 2005). In large basins, according to a plausible argument, flood conditions will result only when rainfall is so widespread and persistent that land cover is irrelevant to flows. Salient empirical data is however scarce. Due to the economic and environmental costs that would be incurred in performing large-scale forest conversion experiments, the only feasible method for evaluating large-scale hydrologic impacts is through the use of macro-scale hydrological models (Bruijnzeel, 2004; Costa, 2005).

In this paper, we present a macro-scale approach to assessing the potential hydrologic changes, and the consequent impacts on human vulnerability, due to a hypothetical but realistic scenario of future tropical forest conversion to agricultural land uses. We hypothesize that where large populations face the potential for marked change in hydrological regime as a consequence of land cover/use change, there will likely be a larger constituency for forest conservation. We highlight regions where this might be the case.

2. The pan-tropics, forest biomes, basins and biodiversity

The first step in this study was to define the geographical scope of the search for synergy between biodiversity and hydrological function. A focus on tropical forests was born out of a comfort level with the state of evidence-based knowledge on the biodiversity and hydrology of the tropics (the humid tropics in particular), as well as a sense of urgency about the need for new knowledge to support tropical biodiversity conservation efforts. It was then essential to identify reliable sources of information for discriminating the nature and status of biodiversity within tropical forest boundaries. Furthermore, in thinking about the hydrological impacts of tropical deforestation, a broader geographic domain than just the forests must be considered, since basins containing significant tropical forest areas may ultimately discharge into distant, non-tropical areas (e.g. the Nile basin). Thus, the extent and characterization of the study domain was determined by the intersection of a number of spatially-explicit variables: the boundaries of tropical forest areas; patterns of biodiversity within those boundaries; and the larger river basins in which tropical forests were found.

2.1. The pan-tropical domain

A review of the limited sources of biodiversity data that spanned the tropics in a consistent manner identified the mapping and characterization of the world's terrestrial ecoregions undertaken by the World Wildlife Fund[1] (WWF, Olson *et al.*, 2001) as most compatible with the needs of the study. Ecoregions are spatial units made up of complex plant, animal, and microorganism communities and the nonliving environment within which these communities function (CBD, 2002). The WWF map and database of global terrestrial ecoregions, was developed as a result of collaboration amongst over 1,000 scientists from around the world, and employs a 3-tier hierarchical classification system. At the highest level are six broad biogeographical *realms* of which four are found in the tropics. Distributed across the realms are 14 *biomes* – generalized global groupings of ecoregions. The biomes include 3 tropical and sub-tropical forest categories: moist forests (19.6 million km^2), dry forest (3.6 million km^2) and, for the sake of completeness, coniferous forests (0.7 million km^2). These three biomes were selected as the principal means of delineating the tropical forest areas of interest to the study. To complete the WWF hierarchy, biomes are further broken down into 828 unique *ecoregions* globally.

The hydrological context of the selected forest biomes was initially defined by delineating the set of river basins across the tropics in which at least one of these biomes was found. The overall pan-tropic boundary was established by overlaying the three WWF tropical forest biomes (converted to 2-minute resolution gridded fields) on to basins delineated from a 30-minute (0.5 decimal degree) simulated topological network. The minimum area unit of the hydrological model was determined by the availability of a validated and coherent grid of water/river flow paths for the pan-tropics (STN-30, Fekete *et al.*, 2001). Figure 1 shows the resulting pan-tropical boundary that initially defined 1,443 basins ranging in size from 5.9 million km^2 (Amazon) to 2,600 km^2 (single grid cell basins). Many of the world's major river basins, such as the Amazon, the Congo, and the Ganges, fell within the general purview of the study, but so too did many less obvious ones, such as the Nile, by virtue of a relatively small area of tropical forest (when expressed as a share of total basin area) in their headwaters.

[1] WWF-International changed its name to the World Wide Fund for Nature then simply to WWF. WWF-US, which published the Terrestrial Ecoregions of the World map, has retained the name World Wildlife Fund.

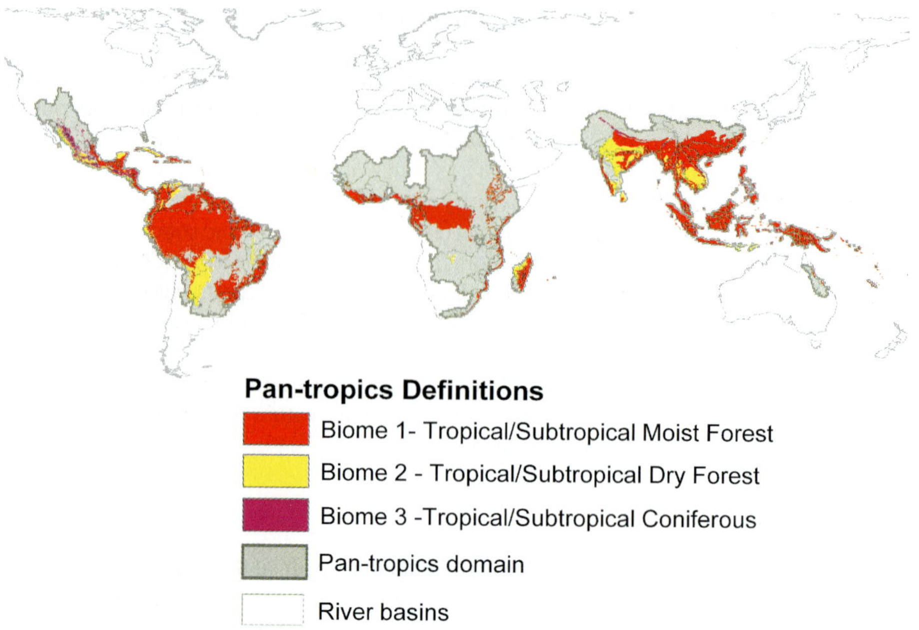

Fig. 1 Tropical forest biomes and the extent of the pan-tropical modeling domain. Forest biomes were extracted from the World Wildlife Fund (WWF) spatial database of terrestrial ecoregions (Olson *et al.*, 2001). The pan-tropical boundary includes river basins within which one or more of the three tropical forest biomes reside

While it was necessary to reduce the number of basins simply on pragmatic grounds, there were also good technical reasons for doing so. Analysis by Fekete *et al.* (2001) suggested that in order to reduce errors associated with representing small basins by a coarse gridded network, a minimum basin size of 30,000 km^2 is recommended. There were some 148 such basins, and these were further narrowed down to 108 focus basins by adopting an additional criterion; that at least 10% of basin precipitation must occur over the tropical forest biomes (Douglas *et al.*, 2005). Of the total area encompassed by this final pan-tropical domain (55 million km^2), the moist forest biome makes up 38%, the dry forest biome 7%, and the coniferous forest biome around 1.3% (Table 1)

2.2. Pan-tropical biodiversity

To obtain an understanding of the characteristics and distribution of biodiversity within tropical forest biomes it was necessary to utilize ecoregion-specific information. Ecoregions are mapped sub-components of biomes, and each ecoregion is associated with information on species richness and endemism (Olson *et al.*, 2001). The WWF defines ecoregions as ‘relatively large units of land containing a distinct assemblage of natural communities and species, with boundaries that approximate the original extent of natural communities prior to major land-use change’. Olson *et al.* argue that WWF ecoregions provide a better reflection of the distribution of species and communities than other global vegetation maps that rely heavily on modeling of specific biophysical features (e.g. Holdridge, 1967; Bailey, 1998; UNESCO, 1969; and Defries *et al.*, 1995). Furthermore, some of these efforts are defined at a scale considered more equivalent to WWF biomes rather than ecoregions (Mace, 2003). The WWF map includes over 800 ecoregions with an average size of approximately 150,000 km^2,

Table 1 Extent and distribution of tropical forest, protected areas, biological diversity, conservation threat and assumed deforestation by the World Wildlife Fund-delineated biomes and realms

Biomes and realms	Total land		Forest area (contemporary)		Protected forest area		Hi-BDI in forest areas		Threatened forest areas		Converted forest areas		Converted high BDI forest areas	
	Area (000 sq km)	Share (%)	Area (000 sq km)	Share (%)	Area (000 sq km)	Share of forest (%)	Area (000 sq km)	Share of forest (%)	Area (000 sq km)	Share of forest (%)	Area (000 sq km)	Share of forest (%)	Area (000 sq km)	Share of forest (%)
Biomes														
Tropical and subtropical moist Broadleaf forest	19,587	38.[illegible]	10,994	73.8	2,179	19.8	9,418	85.7	2,499	22.7	2,409	21.9	2,029	18.5
Tropical and subtropical dry Broadleaf forest	3,619	7.[illegible]	700	4.7	98	14.1	519	74.2	411	58.6	352	50.3	269	38.3
Tropical and subtropical coniferous forest	703	1.[illegible]	276	1.9	29	10.4	241	87.2	219	79.2	199	72.0	181	65.6
Other tropical biomes	31,016	5[illegible]6	2,935	19.7	378	12.9	1,855	63.2	818	27.9				
Total	**54,925**	**100.0**	**14,906**	**27.1**	**2,684**		**12,034**		**3,947**		**2,960**		**2,479**	
Share of forested land (in target biomes)				*100.0*		*18.0*		*80.7*		*26.5*		*19.9*		*16.6*
Realms														
Neotropical	16,923	3[illegible].8	6,933	46.5	1,790	25.8	5,250	75.7	1,490	21.5	983	14.2	684	9.9
Afrotropical	18,781	3[illegible].2	3,553	23.8	431	12.1	2,991	84.2	953	26.8	276	7.8	208	5.8
Indo-Malay	8,356	1[illegible].2	2,881	19.3	260	9.0	2,736	95.0	948	32.9	1,077	37.4	1,026	35.6
Australasia	1,756	[illegible].2	666	4.5	81	12.2	654	98.2	356	53.5	301	45.1	299	44.8
Other realms (within study area)	9,109	1[illegible].6	873	5.9	121	13.9	403	46.1	200	23.0	324	37.1	262	30.1
Total	**54,925**	**100.0**	**14,906**	**27.1**	**2,684**		**12,034**		**3,947**		**2,960**		**2,479**	
Share of forested land				*100.0*		*18.0*		*80.7*		*26.5*		*19.9*		*16.6*

in comparison with the average unit size of over 740,000 km^2 found in the well-known global, biogeographical classification of Udvardy (1975). WWF used many of these coarser datasets to establish the first tiers of their hierarchical classification system – the 8 realms and 14 biomes, and then based the ecoregion delineation on expert opinion and other secondary and more local information (Olson *et al.*, 2001; p. 934). As an example of the WWF hierarchy, in the Tropical and Sub-Tropical Moist Broadleaf Forest *Biome* within the Afro-Tropical *Realm*, there are 30 distinct forest *ecoregions*.

The WWF ecoregion database links each mapped ecoregion unit within a biome to information on location, extent and various measures of biodiversity, as well as conservation status (to be discussed in Section 3). The biodiversity metric of specific relevance to this study was the "*Biological Distinctiveness Index*" (BDI), a scale-dependent attribute of biological richness based on 5 criteria: species richness; endemism; complexity of species distributions; uniqueness and rarity; and geographic uniqueness (e.g. areas that exemplify global rarity of their habitat type). The measures of richness and endemism used as BDI components were assessed for each ecoregion for birds, mammals and plants (Wikramanayake *et al.*, 2002; Dinerstein *et al.*, 1995). The BDI is premised on the assumption that, while all ecoregions are biologically distinct to some degree, some are exceptionally rich, complex or unusual. The WWF ranked ecoregions according to their BDI rating as Globally Outstanding, Regionally Outstanding, Bioregionally Outstanding, and Locally Important. Ecoregions are considered outstanding if they exemplify extraordinary levels of the first 4 criteria or if they meet the criterion for geographic uniqueness (Dinerstein *et al.*, 1995). The BDI metric was derived independently from the level of "threat" to the ecoregion and as such is a "pure" metric of biodiversity (in other words, the BDI of an ecoregion is not higher if that ecoregion is under a greater threat of human exploitation). For the purposes of this study, all ecoregions classified as containing globally or regionally outstanding BDI were, arbitrarily, assumed to exhibit (medium-to-) "high" biodiversity value, and the remaining classes (bioregionally outstanding and locally important) were classified as "low"(-to-medium). Most tropical forest ecoregions fall into the "High BDI" category; over 85% of the area of both moist broadleaf and coniferous tropical forest biomes, and around 75% of the area of the dry broadleaf forest biome (see Table 1). The distribution of ecoregions of High BDI within the WWF tropical forest biomes and tropical realms is shown in Figure 2.

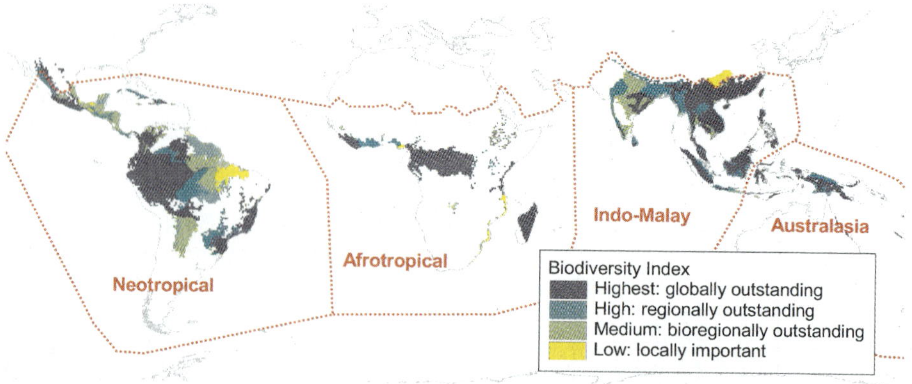

Fig. 2 Biodiversity richness as delineated by the World Wildlife Fund's Biological Distinctiveness Index (BDI) within the tropical forest biomes. The outlines show the major geographic realms that fall within the pan-tropical domain

3. Contemporary land cover and the land cover change scenario

Like other approaches to ecological mapping, WWF biomes and ecoregions provide assessments of untransformed land cover (variously called "potential" or "original" land cover). But a significant amount of transformation has already occurred in the 500 WWF ecoregions that fall within the pan-tropical boundary. A separate assessment of the loss of tropical forests from pre-industrial to contemporary times finds that forest conversion has reached around 23% in the moist broadleaf forest biome, 50% in the dry forerst biome, and 45% for the coniferous forest biome (Douglas *et al.*, 2005). This section describes the process of establishing the contemporary areal extent of remaining forest within the WWF forest biomes, how this contemporary land cover dataset was prepared for use in the hydrological model, and the design and definition of a deforestation scenario (taking the contemporary land use as a baseline).

3.1. Contemporary land cover

From the perspective of this study, potential change in biodiversity and hydrological function are both mediated through change in land cover. Biodiversity impact is assessed through change in forest areas of specific BDI classes. This is a pragmatic approximation that clearly does not address many key dimensions of biodiversity such as habitat fragmentation, species ranges, and sustainability thresholds. Hydrological impact is assessed in a more robust way since more is known (or can be assumed) about the potential above- and below-ground changes brought about by deforestation and how these changes affect hydrological processes. Thus, changes in land cover provide a trigger for changes in relevant parameters in a hydrological model that, in turn, generates changes in predicted river flow (see Section 4). A reliable contemporary land cover dataset, therefore, is necessary for two reasons: first, to calibrate the baseline biodiversity indicator (forest area by BDI class) by establishing to what extent intact forest areas remain within each WWF ecoregion; and second, to provide as reliable as possible an assessment of the actual mix of (all) land cover types within each basin, so that modelled runoff can reasonably be validated against actual gauged basin flows, where these are available. Runoff estimates generated using the contemporary landcover also establish a baseline against which post-deforestation runoff change is assessed.

Several criteria were used to help choose amongst available global land cover dataset options. First was the need to select a land cover data source whose legend would minimize re-classification and harmonization when used in combination with the two other primary spatial data sources: the WWF ecoregion map, which uses a WWF-specific ecoregion legend, and the "TEMVeg" global vegetation map developed by Melillo *et al.* (1993) used in defining land cover shares and associated hydrological parameters for hydrologic modeling. A second criterion was to use a land cover dataset that recognized the heterogeneity of land cover through the use of mosaic classes (i.e., that did not rely solely on a "majority-based" assignment of land cover, but that provided information on land cover sub-categories or shares within a single grid cell). A third criterion, related to the second, was that the land cover dataset be reliable in representing agricultural and urban land uses (since this is a clear indication that transformation of land use/cover has already taken place). In applying these criteria to available land cover datasets that covered the pan-tropical domain, (e.g., GLCCD, 2001; GLC, 2000; Wood *et al.*, 2000), a determination was made that none of them was entirely satisfactory, and that a new composite, contemporary land cover map was needed for the purposes of this study.

The base of the new land cover map was the GLCCD v2 (2001) global 1 km land cover map, integrated with the following components: the global cropland extent (IFPRI, 2002), the global irrigated area map (Döll and Siebert, 2000), global grazing lands (Ramankutty, 2003), and the global night-time lights database (Elvidge *et al.*, 2001). The process of integration is described in Sebastian *et al.* (2003). Although published finally at a coarse resolution (0.5 degree or ~50 km) this database takes into account land cover shares (available in some of the component databases) below 1 km resolution, and is the first attempt to integrate these thematic land cover data sources into one comprehensive land cover product. The database includes, by grid cell, the majority land cover class and the share of each individual land class (e.g. forest and agriculture shares by grid cell). These data were translated into 0.5 degree grid cell shares of the 20 TEMVeg land cover classes before aggregating to the eight general land cover classes required by the hydrological model.

3.2. Scenario for land use change

The trajectory of actual land use and land cover is difficult to project, and is influenced by a very wide range of economic, social, cultural and environmental factors, including land use policies and land tenure arrangements. In its *Terrestrial Ecoregions of the World* database, the WWF developed an indicator termed the *conservation status* to 'estimate the present and future capability of an ecoregion to meet three goals of biodiversity conservation: to maintain viable species populations and communities, sustain ecological processes, and respond effectively to short-and long-term environmental change' (Wikramanayake *et al.*, 2002; p. 41). The conservation status is determined at the landscape level and is based on an interpretation of evidence on loss of original habitat; number and size of habitat blocks; fragmentation/degradation; conversion rate and degree of protection, and is intended to provide a 30 year prediction of future conservation status given current conservation status trajectories. The classification of the conservation status was based on a quantitative assessment using available maps and current land cover data in conjunction with expert opinion on the region, according to the following categories (Dinerstein *et al.*, 1995):

- *Critical* – The remaining intact habitat is restricted to isolated small fragments with low probabilities of persistence over the next 5–10 years without immediate or continuing protection and restoration.
- *Endangered* – The remaining intact habitat is restricted to isolated fragments of varying size (a few large blocks may be present) with medium to low probabilities of persistence over the next 10–15 years without immediate or continuing protection or restoration.
- *Vulnerable* – The remaining intact habitat occurs in habitat blocks ranging from large to small; many intact clusters will likely persist over the next 15–20 years, especially if given adequate protection and moderate restoration.
- *Relatively stable* – Natural communities have been altered in certain areas, causing local declines in exploited populations and disruption of ecosystem processes.
- *Relatively intact* – Natural communities within an ecoregion are largely intact with species, populations, and ecosystem processes occurring within their natural ranges of variation.

WWF's ecoregion dataset includes both a 'snapshot' or current, and a 'global' or future conservation status, according to the above classes. The 'global' status reflects a 30-year prediction of future conservation status created by modifying the current status by estimates of future threat. The threat estimates were determined based on the cumulative impacts of habitat conversion, degradation, wildlife exploitation and exotic species (Ricketts *et al.*, 1999; Dinerstein *et al.*, 1995; Wikramanayake *et al.*, 2002).

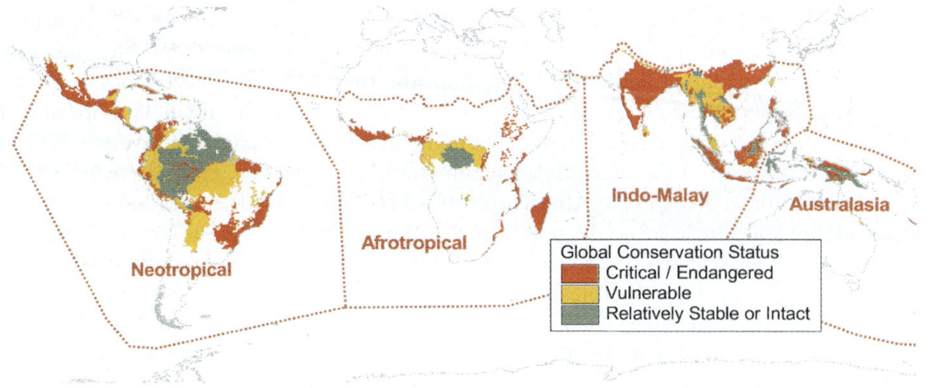

Fig. 3 Deforestation threat as defined by the World Wildlife Fund's Future Conservation Status within the tropical forest biomes. The outlines show the major geographic realms that fall within the pan-tropical domain

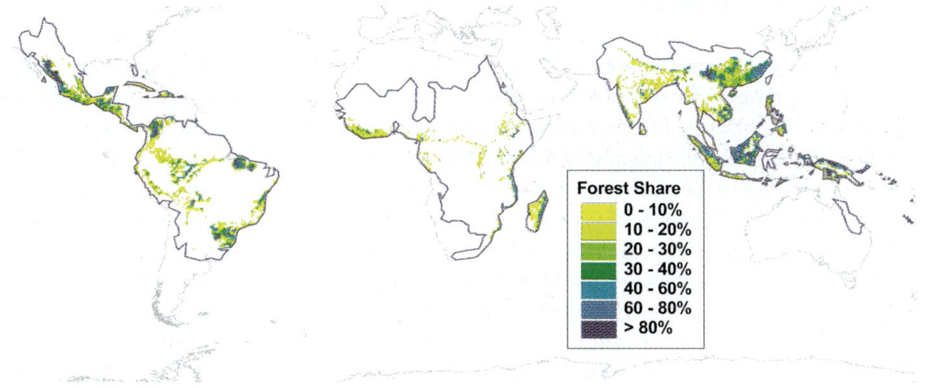

Fig. 4 Hypothetical forest converted (as fraction of grid cell), assuming that all of the most threatened forests are converted to agricultural land use, except for forests under active protection

For the purposes of this study, all ecoregions within the tropical forest biomes classified as *Critical* or *Endangered* according to the WWF's Global Conservation Status were considered to be threatened. This scenario assumes that all area within these ecoregions, *except for existing protected areas*, would be deforested within the overall 30 year time frame of the conservation status (and potentially within the next 15 years). The global protected areas database of the World Conservation Monitoring Centre (WCMC, 1992) was used to define the individual protected areas omitted from the deforestation scenario. Overall, approximately 18% of tropical forest biomes are protected, with the share of protected areas being greatest (26%) in the Neotropical realm (Latin America and the Caribbean), and least (9%) in the Indo-Malay realm (Table 1). Figure 3 shows which ecoregions within the tropical forest biomes were assigned to the most threatened (critical/endangered) categories.

The task in developing the hypothetical deforestation scenario was to identify the most-threatened ecoregions and convert whatever forest cover remained in them to agricultural land cover (the default deforested land use). Figure 4 shows the share of each 0.5 degree pixel within the most-threatened ecoregions that still contains forest, as identified in the contemporary land cover surface. It was these areas of forest that were lost (converted from forest to agriculture) under our hypothetical future land use change scenario.

The total forest area subject to conversion under this scenario is around 3 million km^2 (about 25% of the contemporary forests remaining in the tropical forest biomes) leaving only 9 million of the original (pre-industrial) 29 million km^2 tropical forests intact (Douglas *et al.*, 2005). Over 80% of the converted forest areas, some 2.4 million km^2, lie in the tropical and subtropical moist broadleaf forest biome, but regional differences are large. Around 37% of remaining forest area in the Indo-Malay realm would be converted, but only some 8% of the remaining forest areas in the Afro-Tropical realm (Table 1). After conversion of most-threatened topical forest areas, the proportion of agricultural area in the tropics increased from 13% to an estimated 18%. Interestingly, while this scenario was predicated on threats to biodiversity as a result of deforestation, the resulting 5% increase in agricultural area is within the range of land expansion needed to meet the growth in food demands over the next few decades (Bruinsma, 2003). In other words, while the land use change scenario developed for this study is hypothetical, the possibility of it becoming reality is quite high.

4. The hydrological model

Hydrological modeling was performed using the Water Balance Model (WBM) at a 0.5 degree (30 minute lat *x* long) spatial resolution (Vörösmarty *et al.*, 1998; Fekete *et al.*, 2001; Federer *et al.*, 2003). The WBM simulates monthly soil moisture variations, evapotranspiration, and runoff on single grid cells using biophysical data sets that include climatic drivers as well as vegetation and soil properties. The state variables are determined by interactions among time-varying precipitation, potential evaporation, and soil water content. Soil type and soil texture attributes were derived from the FAO Digital Soil Map of the World (1995; Global Soil Data Task, 2000). The soil type and texture attributes of the dominant soil type in each mapping unit were taken as representative of the entire mapping unit. Potential evapotranspiration (PET) was computed using a modification of the well-known Penman-Monteith surface dependent method (Monteith, 1965; Shuttleworth and Wallace, 1985; Federer *et al.*, 1996) which was found to perform best for global-scale land cover and climate modeling studies (Federer *et al.*, 1996; Vörösmarty *et al.*, 1998). This function requires temperature, net radiation, humidity, wind speed and cover type as inputs. The 30-min gridded datasets of climate variable time series (precipitation, temperature and diurnal temperature range, relative humidity, vapor pressure, and percent cloud cover) were developed by New *et al.* (1998). Operationally, for any time step in which rainfall exceeds the soil moisture deficit (the amount of soil moisture holding capacity currently unreplensihed), the excess is used to augment a rainfall-derived detention pool and to generate runoff. The soil infiltration rate is assumed to equal the rainfall rate; therefore overland flow is not explicitly simulated.

The necessary vegetation characteristics for WBM were derived from land cover inputs developed specifically for this study as described in Section 3. The major influence of vegetation type in WBM is in the computation of evapotranspiration. Broad vegetation classes are characterized using parameter values derived initially from the literature and subsequently refined through model application and validation (Table 2; see Federer *et al.*, 2003 for details). Two important vegetation parameters are leaf conductance and rooting depth. Maximum leaf conductance (GLmax) values in WBM are consistent with maximum leaf conductance observations for tropical forests and crops presented in Schulze *et al.* (1994). A sensitivity analysis showed that a 20% change in GLmax resulted in a change in annual runoff of 1.5% or less over the pan-tropics. Methods for determining appropriate rooting depths for global models can vary quite substantially (Kiedon and Heimann, 1998; Zeng, 2001). To improve the performance of the hydrological model in matching recorded basin flows, the following

Table 2 Vegetation parameters used in the Water Balance Model

Parameter (units)	Conifer forest	Broadleaf forest	Savannah/Pasture	Grassland	Cropland
Albedo (1)	0.14	0.18	0.18	0.2	0.22
Conopy height (m)	25	25	8	0.5	0.3
Max leaf area index (1)	6	6	3	3	3
GLmax[a] (m/s)	0.0053	0.0053	0.0053	0.008	0.011
Leaf width (m)	0.004	0.1	0.03	0.01	0.1
Zero-level roughness height (m)	0.02	0.02	0.02	0.01	0.005
98% root mass depth[b] (m)	1.61	1.01	1.0	1.0	0.49

Notes: [a]GLmax is maximum leaf conductance, the reciprocal of stomatal resistance
[b]Conifer and broadleaf rooting depths interpred from Jackson (1996)
Rooting depths for other cover types were values already established in previous model applications

adjustments were made in vegetation related parameters: forest rooting depths were modified to those published by Jackson *et al.* (1996), as recommended by Federer *et al.* (2003); pastures were assumed to be equivalent to savannahs; closed canopy woodlands were included in the dry forest class; mixed forests were split equally between the moist and dry forest classes; and both dry and moist tropical forests were modeled using the same broadleaf forest parameters. Hence, soil moisture availability rather than differences in vegetation characteristics was the major determinant of runoff generation in dry forests. To represent forest canopy interception (which is not explicitly modeled) the effective monthly rainfall utilized for simulating runoff from forest areas was computed off-line as 80% of observed monthly precipitation. A value of 20% canopy interception was selected because it was within the range of published values for annual canopy interception losses (Jackson, 1975; Calder, 1990; Bruijnzeel, 1990) and because it resulted in an average grid cell increase in annual runoff due the conversion of forest to agricultural land use similar to field observations (Oyebande, 1988; Bonell and Balek, 1993; Bruijnzeel, 1991; 1996). Average rates of wet canopy evaporation from forests can exceed those of shorter vegetation by two to five times (Calder, 1990; Bruijnzeel, 1990); hence interception for other vegetation types was not modeled. Irrigated croplands were simulated as having saturated soils. Irrigation withdrawals are not simulated in the current version of WBM, therefore, runoff from irrigated lands may be overestimated.

A common perception is that deforestation increases human vulnerability to extreme events such as floods. However, the most dramatic hydrologic effects of land use change are often short-lived, and have only been shown to impact smaller magnitude, higher frequency events, since the role of land cover decreases as the magnitude of the event increases (Bruijnzeel, 1996). The hydrologic analysis for this study was limited to modeling changes in the long-term average annual, monthly maximum and monthly minimum runoff as a first step in understanding the impacts of land use change on these events. Runoff was computed by first generating separate runoff estimates for all land cover types independently, and then summing the runoff derived from each cover type within each grid cell according to the share of grid cell area occupied by that cover type (using the land cover shares per pixel as estimated by the prior land cover and land cover change analysis). Increases in annual runoff due to the conversion of forest to agriculture are within the range of values observed from field studies (Oyebande, 1988; Bruijnzeel, 1996, 2004). The model performed well in matching long-term average annual flows in rivers across the pan-tropics (see Appendix A of Douglas *et al.*, 2005 for a more detailed discussion). Annual changes in runoff were obtained by running the hydrologic model with long-term mean annual climate inputs, computed from New *et al.*, (1998), and comparing runoff generated from the contemporary land cover and the hypothetical future land cover datasets. Changes in maximum and minimum monthly runoff were obtained by running the hydrologic model with long-term mean monthly climate inputs and then selecting the maximum and minimum differences in runoff between the two land cover datasets. Annual and monthly river discharge (Q) was computed by accumulation of gridded runoff along a digital river network (STN-30, Fekete *et al.*, 2001). Flow impoundments were not represented in the model, hence the effects of hydroelectric power generation and reservoir siltation, which are important impacts in some pan-tropical basins (e.g., the Parana), were not investigated in this study.

5. Impacts of deforestation on biodiversity and hydrology

5.1. Biodiversity related impacts

Just over 80% of the tropical forest biomes are characterized as having High BDI, while of the contemporary forest areas in other tropical biomes, only around 63% are categorized as High BDI. The Australasia and Indo-Malay realms have the highest share of High BDI, around 98 and 95% respectively, and the Neotropical realm has the lowest, at around 76%. Of note is that forest protection is inversely related to the share of High BDI, e.g. only 9% of the more biodiversity-rich Indo-Malay tropical forest biomes are protected, in comparison to 26% of the Neotropical forest biomes (Table 1). With regard to conversion threat and projected forest conversion, around 26% of the tropical forest biomes are categorized as most-threatened, but this share falls to around 20% after subtracting out existing protected areas. The share of most-threatened forest areas differs considerably by biome. Around 23% of moist broadleaf forest, 59% of dry broadleaf forest, and 79% of coniferous forests are classified as most-threatened. Most-threatened status is most prevalent in Australasia (54%) and least prevalent in the Neotropics (22%). The share of High BDI areas within the converted forest areas also varies significantly by biome and realm. 84, 76 and 91% respectively of the most-threatened forest areas are classified as exhibiting High BDI for the moist broadleaf, dry broadleaf and coniferous forest biomes respectively. From a regional perspective, the share of High BDI areas that are converted are consistent with the overall share of High BDI in the forest areas of the Indo-Malay and Australasia realms, but in the Neotropical and Afrotropical realms they are somewhat less (e.g. a proportionately lesser share of High BDI areas are to be found in the most-threatened forest areas in those regions). The shares of total forest with High BDI and converted area with High BDI are 76 and 70%, respectively, in the Neotropics, and 84 and 75%, respectively, in the Afrotropical realm (Table 1). All the converted forest area in Australasia is classified as High BDI.

5.2. Hydrology related impacts

Figure 5 shows the change in average discharge (ΔQ, in km^3) relative to contemporary discharge (Q, in km^3) for long-term average annual flows (Figure 5a, from Douglas *et al.*, 2005) and long-term average maximum flows (Figure 5b). For annual discharge, the total hydrologic impact of the projected land cover change was less than 5% of contemporary Q, but the impacts were focused in southern China, western Mexico and the Yucatan peninsula, with more localized areas in Paraguay and Bolivia, and in Kenya. For the mean maximum monthly flows, a similar pattern emerged. Figure 5b shows a slight reduction in the area

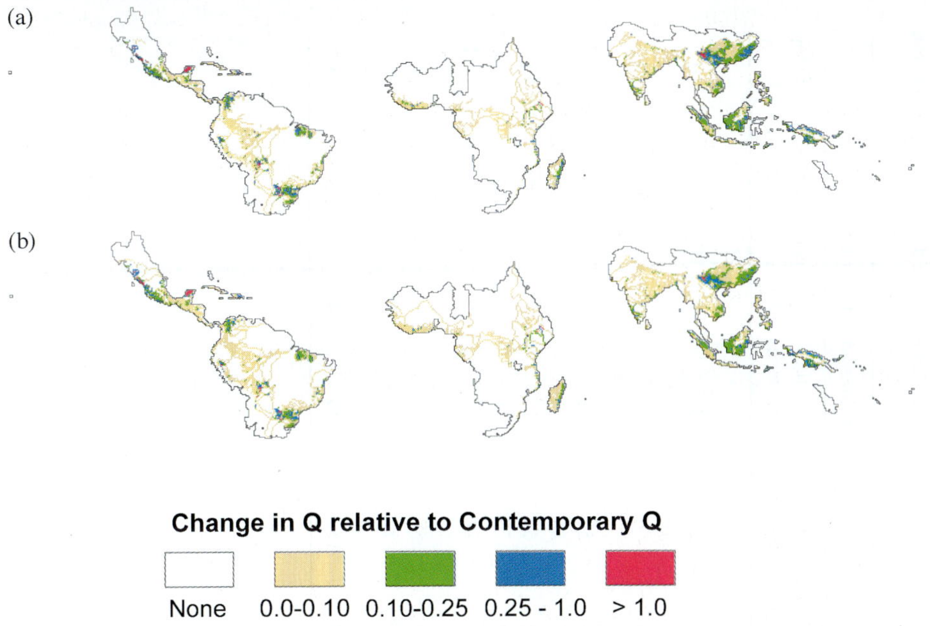

Fig. 5 Proportionate change in (a) mean annual and (b) monthly maximum flows arising from the removal of the most threatened tropical forests areas depicted in Figure t4

where $\Delta Q_{\max}/Q_{\max} \geq 0.25$ in Southeast Asia when compared to Figure 5a, whereas the patterns are essentially identical along the western coast of Central America. Annual rainfall in western coastal Central America ranges between 500 and 1000 mm, while over continental Southeast Asia annual rainfall exceeds 1000 mm and over Indonesia, it exceeds 2000 mm. The spatial differences between Figure 5a and b again support the assertion that deforestation results in greater relative hydrologic impacts in drier climates than in wet.

Table 3 summarizes the annual and maximum monthly discharge changes by biome, realm and basin size. Over the entire pan-tropics, the long-term average maximum monthly discharge increased by 159 km^3/mo or about 3% of the contemporary maximum monthly flows. The majority of this increase (138 km^3/mo) occurred in the moist broadleaf forest biome, but the coniferous forests experienced a greater percentage increase (11%) than either moist or dry forests, both of which had a 5% increase in maximum discharge on average. The Indo-Malay realm had the largest increase in magnitude, but the Australasia realm saw the largest percentage hydrologic impact for annual and maximum monthly increases. Forest conversion on the island of New Guinea accounted for all the hydrologic changes in this realm. While it is generally accepted that forest conversion results in increased annual flows and, as shown in this analysis, in increased high flows, the impact on low (base) flows is more complex and difficult to predict (Bruijnzeel, 1996). Aggregated over the entire pan-tropics, long-term mean minimum monthly flows were projected to increase by about 3%. Our hypothetical forest conversion in the moist and dry tropical forests resulted in a 10 and 2% increase in minimum flows, while a 53% decrease in minimum flows occurred in coniferous forests. However, the decrease in coniferous forest base flow only amounted to 1 km^3/month due to the small extent of the coniferous forest biome within the pan-tropics. Because of limited rainfall and greater seasonal variability in the dry and coniferous forest biomes, increased runoff from forest conversion may have been offset by greater soil moisture deficits, resulting

Table 3 Changes in long-term average annual, monthly maximum and monthly minimum discharge (Q) due to the hypothetical conversion of the most-threatened forests to agricultural land uses (cropland and pasture)

	Contemporary flows			Change in flows after deforestation					
Realms, biomes and basins	Avg Annual (km^3/yr)	Av. Max month (km^3/yr)	Av. Min month (km^3/yr)	ΔQ Annual (km^3/yr)	ΔQ Max (km^3/mon)	ΔQ Min (km^3/mon)	ΔQ Annual (%)	ΔQ Max (%)	ΔQ Min (%)
Change in Q by tropical forest biome									
Trop & Sub-trop Moist Broadleaf Forest	15,608	2,893	319	927	138	31	6	5	10
Trop & Sub-trop Dry Broadleaf Forest	1,052	298	5	63	16	0.1	6	5	2
Trop & Sub-trop Coniferous Forest	242	56	2	19	6	−1	8	11	−53
Change in Q by realm									
Neotropical (Latin America)	10,086	1,916	160	294	50	6	3	3	3
Afrotropical (Africa)	4,510	1,128	33	73	14	1	2	1.2	3
Indo-Malay (E, S, SE Asia)	5,765	1,225	96	411	63	13	7	5	13
Australasia	1,471	198	66	182	20	11	12	10	16
Other Realms (in study area)	873	198	11	49	12	0	6	6	−1.2
Change in Q by basin size									
Very Large Basins (>100,000 km^2)	12,351	2,567	157	325	58	6	3	2	4
Large Basins (>50,000–100,000 km^2)	1,970	360	49	201	29	7.3	10	8	15
Medium Basins (30,000 – 50,000 km^2)	969	164	28	66	9	3.0	7	5	11
Pan Tropics	**22,705**	**4,665**	**366**	**1,008**	**159**	**30**	**4**	**3**	**8**

in less runoff available for base (low) flow. In the mountainous regions of western Mexico and Central America, a combination of low vapor pressure deficits and low modeled leaf conductance used in simulating coniferous forests (see Table 1) may have resulted in an underestimation of ET and an overestimation of runoff from these forests. The conversion of these forests, predominantly to pasture, resulted in the model predicting decreased rather than increased runoff due to deforestation in these areas. The location of these "negative runoff" effects roughly coincide with the locations of tropical montane cloud forests (TMCF) shown in Bruinjzeel (2005, Figure 18.2, pg. 465). Bruinjzeel (2005) and Bruinjzeel and Proctor (1995) note that little is known about the hydrological functioning of TMCF or about the hydrologic effects of converting these forests to cropland or pasture. Measured interception losses in tropical montane environments can be as high as 45% of precipitation (Bruinjzeel, 2005, Table 18.2, pg. 470), which is more than twice the amount that we assumed in our model. This could have led to our anomalous "negative runoff" results in these areas, the magnitude of which ultimately had a negligible impact on our results. Field studies are currently underway to better quantify evapotranspiration and interception processes along with appropriate vegetation parameters for TMCF in Central America (Sampurno Bruinjzeel, World Bank, personal communication, October 13, 2005), which may shed light on the modeled behavior in these areas.

Table 3 also summarizes hydrologic changes by basin size. Out of all 108 focus basins, 99 (92%) showed an increase in long-term mean maximum monthly flows while 8 had no change. For the long-term mean minimum monthly flows, 84 of the 108 focus basins (78%) showed an increase while 16 (15%) showed a decrease. The remaining 8 basins had no change. While the very large basins (basin area >100,000 km^2) had the largest magnitude change in long-term mean annual and maximum flows (325 and 58 km^3, respectively), large basins (area between 50,000 and 100,000 km^2) had the largest change in mean minimum flows (7.3 km^3 or 8%). The number of basins with increasing minimum flows was distributed fairly uniformly across the three size classes (29, 32, 23 respectively), but most of the basins experiencing decreased minimum flows were in the large and very large size classes. In relative terms, however, large basins (basin size between 50,000 and 100,000 km^2) had the largest increases in annual, maximum and minimum flows, followed by medium sized basins (basin size between 30,000 and 50,000 km^2).

5.2.1. *Potential human vulnerability to hydrological change*

A key factor in examining the effects on human welfare of changes in land use and hydrological function is recognition that the populations affected by change are topologically linked to disturbance through river networks. Thus affected populations could be living both in the areas where the land use change takes place and in areas downstream of these changes. The hydrological response can be propagated far downstream of the actual point of disturbance and become intensified or diluted depending on the characteristics of change in the influent tributaries (Douglas *et al.*, 2005). Of the 3.7 billion people who live within the pan-tropic domain, approximately 2.2 billion people – about 1/3 of the world's total – reside within the boundaries of the WWF tropical forest biomes (Table 4) with 1.7 billion (or 70%) living in areas classified as having High BDI. As previously noted, the WWF biomes delineate "potential" tropical forest areas, at least one-third of which have already been deforested (Douglas *et al.*, 2005). Approximately 570 million people (28% of the pan-tropical population) live in contemporary (circa 1992/3) tropical forest areas. Of these, nearly one-half (250 million) live in the most-threatened forest areas, highlighting the intense human pressure on the remaining forest (Table 4). Densely populated urban areas could be especially vulnerable to the effects

Table 4 Distribution of forest and population by biome, realm and basin size

Realms, biomes and basins	Total area (000 sq km)	Totol population (M. persons)	Total population in contemporary Forest area (M. persons)	Population in deforested pixels (M. persons)	Flood plain population (M. persons)	Population densities: Total pop/Total area person km^{-2}	Population densities: Forest pop/Forest Area person km^{-2}	Population densities: Floodplain pop/Total Area person km^{-2}
By tropical forest biome								
Trop & Sub-trop moist broadleaf forest	19,587	1,736	337	211	453	89	31	
Trop & Sub-trop dry broadleaf forest	3,619	406	48	26	55	112	68	
Trop & Sub-trop coniferous forest	703	65	22	13	1	93	78	
By Realm								
Neotropical (Latin America)	16,923	433	71	41	60	26	10	
Afrotropical (Africa)	18,781	625	72	14	90	33	20	
Indo-Malay (E, S, SE Asia)	8,356	1,962	299	162	502	235	104	
Australasia	1,756	25	11	3	1.1	14	16	
Other realms (in study area)	9,109	637	116	30	168	70	133	
By basin size (108 focus basins only)								
Very large basins (>100,000 km^2)	25498	1804	299	108	461	71	30	18
Large basins (>50,000–100,000 km^2)	2890	264	44	30	69	91	40	24
Medium basins (30,000–50,000 km^2)	1133	87	13	7	45	77	28	40
Pan tropics	**54,925**	**3,682**	**568**	**250**	**821**	**67**	**38**	

of hydrologic changes when located in floodplains along major rivers. At the time of this study, the Global Rural Urban Mapping Project (GRUMP, CIESIN, 2005) had not yet been completed and it was not possible to break the pan-tropical population into rural and urban inhabitants. However, according to this dataset (now completed), approximately 35% of the pan-tropical population lives in urban areas.

6. Implications – a strategic perspective

Douglas *et al.* (2005) used some elements of the analyses described above to identify "hydrological hotspots" where the following conditions applied at the pixel level: forest area within a tropical forest biome, high BDI (WWF globally or regionally outstanding), most-threatened (WWF critical or endangered conservation status), and ≥25% projected increase in annual runoff if the forest areas are converted to agriculture. Categories of hotspots for pixels that met these criteria were then assigned on the basis of floodplain population. About 104 million people (roughly 40% of the total number of people living within the most-threatened forest areas) are deemed to be potentially at risk from high levels of hydrological response to deforestation. More than three-quarters of these people (80 million) live on floodplains within or downstream of these highly responsive areas, which makes them particularly vulnerable to both immediate and long-term changes in hydrologic regime. Some areas with elevated risk of hydrological change and biodiversity loss according to this approach are found in east and southeast Asia, in particular in the Zhujiang, Menjiang, Chang Jiang, Fuchun Jiang, Hanjiang, Menjiang and Hong basins of southern China. Other areas include western Mexico where a series of smaller watersheds along the Pacific coast were highlighted as having increased vulnerability in terms of hydrology, biodiversity and populations. In South America, the Parana basin which covers parts of Argentina, Paraguay and southern Brazil also contains several hotspot areas. The Parana basin was targeted by the land use change scenario primarily due to the lack of protection and the degree to which forest areas are currently deemed to be under threat.

The above analysis can be extended to examine a mix of possible biodiversity and hydrological outcomes from deforestation at the basin scale, and the potential strategic implications of each. To do this, the 108 focus basins were categorized according to:

(a) The proportion of High-BDI forest areas that would be lost, where "High BDI" areas are those originally occupied by globally or regionally outstanding forests (see Figure 2). Basins where this ratio exceeded 0.75 (the mean across basins) were categorized as "higher BDI loss"
(b) The ratio of threatened population to total forest area converted (TP/FC), where threatened populations are those with within "hydrologic hotspots", grid cells where the change in Q due to deforestation relative to contemporary $Q(\Delta Q/Q)$ was ≥25%. Basins where TP/FC >40 people/km^2 (the mean across basins) were categorized as 'higher potential downstream impact'.

The result is four categories based on relative biodiversity and hydrological impacts:

Category 1 (Low B-Low H): lower biodiversity loss, less population affected
Category 2 (Low B–High H): lower biodiversity loss; more population affected
Category 3 (High B–Low H): higher biodiversity loss; less population affected
Category 4 (High B–High H) higher biodiversity loss; more population affected

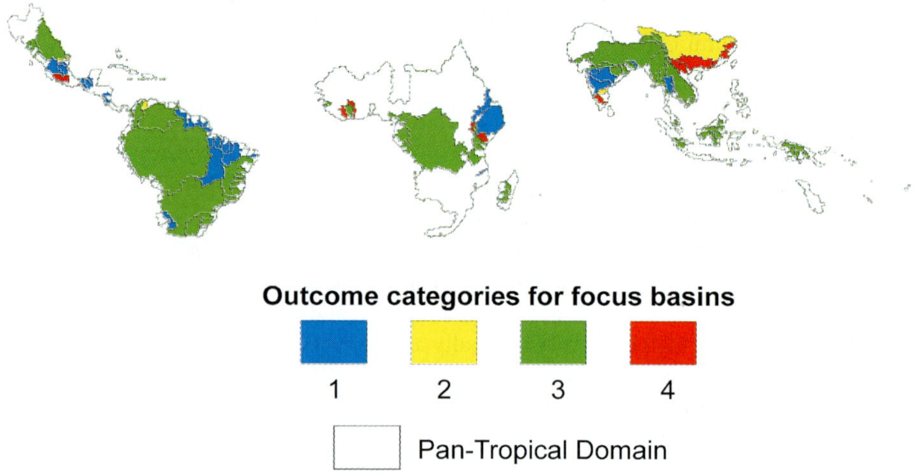

Fig. 6 Outcome categories (as defined in text) assigned to pan-tropical focus basins

Most basins (60 out of 108, or 56%), were classified as *High B-Low H* including the Ganges, Parana, Mekong, Amazon, Zaire, Orinoco, Rio Grande, Sao Francisco and Magdalena, basins. This category represents 59% of the total focus basin area. Thirty-one percent of focus basins were in the *Low B-Low H* (e.g., Krishna, Jubba, Godavari, Chao Phraya) and 4% were in the *Low B-High H* (e.g., Chang Jiang, Penner) categories. Ten percent of the focus basins were classified as *High B-High H*, including the Zhujiang and Hong rivers. Figure 6 shows the spatial distribution of basins by outcome category. Most of the basins in the Neotropics were classified as category 1 or 3, due to relatively lower population densities, whereas in the Indo-Malay region, all categories are represented. These rough categorizations can be used to assess the feasibility of devising strategies, assembling constituencies and raising funds for forest conservation. Support for conservation depends on beneficiaries' perception of its benefits relative to the costs of organizing, negotiating with forest dwellers, prohibiting illegal deforestation, and so on. On the hydrological side, forest dwellers' perceived benefit/cost ratio for supporting conservation may be proportional to the ratio of threatened population/converted forest area (TP/FC). For instance, potentially affected populations may be willing to pay \$10/person/year to avert a 25% increase in annual flows, to pick an arbitrary but plausible number. If TP/FC $>$ 40, then forest conservation may be justified, and possible, if costs of conserving the forest are less than \$400/km^2/year. Mobilizing conservation support becomes more feasible with higher TP/FC, wealthier populations, and lower opportunity costs or implementation costs of forest maintenance. There are 13 basins with TP/FC $>$ 40, and 9 with TP/FC $>$ 100. The latter have a combined population of 42 million and a combined at-risk forest area of 229,000 km^2. In basins with the highest TP/FC, domestic hydrological concerns might play a leading role in driving forest conservation, with biodiversity conservation as a side benefit. Conversely, in category 1 and 3 basins, interest in maintaining globally important biodiversity will probably play the most important role. (Note however that there may well be important domestic benefits of biodiversity conservation in the lower-BDI forests.) In many ways category 4 (High Biodiversity, High Hydrological Impact) is the most interesting. In principle, conservation of these forests could be supported and financed either by biodiversity or hydrology beneficiaries. In practice, it is likely that an alliance between these two interests will be important to achieving social and financial support for conservation.

7. Concluding remarks

This study was predicated on the existence of significant tracts of tropical forest that provide both havens of biodiversity richness and socially beneficial watershed services. The goal was to identify the location and extent of such tracts within the tropics globally and to generate evidence of the nature and scale of the biodiversity and the hydrological services they deliver. Land cover surfaces were generated to represent two "snap shots" in time; a contemporary view, based on a combination of existing evidence representing the state of land cover/land use in the mid-1990s; and a hypothetical future land cover, representing conversion of the most-threatened tropical forest tracts to agriculture. The biodiversity and hydrological impacts and potential human threats of the hypothesized land cover changes were examined, some hydrological hotspots identified, and a prototype biodiversity conservation strategy schema developed. These results appear promising and point to the validity of future work in two main areas; further validation and elaboration of the empirical results in terms of hydrological hotspots and improved typologies of biodiversity conservation strategies, as well as the need for continued improvement in the underlying data and analytical approaches employed.

With regard to the measures used in the assessment, there are clearly some weaknesses. First is the disconnect between a relatively rich and consistent global characterization of biological distinctiveness developed by WWF for forest biomes and ecoregions that may in reality have already been significantly degraded – at least according to the contemporary land cover evidence. Second, the biodiversity grouping schema used here that treats only areas of global and regional distinctiveness as being of high biodiversity value gives perhaps insufficient weight to other, still important categories of biodiversity. But the approach does highlight where major challenges will likely be faced from a regional and pan-tropic perspective. Third, a better understanding of the actual magnitude of deforestation threat would represent an improvement over the simple threat ratings used here, and might significantly change the spatial pattern of prioritization. Finally, there are also shortcomings in the resolution (both temporally and spatially) and completeness of the climate, hydrological, and basin scale components of assessing (changes in) river flow. Ideally the hydrological modeling would have been capable of examining the impacts of deforestation on extreme runoff events (particularly high flows) at a higher temporal resolution than one month, and over a lengthy time series rather than through the use of a long-term average climatologies. Some improvements were made in the hydrological process model in terms of representation of interception storage in forest areas, but runoff modeling could still be improved through better representation of infiltration, delayed base-flow storage, and river flow routing. But there are also pragmatic analytical constraints when running a global hydrological model with multiple land cover types represented in each pixel that force analytical tradeoffs to be made. Work on developing a downstream indicator of human vulnerability is still in progress.

Acknowledgements This work was supported [in part] by the Bank-Netherlands Partnership Program. The findings and interpretations are the authors' and do not necessarily reflect the views of the World Bank, its Executive Board of Directors, or the countries they represent. We gratefully acknowledge the insightful comments of two anomymous reviewers; incorporation of their comments made this a more complete and balanced paper.

References

Bailey, RG (1998) Ecoregions: the ecosystem geography of oceans and continents. Springer-Verlag, New York

Bonell M, Balek J (1993) Recent scientific developments and research needs in hydrological processes. In: Bonell M, Hufschmidt M, Gladwell J (eds) Hydrology and water management in the humid tropics. Cambridge University Press, Cambridge, UK

Bruinjzeel, LA (1990) Hydrology of moist tropical forests and effects of conversion: a state of knowledge review. UNESCO, International Hydrologic Programme, Paris, p 224

Bruijnzeel LA (1991) Hydrological impacts of tropical forest conversion. Nat Resour 27(2):36–46

Bruijnzeel LA (1996) Chapter 2 – Predicting the hydrological impacts of land cover transformation in the humid tropics: the need for more research. In: Gash JHC , Nobre CA (eds) Amazonian deforestation and climate. John Wiley & Sons

Bruijnzeel LA (2004) Hydrological functions of tropical forests: not seeing the soil for the trees? Agriculture, ecosystems and the environment. Agr Ecosyst Environ 104:185–228

Bruijnzeel LA (2005) Tropical montane cloud forest: a unique hydrological case. In: Bonell M, Bruijnzeel LA (eds) Forests, water and people in the humid tropics: Past, present and future hydrological research for integrated land and water management. UNESCO International Hydrology Series, Cambridge University Press, U.K

Bruijnzeel LA, Bonell M, Gilmour DA, Lamb D (2005) Forests, water and people in the humid tropics: an emerging view. In: Bonell M, Bruijnzeel LA (eds) Forests, water and people in the humid tropics: Past, present and future hydrological research for integrated land and water management. UNESCO International Hydrology Series, Cambridge University Press, U.K

Bruinjzeel LA, Proctor J (1995) Hydrology and biogeochemistry of tropical montane cloud forests: what do we really know? In: Hamilton LS, Juvik JO, Scatena FN (eds) Tropical montane cloud forests, ecological studies 110. Springer Verlag, New York

Bruinsma J (ed) (2003) World agriculture towards 2015/2030. Rome: Food and Agriculture Organization

Calder IR (1990) Evaporation in the uplands. Wiley, New York, p 148

Calder IR (2005) The blue revolution: land use and integrated water resources management. Earthscan, London

Convention on Biological Diversity [CBD] (2002) Text of convention on biological diversity: Article 2 (1992). Available on-line at www.biodiv.org, (accessed October 20, 2005)

Chomitz K, Kumari K (2005) The domestic benefits of tropical forest preservation: a critical review emphasizing hydrological functions. World Bank Res Obs 13(1):13–35

CIESIN (2005) Global Urban-Rural Mapping Project, Center for International Earth Science Information Network (CIESIN), Columbia University, http://beta.sedac.ciesin.columbia.edu/gpw/index.jsp; accessed online on September 8, 2005

Costa MH (2005) Large-scale hydrological impacts of tropical forest conversion. In: Bonell M, Bruijnzeel LA (eds) Forests, water and people in the humid tropics: past, present and future hydrological research for integrated land and water management. UNESCO International Hydrology Series, Cambridge University Press, U.K

Defries R, Hansen M, Townshend J (1995) Global discrimination of land cover types from metrics derived from AVHRR pathfinder data. Remote Sensing Environ 5–4:209–222

Dinerstein E, Olson DM, Graham DJ, Webster AL, Primm SA, Bookbinder MP, Ledec G (1995) A conservation assessment of the terrestrial ecoregions of Latin America and the Caribbean. World, Washington, D.C.

Döll P, Siebert S (2000) A digital global map of irrigated areas. ICID J 49(2):55–66

Douglas EM, Sebastian K, Vörösmarty CJ, Wood S (2005) The role of tropical forests in supporting biodiversity and hydrologic integrity. World Bank Policy Research Working Paper No. 3635, Social Science Research Network, Washington, DC, 2005. Available on-line at http://econ.worldbank.org (accessed October 20, 2005)

Dudley N, Stolton S (2003) Running pure. World Bank/WWF Alliance

Elvidge CD, Imhoff ML, Baugh KE, Hobson VR, Nelson I, Safran J, Dietz JB, Tuttle BT (2001) Nighttime lights of the world: 1994–1995. ISPRS J Photogrammetry Remote Sensing 56:81–99

Federer CA, Vörösmarty C, Fekete B (2003) Sensitivity of annual evaporation to soil and root properties in two models of contrasting complexity. J. Hydrometeorology 4:1276–1290

Federer CA, Vörösmarty C, Fekete B (1996) Intercomparison of methods for calculating potential evaporation in regional and global water balance models. Water Resour Res 32(7):2315–2321

Fekete BM, Vörösmarty CJ, Lammers RB (2001) Scaling gridded river networks for macro scale hydrology: development, analysis, and control of error. Water Resour Res 3(77):1955–1967

FAO/CIFOR (2005) Forests and floods: drowning in fiction or thriving on the facts? Bangkok, FAO. RAP Publication 2005/03

Food and Agriculture Organization of the United Nations (FAO) (2001) Global Forest Resources Assessment 2000: Main Report, FAO Forestry Paper 140, Rome, Italy

Food and Agriculture Organization of the United Nations (FAO) (1995) Digital Soil Map of the World, Version 3.5. FAO, Rome, Italy

Global Land Cover [GLC] (2000) Available on-line at http://www-gvm.jrc.it/glc2000/. Accessed on October 20, 2005

Global Land Cover Characteristics Database. Version 2.0 [GLCCD] (2001) Available online at: http://edcdaac.usgs.gov/glcc/glcc.html

Global Soil Data Task (2000) Global Soil Data Products CD-ROM (IGBP-DIS). CD-ROM. International Geosphere-Biosphere Programme, Data and Information System, Potsdam, Germany. Available from Oak Ridge National Laboratory Distributed Active Archive Center, Oak Ridge, Tennessee, U.S.A. [http://www.daac.ornl.gov]

Holdridge LR (1967) Life zone ecology. Tropical Science Center, San José (Costa Rica)

International Food Policy Research Institute (2002) Global Agricultural Extent v2.0. Available online at: http://www.asb.cgiar.org/BNPP/phase2/bnpp phase2 datasets.htm

Jackson IJ (1975) Relationship between rainfall parameters and interception by tropical forests. J Hydrol 24:215–238

Jackson RB, Canadell J, Ehleringer JR, Mooney HA, Sala OE, Schulze ED (1996) A global analysis of root distributions for terrestrial biomes. Oecologia 108:389–411

Kiedon A, Heimann M (1998) A method of determining rooting depth from a terrestrial biosphere model and its impacts on the global water and carbon cycles. Global Change Biol 4(3):275–286

LandScan (2002) LandScan Global Population Database, Oak Ridge National Laboratory, Oak Ridge, TN. Available at http://www.ornl.gov.gist/

Mace G (2003) Personal Communication related to work on the Millennium Ecosystem Assessment

Millennium Ecosystem Assessment (2005) Ecosystems and human well-being: synthesis. Island Press, Washington DC

Melillo JM, McGuire AD, Kickligher DW, Moore B, Vörösmarty CJ, Schloss AL (1993) Global climate change and terrestrial net primary production. Nature 363:234–240

Monteith JL (1965) Evaporation and environment. In: The state and movement of water in living organisms. In: Proceedings of the 19th Symposium of the Society of Experimental Biology. Cambridge University Press, Cambridge, UK, pp 205–233

New M, Hulme M, Jones P (1998) Representing twentieth century space-time climate variability, Part II: development of 1901–1996 monthly grids. J Clim 13:2217–2238

Olson DM, Dinerstein E, Wikramanayake ED, Burgess ND, Powell GVN, Underwood EC, D'Amico JA, Itoua I, Strand HE, Morrison JC, Loucks CJ, Allnutt TF, Ricketts TH, Kura Y, Lamoreux JF, Wettengel WW, Hedao P, Kassem KR (2001) Terrestrial ecoregions of the world: a new map of life on earth. BioScience 51(11):935–938

Oyebande L (1988) Effects of tropical forest on water yield. In: Reynolds E, Thompson F (eds) Forest, climate, and hydrology: regional impacts. The United Nations University, Kefford Press, Singapore

Ramankutty N (2003) Global grazing lands dataset. Center for sustainability and the global environment (SAGE). University of Wisconsin, Madison. Data made available through personal communication

Ricketts TH, Dinerstein E, Olson DM, Loucks CJ, Eichbaum W (1999) Terrestrial ecoregions of North America: a conservation assessment. Island Press, Washington, D.C

Schulze ED, Kelliher FM, Korner C, Lloyd J, Leuning R (1994) Relationships among maximum stomatal conductance, ecosystem surface conductance, and carbon assimilation rate and plant nitrogen nutrition: a global ecology scaling exercise. Annu Rev Ecol Syst 25:629–660

Sebastian K, Douglas E, Wood S, Vörösmarty C (2003) Functional value of biodiversity: pantropic/meso-scale analysis and synthesis – Phase II Final Report. (unpublished). Washington, D.C.: World Bank. http://www.asb.cgiar.org/BNPP/phase2/ifpri/techreport/Act1_technical_report.pdf

Shuttleworth JW, Wallace JS (1985) Evaporation from sparse crops: an energy combination theory. Q J R Meteorol Soc 111:839–855

Tomich, TP, Chomitz KM, Francisco H, Izac AMN, Murdiyarso D, Ratner BD, Thomas DE, van Noordwijk M (2004) Policy analysis and environmental problems at different scales: asking the right questions. Agric Ecosyst Envir 104:5–18

Udvardy MDE (1975) A classification of the biogeographical provinces of the world. Morges (Switzerland): International Union of Conservation of Nature and Natural Resources. IUCN Occasional Paper no. 18

United Nations Educational, Scientific and Cultural Organization [UNESCO] (1969) A framework for a classification of world vegetation. UNESCO, Paris

Vörösmarty CJ, Federer CA, Schloss AL (1998) Potential evaporation functions compared on US watersheds: possible implications for global-scale water balance and terrestrial ecosystem modeling. J Hydrol 207:147–169

World Conservation Monitoring Centre [WCMC] (1992) Global biodiversity: status of the earth's living resources. Chapman and Hall, London

Wikramanayake E, Dinerstein E, Loucks CJ, Olson DM, Morrison J, Lamoreaux J, et al (2000) Terrestrial ecoregions of the Indo-Pacific: a conservation assessment. Island Press, Washington, DC, p. 643

Wood S, Sebastian K, Scherr SJ (2000) Pilot analysis of global ecosystems: agroecosystems. World Resources Institute/International Food Policy Research Institute. Washington D.C

World Bank (2002) Sustainable development in a dynamic world: transforming institution, growth, and the quality of life. Oxford University Press, New York. World Development Report, Washington, DC

Zeng XB (2001) Global vegetation root distribution for land modeling. J Hydrometeorol 2(5):525–530

Water Resour Manage (2007) 21:233–247
DOI 10.1007/s11269-006-9051-1

ORIGINAL ARTICLE

Towards better water security in North China

Jun Xia · Lu Zhang · Changming Liu · Jingjie Yu

Received: 8 June 2005 / Accepted: 28 October 2005

Abstract Water shortages and related environmental degradation in North China are major issues facing the country. As runoff from the mountainous parts of the region steadily decrease and water resources become overcommitted, serious water and environmental problems have resulted. These include drying-up of rivers, decline in groundwater levels, degradation of lakes and wetlands, and water pollution. Thus, 4000 km of the lower reaches of the Hai River – some 40% of its length – has experienced zero flows and, as result, parts of this river have become an ephemeral stream. The area of wetland within the Basin has decreased from 10,000 km^2 at the beginning of 1950s to 1,000 km^2 at present. Over-extraction of groundwater occurs beneath 70% of the North China Plain, with the total groundwater over-extraction estimated at 90 billion m^3. Thus, problems of water shortage and related environmental issues in North China have become the most significant limiting factors affecting sustainable development in this important region of China.

This paper addresses the water security issues facing North China in the 21st Century using the Hai River basin as an example. We describe hydrologic cycles under changing environments, water-saving agriculture, assessment of water resource security, and efforts towards achieving integrated catchment management.

Keywords Water security · North China · Integrated catchment management · Water use efficiency

---

J. Xia (✉)
Key Laboratory of Water Cycle & Related Land Surface Processes, Institute of Geographical Sciences and Natural Resources Research, Chinese Academy of Sciences, Beijing, 100101 China;
e-mail: xiaj@igsnrr.ac.cn; State Key Laboratory of Water Resources & Hydropower Engineering Sciences, Wuhan University, Wuhan, 430072, China

L. Zhang
CSIRO Land & Water, Christian Laboratory, Canberra, ACT, Australia

C. Liu · J. Yu
Key Laboratory of Water Cycle & Related Land Surface Processes, Institute of Geographical Sciences and Natural Resources Research, Chinese Academy of Sciences, Beijing, 100101 China

1. Water security problems in North China

North China, covering an area of more than 1.5 million square kilometers, is one of China's six administration regions, and plays an important economic role (Fig. 1). In 2000, the population of the North China Plain was 437 million (35% of the nation), GDP was ¥ 3130 billion (32% of the whole), and irrigated agriculture occupied 23 million ha (42%) (Chen, 2002). Whereas the Hai River Basin, the major river basin in the region, has an mean annual water yield of about 42 billion m^3, water resources per capita is only 305 m^3 less than one-seventh of the Chinese average and 1/24th of the world's (Figure 2). The region is currently experiencing serious environmental degradation: groundwater depression, drying up of rivers, reduced wetlands, and pollution of surface water and groundwater. These problems are associated with rapid economical development which has led to excessive exploitation of water resources and reduction of runoff (Chen and Xia, 1999; Chen, 1994; Xia and Tackeuchi, 1999; Xia and Tan, 2002). The result is that water security in North China is severely threatened.

Water resources issues in North China have received considerable attention from the Chinese government, and there has been notable progress (Liu and Wei, 1989; Liu and Yu, 2001; Chen, 1985). However, due to the nature and complexity of the situation, water shortages remain a major problem. Apart from social and economic factors, the biophysical aspects of the hydrological cycle, and its ecosystem interactions, require better understanding in order to provide a sound scientific basis for policy development.

1.1. Runoff reduction in mountainous areas

The mountainous areas in the North-west are the source of most of North China's water supply. However, runoff has significantly decreased over the last 30 years, leading to serious water shortages on the North China Plain and its major cities (Chen, 1999). For example, the average annual inflow to the Guanting Reservoir, one of the major

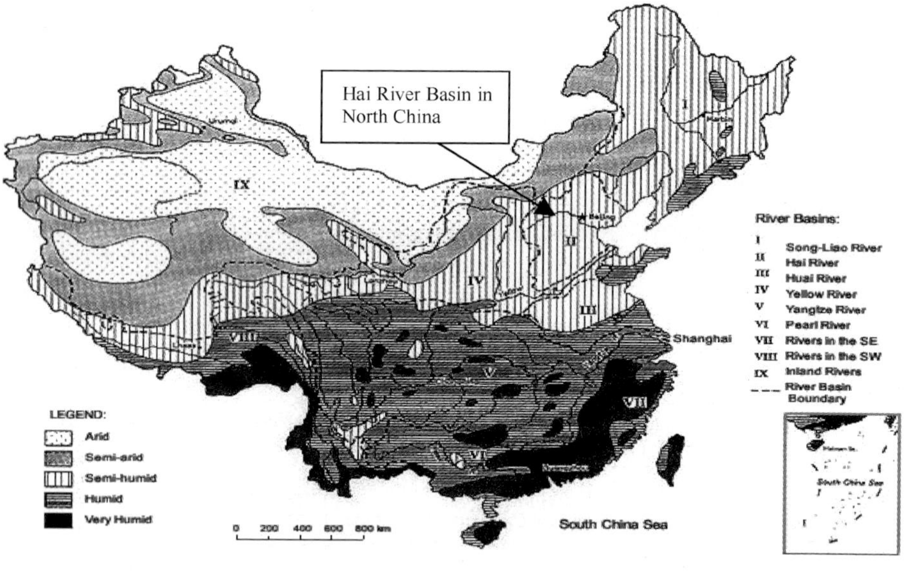

Fig. 1 Major river basins, water resources regions, and availability of water resources in China (Source: Xia and Chen, 2001)

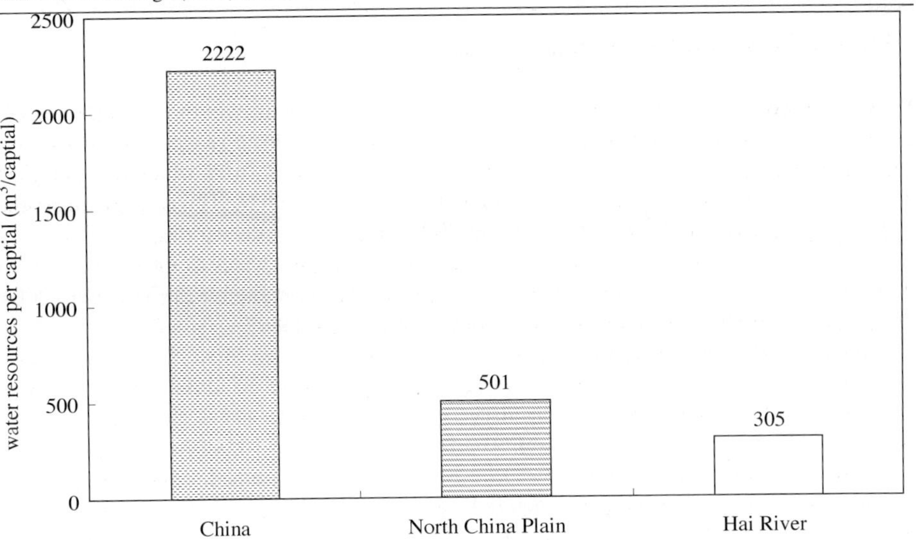

Fig. 2 Water resources per capita in China, showing low values in the Hai River basin of North China (Source: Qian *et al.*, 2001)

water supplies for Beijing, decreased from 1.13 billion m^3 in 1955–1984 to 0.27 billion m^3 in 1985–1995, a 75% reduction (Wang *et al.*, 2000). The reduction is chiefly due to land-use change (e.g. dams and afforestation), since average rainfall has stayed about the same. Similarly, inflow to the Miyun reservoir near Beijing shows the same downward trend. Compared with the period of 1960–1979, the average inflow during the period of 1980–1997 had decreased by 0.4 billion m^3 (Wang *et al.*, 2000). Reductions in runoff from the mountains have put tremendous pressure on the ecological environment (Liu and Wei, 1989) and on the social, economic, and development fabric of the region (Liu and Wei, 1989).

In seeking a solution to this problem, understanding the hydrological processes in the mountains, particularly runoff mechanisms, is a prerequisite. Research will need to focus on the effect of large land use changes, quantifying the impact and looking at how engineering work may have reduced runoff. This will provide a scientific basis for future water resources planning and management decisions.

1.2. Groundwater depression around major cities

Large parts of North China rely on groundwater as a major source of water supply. With population growth and economic development, groundwater has been over-exploited, leading to a steady decline in the water table. It has been estimated that 90,000 km^2, 70% of the North China Plain, has been affected by groundwater over-exploitation (Liu and Yu, 2001). As result, severely lowered water tables now surround the major cities in the region – Beijing, Shijiazhuang, Baoding, Xingtai, Handan, and Tangshan. Exhaustion of groundwater directly threatens the security of water resources in the North China Plain and has also led to serious land subsidence and seawater intrusion.

Nevertheless, over-extraction continues, and if population and economic growth in the region is to continue, ways of increasing recharge to the groundwater system and of preventing groundwater pollution will need to be found.

1.3. Water pollution and ecosystem degradation

Pollution of surface water presents a serious threat to water security in the region, especially in the Hai river basin. Water pollution has grown from some local reaches 20 years ago to the whole basin, spreading from downstream to middle reaches, from urban to countryside, and from ground surface to subsurface. Waste water discharge in the Hai river basin reached 6 billion ton's in recent years (Wang *et al.*, 2000). Because of deteriorating water quality, Guanting reservoir has, since 1997, ceased supplying domestic water to Beijing. Each year, more than 2 billion m^3 of sewage water is used for irrigation in the basin (Xia *et al.*, 2004), causing contamination of shallow groundwater, soil, and crops.

Damage to the riverine environment has also become serious. Most of the watercourses in the middle reaches of rivers have virtually dried up and are no longer able to recharge groundwater systems, instead they discharge sediment and salt. These changes have led to land degradation and wetland reduction. Among a total length of 10,000 km of watercourses, some 4000 km has dried up (Wang *et al.*, 2000), and 194 natural lakes and depressions (amounting to 6.67 km^2) have disappeared (Wang *et al.*, 2000). Stream discharge to the ocean dwindled from an annual mean of 24 billion m^3 in the 1950s to 1 billion m^3 in 2001 (Wang *et al.*, 2000). As a result, the ecosystem of the Hai river basin has drastically changed. The river's mouth has become choked with silt and loaded with salt, the natural ecological balance has been destroyed, and many estuarine organisms have become extinct (Xia *et al.*, 2004).

These problems have forced us towards improving knowledge of how water pollution affects the ecology of this environment, and how human activities – like the large hydraulic engineering works undertaken in the 1960s in the Hai river basin – impact on the ecosystem. The effect of natural changes, such as drought, also needs to be taken into account. Answers to the problems are compounded by the inherent conflicts between flood control, water shortages, and ecological protection. All these issues, which call for urgent attention, are generating intense discussion at present (Xia *et al.*, 2004; Chen *et al.*, 1999; Huang and Xia, 2001; Liu and Chen, 2000p; MWR, 1997; Qian, 1991).

1.4. Changed water consumption patterns

Because of its key place in the China's economy, and its special climate and geography, the availability of water resources in North China is becoming critical, especially in the Hai basin which has seen continuous rapid growth in economy and population over the last 20 years. Industrial growth in Beijing, Tianjin, and some megalopolises has plateaued, but most other areas in this basin are expected to sustain growth until 2030, and their demand for water – for homes, industry, and the environment – will increase accordingly (Xia *et al.*, 2004; Jia and Zhang, 2003). Unless measures are taken, there will be a serious conflict between supply and demand, and ultimately the environment will suffer (Qian, 1992; Qian and Zhang, 2001).

To meet the shortfall and give the environment a chance, one option is to find some way of transferring water from further south (Xia *et al.*, 2004; MWR, 1987), but in any case a rational way of allocating water resources is needed. Water allocation policies should take into account projected growth scenarios, but there is no doubt that reasonable limits to growth will be required. Moves towards a sustainable life-style will help, as will water-saving measures and concern for the environment (Xia *et al.*, 2004). The problem is one of reconciling the conflicting needs of different end-users.

The answer is difficult because as yet we are not even clear on what sort of extra water-supply infrastructure will be needed, and predicting demand for water, particularly that required for the environment, is not easy (Xia *et al.*, 2004).

1.5. Seeking a way forward: Integrated catchment management

The Chinese government has made much effort to address the water shortages and related environmental problems in North China (Qian, 1992; Qian and Zhang, 2001; World Bank, 1997). So far as water resources management is concerned, its objective for the 21st Century is to implement a strategy of developing the region socially and economically while protecting the environment (SPC, 1995; Xia and Chen, 2001; Xia *et al.*, 2001; Xia, 1997).

In scientific terms, implementing such a policy requires tools of integrated catchment management. Such a tool should be able to reasonably quantify the carrying capacity of a basin. It should be able to objectively weigh up the options between saving water and transferring it from elsewhere. An objective way of balancing the needs of the Hai river basin in terms of development and cost to the environment would be highly valuable.

The ideal model should be able to determine how groundwater tables interact with environmental water requirements; it should be able to balance the benefits of increased groundwater recharge, controlled land subsidence, and prevention of saline incursions against the cost of water transfers and implementing water-saving measures. Ultimately, of course, science needs to offer a way towards a feasible and permanent solution. Can we overcome water shortages and also repair environmental damage? All these challenging questions call for intense study and earnest consideration.

2. Towards scientific solutions

The Chinese Academy of Sciences (CAS) occupies a key position in organising the nation's fundamental and applied research. Since the 1960s, CAS has promoted research in water resources, agriculture, and ecology, with experimental expertise built up at the Yucheng Synthesised Agricultural Experiment Station, the Luancheng Agricultural-Ecological Experiment Station, and elsewhere (Liu and He, 1996; Zuo *et al.*, 1985). One key project, a study on integrated waste-water disposal and the development of a medium-to-low intensity production region in the Huang-Huai-Hai Plain, was initiated by CAS in the 1980s. Again, during the Ninth Five-Year Plan, another key project – on variation and allocation of water resources in North China – was carried out, along with some related projects.

More recently, in 2001, a Knowledge Innovation Project involving research on hydrological cycles and water security in North China, was funded by CAS. The project's chief goal was to scientifically examine three important issues: hydrological cycles under changing environments, how farmland consumed water and the potential for water-saving agricultural practices, and water security under natural change and human activity (Xia *et al.*, 2004; Plate, 1993). A diversity of scientific problems were tackled, including: (1) the interaction between land use change and the water cycle, and modelling it under changing environments using a distributed hydrological model; (2) how water-saving measures reduce evapotranspiration under farmland system, and an evaluation of its potential to save water; (3) establishing a system of water-security assessment that takes account of the principles of sustainable development at the same time as the economic law of supply and demand in regard to water. These three basic themes are set out in more detail below.

2.1. Study of the hydrological cycle

Here the work mainly centres on water-use problems involving the interplay between, firstly, the environment versus urbanisation and, secondly, the environment versus agriculture (WWC, 2000; Rodda, 1995; Kundzewicz, 1997). Experimental stations have been established in two typical basins: the Ziya basin of the Hai River agricultural plain and the Chaobai basin, which forms the catchment of the Miyun reservoir. Hydrological characteristics and their variations were identified, allowing refinements to be made to a distributed hydrologic model, both under natural conditions and human activity. The effects of a decrease in the reservoir inflow and space - time variations in soil water and groundwater were analysed (Xia *et al.*, 2004). The aim was to provide a scientific foundation for understanding changes in water resources, for measuring and calculating water resource availability, and for assessing how much water was required for the environment. This work provided a foundation for tackling the broader issue of water security.

2.2. Ways of reducing water consumption

Studies at the Luancheng and Yucheng Agricultural-Ecological Experimental Stations have focused on solving common problems encountered by farmers in the region (Zhang *et al.*, 2005; Luo *et al.*, 2000). In brief, they have looked at the physiological and environmental aspects of water consumption on farms and how water can be saved by considering water-energy transfer in soils and plants. Irrigation is a prime user of water, and work has shown ways of saving water on farms by using more-efficient irrigation techniques. The main thrust is to tailor water-saving technology and practices appropriate to each region, thereby improving water-use efficiency and slowing, or halting, the continuing downwards trend in the groundwater table as groundwater is the source for most irrigation in the region.

2.3. Water allocation and water security

This program aims to promote allocation of water for the benefit of the environment. The idea is that water should be allocated under the sustainable development principle. To do this, we will need to be able to predict trends in the water availability and water supply over the next 30 years. The effect of industrialisation, urbanisation, and new water rights and water-pricing systems will also need to be taken into account (Falkenmark, 1997).

Another part of the work is a wide-ranging investigation of the water-saving options. This ties in with building a theoretical model that could allow us to assess, and put a number on, a region's water security or vulnerability. Thinking of new ways to improve water security is a recurring theme.

This project will expand our knowledge and, at the same time, suggest innovative ways of regulating the soil - plant water interaction. It will improve water security for both humans and the environment. Hopefully, it will improve the skill and expertise of scientists for the benefit of the nation; such an aggregation of experts could help in achieving other national strategic targets such as sustainable development and environmental protection.

3. Examples of research achievements

In this section we will look at some successful research that has had a marked effect on improving our understanding of how water passes through the soil - plant-air interface.

The approaches used have ranged from practical field experiments to physical theories and complex computer models.

3.1. Basic studies of the hydrological cycle

A detailed knowledge of the water cycle on North China farmland has accumulated over the years thanks to focused experimental catchments and dedicated research stations shown in Figure 3. These typical basins include the Chaobai river basin, a mountainous region including the Dongtaigou, Xiaochangyu, and Huaishahe catchment; the Chongling experimental basin in the hilly Daqing river region; and the Ziya river basin (the Shimen catchment, the Luancheng station, and the Nanpi station) on the agricultural plains. Hydrological data has been acquired by a range of techniques that include isotope tracers.

Taking the Dongtaigou experimental catchment as an example, the impact of land use change, especially the soil and water conservation work, on the reservoir's inflow has been analysed using comparative studies on two small catchments (Xia *et al.*, 2004). During one storm event in 2003, 22 mm of rain fell during 0.45 h (the average rainfall intensity was 50 mm/h, and the maximum rain intensity was 1.8 mm/min). Comparing the two catchments, the runoff depth of the natural catchment without water conservation work was 18.5 mm, but none in the catchment with conservation. In the untreated catchment, sediment and litter scoured by water was about 1.5 tonnes with an erosion modulus of 8.8 t/km^2. In the controlled catchment, there was no runoff and no erosion.

Work on groundwater wells at the Luancheng and Nanpi Experimental Stations has demonstrated that the rainfall-infiltration process, and associated variations in soil water potential, are relatively complex within 2 m of the surface (Xia *et al.*, 2004). Strong variations are evident, with the soil water potential ranging from −400 mm to −1200 mm, and the soil moisture content varying from 10 to 39%. Soil water levels diminish progressively as the

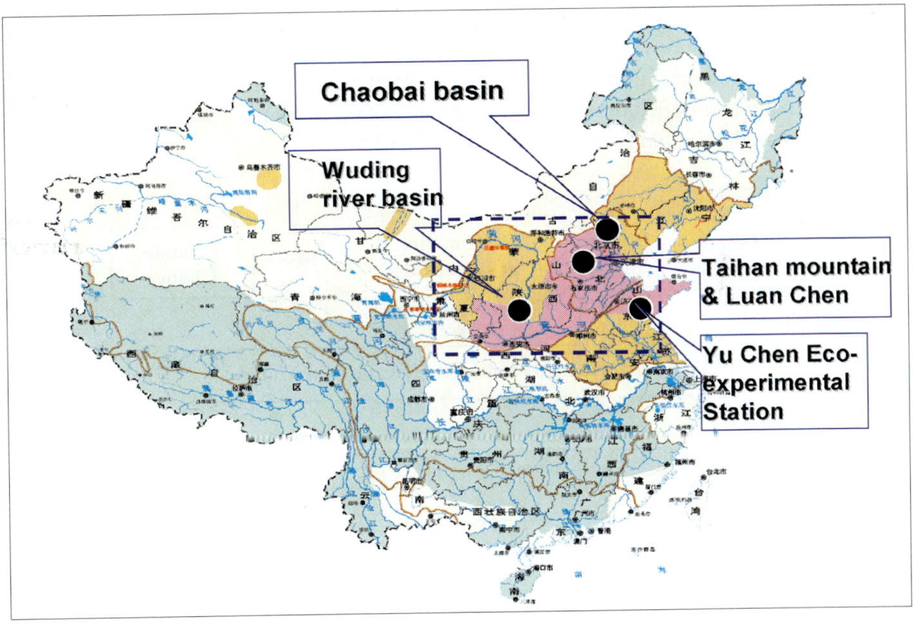

Fig. 3 Experimental catchments of Chinese Academy of Sciences in North China

depth increases to 3 m and beyond; this shows that soil water plays a key role in water transfer and crop water consumption down to significant depths. Understanding the movements of soil water remains a key problem in studies on water resources of the North China Plain.

3.2. Distributed hydrological models

Much research has been devoted to developing an accurate distributed hydrological model. We have developed hydrological non-linear system theory, and put forward a Distributed Time Variant-Gain Model (DTVGM), which has been applied to the Chaobai river basin (Xia, 2002; Xia *et al.*, 2003, 2005; Wang *et al.*, 2002). DTVGM is based on GIS/DEM, and abstracts the land surface gradient, flow direction, flow pathway, river network, basin boundary, and land cover. It extends the lumped non-linear Time Variant Gain Model (TVGM) to the distributed hydrologic modelling on raster grids provided by DEM, and conducts successive flow routing between adjacent concentration grids. Outputs of the model included the time - space distribution of the hydrologic elements and the flow processes of the basin's outlets.

DTVGM is a research tool integrating a hydrological understanding with systems theory, giving a better theoretical base and applicability. It can numerically simulate and analyse the impacts of land use change and water conservation work on reservoir inflow, and the responses of hydrologic processes to variations in precipitation. Lastly, we have verified a groundwater-surface water integration system model (SWAT/MODFLOW) for the Plain region and a large-scale land surface hydrologic model in the Hai River basin (Fig. 4).

3.3. Field experiments on agricultural water use efficiency

At Yucheng and Luancheng experimental stations, we have used closely monitored experimental plots to observe crop - water relationships. Researchers from have built up a spatial database of water diversions, meteorology, soils, and crops for the Panzhuang irrigated area in the lower reaches of the Yellow River. Using this experimental base, studies on crop water deficits based on the canopy temperature have been carried out. Minimum canopy resistances have been measured in different growth stages of North China winter wheat and a diagnostic index system of water deficit developed (Luo *et al.*, 2000). In terms of water uptake, we have put forward a model that takes into account root distribution.

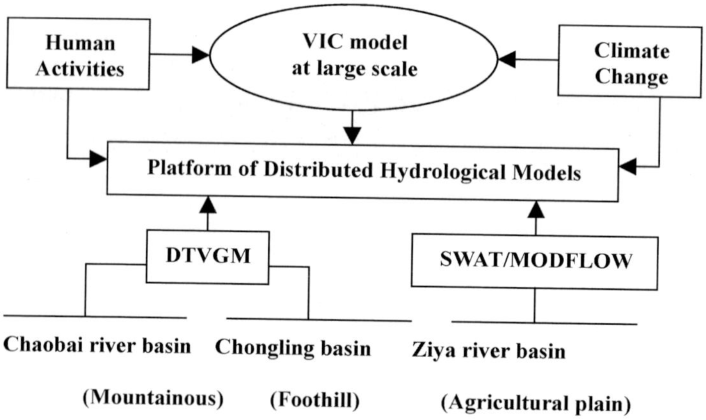

Fig. 4 Integrated system of hydrologic models at various scales in North China

Improving agricultural water use efficiency is important, and in this respect we have looked at a physiological/ecological approaches. Choice of species is a key factor, because, as Fig. 5 shows, water-use efficiencies (WUEs) can vary by 15–20% depending on species (Sun *et al.*, 2003). We established a correlation between WUEs of winter wheat and five other factors: photosynthetic rate, specific leaf weight of functional leaves, transpiration rate of functional leaves, wax content of flag leaf, and harvest ratio.

Turning to irrigation, we have looked at a novel scheme whereby the irrigation supply is regulated to keep a crop in water deficit. Under such conditions, we found that the best times for applying water to winter wheat were the turning-green stage and the later 'grain-filling stage. As a result, moderate levels of regulated water deficits do not reduce yields; at the same time, the optimised irrigation system could reduce water use by 60–120 mm, increasing WUE by 8–13%.

Another practice aimed at saving water in farming systems is straw mulching (Chilima *et al.*, 2002; Wang *et al.*, 2001). Spreading straw on the soil has clear effects in conserving water, reducing surface evaporation by 20–35 mm and increasing the farm crop WUE by 8–13%. Covering winter wheat and summer maize with whole straw reduced irrigation by 60–80 mm in a normal year (Table 1). Tillage is reduced or eliminated, saving water and energy, but it needs appropriate field management practices (Subhani *et al.*, 2000; Wang *et al.*, 2004).

Five different of water saving irrigation regimes have been investigated in the well irrigation district, each using a different pattern of water distribution and application (see Chen *et al.*, 2004 for details). These regimes have had widespread application, and can reduce water use by 60–150 mm and increase WUE by 20–30%. The key results of the research are

Table 1 Impact of different farming practices on the water use efficiency of winter wheat (Zhang *et al.*, 2005)

Treatment	Soil water consumption (mm)	Rainfall amount (mm)	Irrigation amount (mm)	Gross water consumption (mm)	Yield (kg/ha)	WUE (kg/mm/ha)
Deep plowing	73.76	133.8	222.8	430.4	4789	11.1
Rotary tillage	35.34	133.8	222.8	391.9	5775	14.7
Harrowing	46.54	133.8	147.8	328.1	3986	12.2
Zero tillage	−47.11	133.8	147.8	234.5	3539	15.2

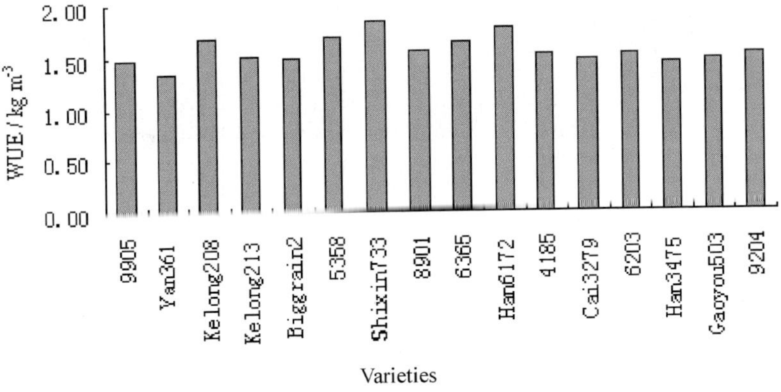

Fig. 5 WUE (water use efficiencies) for different varieties of winter wheat (Zhang *et al.*, 2005)

set out in Figure 6, which shows that the potential for saving water in agricultural production is high.

The amount of water saved could reach 8 billion m^3 if the techniques were fully implemented in agriculture while maintaining the current gross food production. Other important ways of saving water include adjusting and optimizing planting pattern, reducing the planting area of thirsty crops, and increasing it for crops that don't consume so much.

3.4. Assessing water security

North China has reached a stage where its water supply, in the long term, is not assured (Jia and Zhang, 2003; Zhang, 1999; Zhang and Zhang, 1995). Water security means the ability to supply water, according to a specified quality, to homes and industry under conditions satisfactory to the environment and at an acceptable price. The definition of water security includes: (a) population-wide security, that is, everyone can obtain secure water for domestic use; (b) economic security, namely water resources can satisfy the normal requirements of economic development; (c) ecological security, namely water resources can meet the lowest water demands of ecosystems without causing damage.

Ways of overcoming a lack of security include: public education, controlling water supply and demand, and using water resources so as to satisfy only reasonable socio-economic requirements. In terms of broadening water supply sources, besides the proposed South–North Water Transfer project, alternatives include sewage and rainwater utilization, and even seawater utilization; all of these are worthy of attention, and should be regarded as a strategic measure to ensure North China's water security.

To reduce expenditure, specific ways of limiting water resources demand are: (i) increasing water use efficiency and decreasing water use quota across all industries; (ii) reducing high water-consuming industries and enhancing low water-consuming ones, in this way improving the water use efficiency of the whole economic system; (iii) domestically, reducing water wastage by encouraging good water use habits; and (iv) decreasing water use in the home by popularizing water-saving techniques. Generally speaking, preventing water pollution is the most essential step for ensuring North China's water security. Because the normal water supply of the North China region is relatively small, not even the proposed South–North Water Transfer Project will remedy that. Therefore, the cornerstone of overcoming North

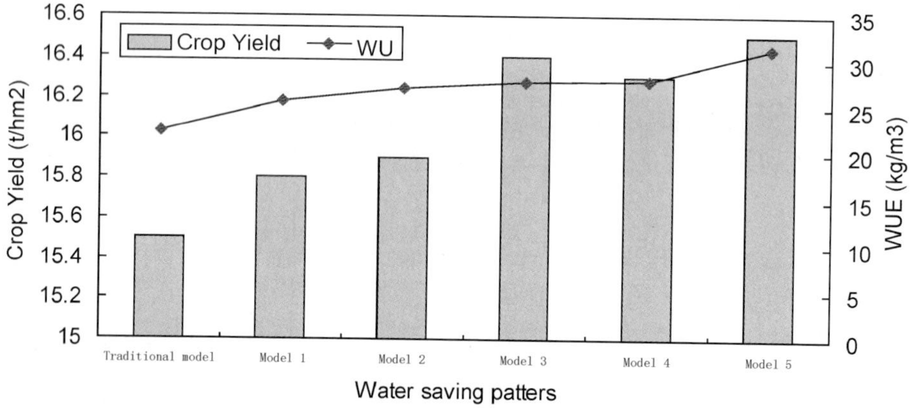

Fig. 6 Comparison of crop yields and WUE (water use efficiencies) for different irrigation regimes (Zhang *et al.*, 2005)

China's water security problem lies in sophisticated water management techniques and in finding new non-conventional water sources (Zhang *et al.*, 1992).

4. Challenges and opportunities

Water resources underlie the basic need of people's livelihood and the national welfare; without water, there can be no economic development. Understanding the hydrological cycle will provide a scientific basis for efficient water resources management and allow us to deal with temporal and spatial variations in water resources.

For a long time, conflicts existed between high-intensity human activity, economic development, and the environment. From a strategic point of view, the principle of sustainable development provides a potential avenue for solving these conflicts. Insight into interactions between the environment and industrialisation may provide important clues into how sustainable development patterns for North China can be achieved. In the next decade, we anticipate progress in the following areas: (a) understanding of the hydrological cycle; (b) allocation of water between the environment and socio-economic uses; (c) integrated catchment modelling.

As a rapidly developing region, North China plays a vital role in China's economic and social well-being. Of the serious environmental problems facing the region, water security has become a key issue in maintaining sustainable development. This presents challenges for both hydrologists and policy makers. The complex nature of the problems requires lateral thinking and multi-disciplinary research, including social and economic research. It is encouraging that both the central and local governments are committed to improving living standards in the region while protecting its environment.

To achieve the goal of economic development while improving environmental quality, in our estimation a research agenda along the following lines rates highly.

4.1. Studies of catchment-scale ecohydrological processes

Ecohydrology is an import component of catchmnet hydrological processes. Vegetation cover is closely related to water availability, especially in the arid and semi-arid region of northern China. Water is the limiting factor in vegetation growth. At present, drying up of rivers in the region is the direct result of the water conservation works in its catchments. Vegetation can influence water yield and hence affect water security (Zhang *et al.*, 2001). However, there is limited research on the interactions between hydrological and ecohydological processes in North China. We need to better understand the interactions in the soil - vegetation - atmosphere system; we also need to be able to quantify catchment-scale evapotranspiration and its change under future climate and land use change scenarios. Study of catchment-scale eco-hydrological processes will improve our understanding of vegetation - water relationships in various climatic zones and provide a scientific basis for land use change planning.

4.2. Water use efficiency and alternative water resources

North China is a region with limited water resources. As the local economy rapidly develops, water shortages are becoming critical. Among the various sectors, agriculture is the biggest water consumer and its water use efficiency is currently low. This issue is an important one in developing a national long-term science and technology plan. The key is to improve agricultural water use efficiency through advanced irrigation techniques and more efficient use of rain. It is perhaps equally important to consider alternative water resources such as sewage and seawater. This will require highly efficient technologies for processing seawater if users are to have access to an affordable water supply. There are successful examples of using sewage water for irrigation; however, to avoid soil and groundwater pollution, extreme caution is needed.

4.3. Integrated catchment management

The catchment is the basic unit of the natural water cycle; it is also the basic system in terms of water collection and use. Protecting water quality and rationally allocating it to users are issues that must also be decided on a regional scale. This is where integrated catchment management plays an important role, and this approach is the focal point of much current international hydrological research. Here, science is attempting to provide a rational basis for gauging catchment responses to human activities, regulations, and climate change. Research can help through investigating catchment hydro-ecological processes, physical aspects of sedimentation in the water cycle, the chemistry of point and non-point pollution sources, and the mechanism and control of soil erosion and transport processes.

Catchment hydrological models provide a tool for simulating and analysing catchments. The starting point is an acknowledgement that water resources management in China greatly lags behind that of developed countries. Water shortages, water pollution, flood damage, and soil and water erosion are still very serious, particularly in the Northern part of China. Integrated catchment management provides a valuable way of exploring options – how to make water resources sustainable, how to assure water supply, and how to reduce erosion. It offers an efficient approach to solving water problems in China.

Integrated catchment management models can analyse various scenarios of water demand, and provide a scientific basis on which the government can make decisions to overcome water problems. The current main research effort is along the following lines. First, the catchment distributed hydrological models (see Section 3.2) are being used to look at the effect of climate change and land use/cover changes. Simulation models are also being applied to the interaction between hydrology, ecology, and the aquatic environment; an important aspect here is the ability to consider non-point sources of pollution. Integrated catchment management approaches can investigate the optimum way of dividing a given water resource between industry, households, and the environment.

Simulation models are also being used to look at the relevant factors affecting a decision to develop future water resources, and how best to use that water. They can forecast the hydrological effects of land use change, particularly urbanisation, and the quantity and quality of water resources. For example, what is the impact of a structural readjustment in a region's industrial mix and layout, and what is the effect of a water price policy on water demand? These models can provide important background information to decision-makers.

'Clean and green' is an increasingly favoured approach internationally, and the predicted outcomes of management measures – such as clean production and the growing implementation of water-saving techniques – are of great value to the administrator in forecasting water demand. Alternative sources of water – brackish water, seawater, recycled sewage, and the like – are attracting considerable interest, and assessing their potential to overcome water shortages has become a major research theme (Xia *et al.*, 2004.).

4.4. Legislation and market-based approaches for water resources management

Management of water resources requires not only understanding of the biophysical processes involved but also of social and economic constraints. In planning regional development, it is important to consider water security as it will significantly affect the final outcome. North China is a water-limited region and it is unsustainable to develop industries with high water consumption. Currently, there is no system-wide approach for estimating water consumption and the associated economic returns from various industries. This calls for research on water requirements and consumption of various industrial, domestic, and ecological systems. To achieve the goal of sustainable development, it is essential to introduce a market-based approach to water resources management. This may include appropriate water allocations and entitlements for different industries and sectors; water trading should be encouraged as it tends to relieve pressure on limited water resources. While the market-based approach can work in managing domestic, industrial, and agricultural water consumption, the responsibility of securing environmental water requirements lies with the government, and it needs to be regulated through legislation.

5. Conclusion

Despite the daunting problems facing North China, there are a number of different avenues where research has contributed towards improving the situation. In particular, we have highlighted the role of field experiments in improving water use efficiencies and of computer models that offer insight into catchment-scale hydrological processes. Water security continues to remain the outstanding problem, however, and no one solution is likely to overcome it. It will require a concerted effort from scientists from a range of disciplines and catchment manager before the threat of severe water shortages – and the associated threat to the environment – can be alleviated.

Acknowledgements The work described in this paper was supported by the Knowledge Innovation Project of CAS (KZCX2-SW-317/CX10G-E01-08) & Natural Sciences Foundation of China (50279049 /50239050). Lu Zhang would like to thank the Chinese Academy of Sciences for the Outstanding Overseas Chinese Scholars Fund.

References

Chen SY, Zhang XY, Hu CS, Pei D, Liu MY (2004) Comprehensive water-saving models in high yielding region of Hebei Plain. Chin J Eco-Agric 12(1):148–151(in Chinese)

Chen JQ (1994) Water resources in China. In: Qian ZY (ed) Water resources development in China. China Water and Power Press (in Chinese), Beijing, pp 1–42

Chen JQ, Xia J (1999) Facing the challenge: barriers to sustainable water resources development in China. Hydrol Sci J 44(4):507–516

Chen XD, Li XM, Lu GQ (1999) Drying up courses of lower Huanghe river and variable characteristic of water resources, research and practice on sustainable management of water resources. Wuhan Mapping University of Science and Technology Press, pp 23–28

Chen ZK (1985) China's water resources and its utilization. GeoJournal 10(2):167–171

Chen ZK (2002) The serious impact of the lasting drought and water shortage on eco-environment of the North China Plain, In; Proceedings of the sixth academician conference of chinese academy of engineering (in Chinese), pp 90–97

Chilima J, Huang CY, Wu CF (2002) Maize (*Zea mays* L.) and cotton (*Gossypium hirsutum* L.) straw decomposition in soil: effect of straw placement, mineral nitrogen and tillage. Pedosphere 12(4):353–364

Falkenmark M (1997) Society's interaction with the water cycle: a conceptual framework for a more holistic approach. Hydrol Sci J 42(4):451–466

Huang GH, Xia J (2001) Barriers to sustainable water quality management. J Environ Manage 61(1):1–23

Jia SF, Zhang SF (2003) Water resources security appraisement of Hai basin. Prog Geogr 22(4):379–387 (in Chinese)

Kundzewicz ZW (1997) Water resources for sustainable development. Hydrol Sci J 42(4):467–480
Liu CM, Cheng L (2000) Analysis on runoff series with special reference to drying up courses of lower Huanghe River. Acta Geographica Sinica 55(3):257–265 (in Chinese)
Liu CM, He XW (1996) China's water problems and strategies in the 21st century (in Chinese). Science Press, Beijing, China, p 199
Liu CM, Yu JJ (2001) Groundwater exploitation and its impact on the environment in the north china plain. Water Int 26(2):265–272
Liu CM, Wei ZY (1989) Agricultural hydrology and water resources in the north china plain. Science Press (in Chinese), Beijing, China, p 236
Loucks DP (1997) Quantifying trends in system sustainability. Hydrol Sci J 42(4):513–530
Luo Y, Yu Q, Ou YZ, Tang DY, Xie XQ (2000) The evaluation of water uptake models by using precise field observation data. J Hydraulic Eng 4:73–80 (in Chinese)
MWR (Ministry of Water Resources) (1987) Water resources assessment for China (in Chinese). Chinese Water and Power Press, Beijing, China, pp 15–46
MWR (Ministry of Water Resources) (1997) Quality assessment of water resources for China (in Chinese). China Science and Technology Press, Beijing, China, pp 18–42
Plate EJ (1993) Sustainable development of water resources: a challenge to science and engineering. Water Int 18(2):84–93
Qian ZY (1991) Water resources in China (in Chinese). Chinese Water and Power Press, Beijing, China, pp 1–3
Qian ZY (1992) Water resources development in China: history, status quo and prospect. Hehai University Press, Nanjing (in Chinese), pp 2–3
Qian ZY, Zhang GD (2001) Strategic studies of water resources for sustainable development in China, vol. 1 (in Chinese). Chinese Water Resources and Hydropower Press, Beijing, China, pp 1–10
Rodda J (1995) Whither world water? Water Resour Bull 31(2):1–7
Simonovic SP (1996) Decision support systems for sustainable management of water resources: 1, general principles. Water Int 21(4):223–232
SPC (State Planning Commission) (1995) Report on China's population, resources and environment. China Environmental Sciences Press, Beijing, China (in Chinese), p 180
Subhani A, Liao M, Huang CY, Xie ZM (2000) Effects of some management practices on electron transport system (ETS) activity in paddy soil. Pedosphere 10(3):257–264
Sun HY, Liu CM, Zhang YQ (2003) Effects of water stress in different growth stage on water consumption and yield in winter wheat. J Irrigation Drainage 22(2):13–16 (in Chinese)
Wang GS, Xia J, Tan G, Lv AF (2002) A research on Distributed Time Variant Gain Model: a case study on Chaohe river basin. Prog Geogr 21(6):573–582 (in Chinese)
Wang HX, Zhang L, Dawes WR, Liu CM (2001) Improving water efficiency of irrigated crops in the North China Plain – measurements and modelling. Agric Water Manage 48:151–167
Wang XY, Xie HT, Liang WJ, Wen DZ (2004) Rice yield and water use as affected by soil management practices. Pedosphere 14(3):331–337
Wang ZM, Ren XS, Guo HY (2000) Hai water resources facing the 21st century. Tianjin Science and Technology Press, Tianjin (in Chinese), China
World Bank (1997) Clear water and blue skies: China's environment in the new century. World Bank, Washington DC, USA, p 114
WWC (World Water Council) (2000) World Water Vision 2025. Earthscan Publications Ltd., London, United Kingdom, p 108
Xia J (1997) Advances in management of sustainable water resources system. Adv Water Sci 8(4):370–376 (in Chinese)
Xia J (2002) Hydrologic nonlinear system theory and approach. Wuhan University Press (in Chinese), pp 8–16
Xia J, David Chen YQ (2001) Water problems and opportunities in hydrological sciences in China. Hydrol Sci J 46(6):907–922
Xia J, Tackeuchi K (1999) Barriers to sustainable management of water quantity and quality. Guest editors for Special Issue. Hydrol Sci J 44 (4):503–505
Xia J, Tan G (2002) Hydrological science towards global change: progress and challenge. Resour Sci 24(3):1–7 (in Chinese)
Xia J, Liu MY, Jia SF, Song XF, Luo Y, Zhang SF (2004) Water security problem and research perspective in North China. J Nat Resour 19(5):550–560 (in Chinese)
Xia J, Wang GS, Tan G (2003) A distributed hydrological model applied to Heihe mountainous basin in western China. IAHS Publication no. 282, pp 268–274

Xia J, Wang GS, Tan G, Ye AZ, Huang GH (2005) Development of distributed time-variant gain model for nonlinear hydrological system. Sciences in china, Ser D Earth Sci 48(6):713–723

Xia J, Zuo QT, Pang JW (2001) Enlightenment on sustainable management of water resources from past practices for the Bositeng Lake basin in Xinjiang, China, IAHS Pub. No. 268 (Regional Management of Water Resources), UK, pp 41–48

Zhang GL (1999) Water supply and demand in china in the 21st century (in Chinese). Chinese Water Resources and Hydropower Press, Beijing, China, pp 1–4

Zhang L, Dawes WR, Walker GR (2001) The response of mean annual evapotranspiration to vegetation changes at catchment scale. Water Resour Res 37:701–708

Zhang QS, Zhang X (1995) Water issues and sustainable social development in China. Water Int 20:122–128

Zhang ZZ, Chen BX, Chen ZK (1992) Challenges to and opportunities for developments of China's water resources in the 21st century. Water Int 17:21–27

Zhang XY, Che SY, Liu MY, Pei D, Sun HY (2005) Improved water use efficiency associated with cultivars and agronomic management in the North China plain. Agron J 97:783–790

Zuo DK (1985) Proceedings of water balance and south to North water transferring in North China plain (in Chinese). Science Press, Beijing, p 303

Water Resour Manage (2007) 21:249–267
DOI 10.1007/s11269-006-9052-0

ORIGINAL ARTICLE

Towards transition management of European water resources

Rutger van der Brugge · Jan Rotmans

Received: 20 April 2006

Abstract Global change fundamentally changes the nature of water-related problems. We will illustrate this by showing how perceptions of the water-problems in the Netherlands have shifted in the past four decades. The nature of water-related problems changed from a technical problem to a so-called 'persistent' problem, characterized by plurality, uncertainty and complexity. Although integrated water resource management (IWRM) has been advocated to cope with this type of problem, the complexity of the transition process towards such a water management regime is often underestimated. Therefore, transition management is needed in the water sector. Transition management theory is presented and applied to the Dutch case. Transition management strategies are suggested that would reinforce this transition. Comparison between the European Water Framework Directive (WFD) and transition management indicates that the Common Implementation Strategy (CIS) in its current form is not sufficiently stimulating an innovation climate.

Keywords Transition management · Integrated water resource management (IRWM) · European Water Framework Directive (WFD)

Introduction

Our world is rapidly changing. Human induced climate change is expected to have profound consequences on large parts of the world. Especially the impacts on large aquatic systems – upon which many people depend - pose a considerable threat to current and future generations. However, due to the high complexity of this problem, that is the many interactions between atmospheric processes, hydrological processes and ecosystem processes, it is impossible to calculate local impacts accurately. Bearing this in mind, water managers are now analyzing the possibilities to create adaptive water systems that are more resilient to extreme impacts.

R. Van der Brugge · J. Rotmans
D.R.I.F.T – Dutch Research Institute For Transitions, Erasmus University Rotterdam, Faculty of Social Sciences, P.O. Box 1738, 3000 DR Rotterdam, The Netherlands
e-mails: VanderBrugge@fsw.eur.nl; Rotmans@fsw.eur.nl

Global change not only affects the water management strategies, but also changes the nature of water-related problems. The interconnectedness of different water functions reflects the increasing complexity of our modern society. Economic functions for navigation, agriculture and energy supply, ecological functions for sustaining ecosystems, and social functions in terms of safety and drinking water supplies are without exception of critical importance to our modern society. Reinforcing one particular function may have adverse effects on others as a result of (often unknown) feedbacks. Policy-makers, as well as water managers cannot afford to be ignorant about societal functions 'outside their own box'. Moreover, different types of values that can be attributed to water complicate the decision-making process itself. How does one attribute weights to the economic values of water (utility value of water by using pricing mechanisms) the ecological values (water regulation services for ecosystem sustainability) and the cultural values and emotional meaning of water? Modern water management is thus confronted with a complex, uncertain and pluralistic problem.

Integrated water resource management (IWRM) has been advocated to cope with this type of problem. In this respect, the European Union Water Framework Directive (2000/60/EC) is considered to be an important landmark. The Water Framework Directive (WFD) has been put forward as a legislative framework to guarantee the 'good quality of all waters in Europe'. Besides the scale to which it applies, which is unprecedented (and that alone makes it worthwhile to analyze its implementation), the Directive introduced two important new aspects: the river basin approach and stakeholder involvement. The first aspect is of course a prerequisite for the integration of water management strategies up- and downstream. The latter addresses the plurality of objectives and values by involving stakeholders in the river basin management plan. It thereby addresses and confronts many of the difficulties of integrated trans-boundary water management. The Directive thus is a strong impetus for integration and cooperation in European water management. However, one of the weak points is that the WFD does not address water quantity issues, which in terms of climate change is indeed a very serious omission.

Although the WFD thus addresses important first steps in the transition process, it does not sufficiently address the transition process itself. Water quantity issues ask for even more fundamental rethink about water management because these issues presuppose a very tight relation with spatial planning, e.g. spatial planning policies have to be incorporated in the river basin management plan. However, conditions have to be created in which this can be done and in which innovations can be developed. For example, in the Netherlands one clearly sees this process unfolding. Having a large coastal zone and embracing the river delta of two large European rivers, its water managers are particularly forced to make adaptations in the water system in order to deal with sea-level rise and extreme river discharges as a result of climate change. Much effort goes into the institutional integration of the policy fields of spatial planning and water management, but the actual implementation is very difficult.

Even though we focus in this article on the Dutch case, understanding the particular case of the Netherlands may provide insights into the general nature of the transition processes in the EU-Member States and European water management as a whole. We also try to present a more general analytical framework based on the theory of transitions and transition management (Rotmans *et al.*, 2000) and show that the problems that now arise in the river basins are of a more general nature, inherently related to transition processes. We will apply the basic principles underlying transition management to EU water management and from this perspective reflect on the WFD and explore to what extent these principles can be used within the context of the WFD.

The nature of the Dutch water problems

Although approximately 60% of the Dutch soil is beneath sea level, an intricate web of dams, channels, pumps and polders keeps the inhabitants' feet dry. However, full attention is required twenty-four-seven. Nonetheless, the water problems are manifold. In 1993 and 1995 it shocked the nation when the rivers Meuse and Rhine almost flooded and thousands of people had to be evacuated. After this major but rather surprising incident more problems followed, amongst others large agricultural damages in the western part of Holland due to high water levels in 1998 and the 2003 flooding in Wilnis. Besides water abundance, other problems related to water have been drought, industrial cooling water and water pollution. Although, these water-related problems may seem to be singular, they are in fact interrelated and connected.

These water-related problems are specific manifestations of a deeper-lying, more fundamental problem. The growing economic development, increasing population and changing life-styles of the 20th century have led to pressures on the water system through increasing spatial claims from agriculture, industry, traffic, housing and infrastructure. In the attempt to meet these increasing societal demands, water engineers fabricated a skillfully branched water system consisting of canals, dikes and polders. The human controlled water system resulted in altered hydrological cycles and morphological changes. Its slow development has put increasing pressure on space, hydrology, morphology, subsidence of soil and capacity to retain water. Although these interventions were absolutely necessary, the negative consequences that stayed concealed for a long time are now surfacing. Unless these pressures are released, in the future both the magnitude and frequency of water related problems will increase.

Viewed from this angle, the water problems are local manifestations with similar underlying (autonomous) driving forces. They are so-called *persistent problems* and can be characterized by on the one hand the complexity of the interactions of broad societal trends and physical (natural) processes (such as climate change), which gives rise to structural uncertainty and on the other hand by the involvement of many stakeholders with different but plausible perspectives, which leads to problems of management and governance (Dirven *et al.*, 2002). The persistence itself is caused by the strong interconnectivity between water institutions, management structures, routines and techniques. This results in a tight and well-organized water management regime, involving policy departments, regional water management boards, engineering offices, scholars and jurists. Due to its interdependencies and internal logic regime structures are often difficult to change and radical innovations have trouble breaking through.

What does this mean for modern water management practice? First, it means that society has to reconsider the way water functions are weighted and prioritized. Secondly, water management decision-making should be based on balanced integration frameworks and address different scale levels both in space and time. Thirdly, the current institutional arrangement, which is a barrier for the first two, needs to be changed. Transition management is one of the approaches proposed in the literature to deal with persistent problems. In the following sections we will present transition theory and apply it to the observed shift in Dutch water management (Van der Brugge *et al.*, 2005).

Transition theory

Rotmans *et al.* (2000) have introduced the concepts of transition and transition management as new integrative approach in the field of sustainability and governance in order to deal

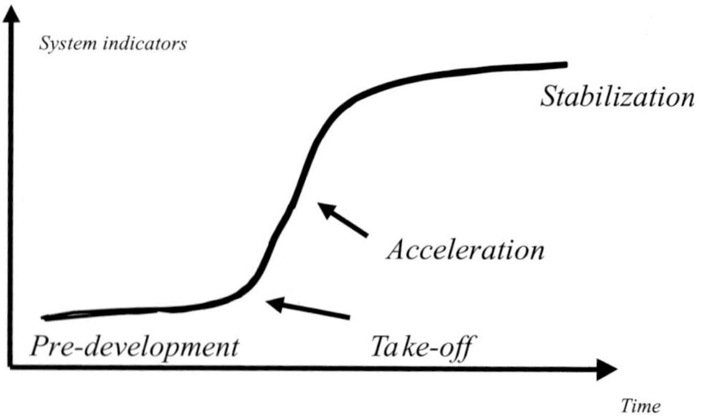

Fig. 1 A transition is the shift between two dynamic equilibria that can be described by a set of system indicators. In the transition process, four phases can be distinguished. In the predevelopment these indicators change only marginally. In the take-off and acceleration phase the indicators change with increasing speed. In the stabilization a new equilibrium is reached (Rotmans *et al.*, 2002)

with persistent problems. The transition framework offers analytical tools for structuring and explaining the dynamic behavior of societal systems, such as the transport sector, energy supply and agriculture, or water management. Transition management attempts to influence, facilitate, stimulate and organize processes that contribute to the transition. They define a transition as 'a continuous process of societal change, whereby the structure of society (or a subsystem of society) fundamentally changes and has the following characteristics:

- It concerns large scale technological, economical, ecological, socio-cultural and institutional developments that influence and reinforce each other;
- It is a long term process that covers at least one generation (25 years);
- There are interactions between different scale levels (niche, regime, landscape).

The transition framework is rooted in complex adaptive systems (CAS) theory (Holland, 1995; Kauffman, 1995; Prigogine and Stengers, 1984) post-normal science (Ravetz, 1999), integrating concepts from governance (Sabatier and Jenkins-Smith, 1999), evolutionary economics (Arthur, 1988; Nelson and Winter, 1982), innovation studies (Smits and Kuhlmann, 2004) and technological transitions (Geels and Kemp, 2000). Transitions are often illustrated as S-shaped curves (Figure 1). Although this is a very simple aggregated curve, the underlying transition dynamics are complex interaction processes between markets, networks, institutions, technologies, policies, individual behavior and autonomous trends in the economic, ecological, socio-cultural and institutional domain. From a CAS perspective, transitions are system transformations between two temporal (dynamic) equilibrium states (attractors). In between there is a period of rapid change during which the system undergoes irreversible re-organization (Rotmans, 1994).

The transition framework comprises three analytical tools in order to recognize and explain transition patterns. These are:

(1) *Multi-phase concept*: a transition is a sequence of the following four phases: predevelopment, take-off, acceleration and stabilization (Figure 1).
(2) *Multi-level concept*: a transition is the result of interacting developments at macro-, meso- and micro-level (Figure 2).

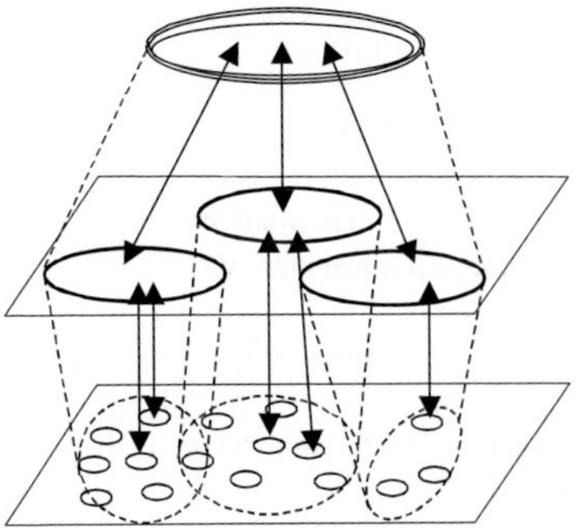

Fig. 2 Multi-level concept is based on (Geels and Kemp, 2000). Developments at the macro-level correspond to slow broad societal trends. Dynamics at the meso-level are determined by the regime. The regime is the dominant pattern of actors, artifacts and structures in the social system. At the micro-level, individual persons, organizations, or innovations are distinguished

(3) *Multi-change concept*: during a transition new structures emerge and existing structures are broken down.

According to Rotmans *et al.* (2000) the general pattern of evolution during the four phases is the following. In the pre-development phase, the system dynamics do not change visibly but stocks are slowly changing. In the take-off phase, the structure of the system begins to change as the result of (1) the emergence of innovations and (2) destabilization of the existing regime. In the acceleration phase, structural transformation of the system takes place. New patterns of system dynamics emerge as a result of accumulation of socio-cultural, economic, ecological and institutional changes innovations that reinforce each other. In the stabilization phase the new pattern of system dynamics reaches a new dynamic equilibrium.

Transitions only unfold when developments at the macro level, meso-level and micro-level 'move into the same direction' (Geels and Kemp, 2000; Rotmans *et al.*, 2000). Geels and Kemp have developed this multi-level perspective to describe and explain transitions in large technological systems and later applied it to sustainability transitions (Figure 2). At the macro-level the *'landscape'* is determined by changes in the macro-economy, politics, population dynamics, natural environment, culture and world views. This level responds to relative slow trends and large-scale developments that canalize the evolution of a particular societal system, although catastrophic events, such as tsunamis or terrorist attacks can almost instantly change (parts of) the (political) landscape. At the meso-level there are patterns of institutions, artifacts, rules and norms assembled and maintained to perform economic and social activities which is referred to as the '*regime*' (Berkhout, 2003) At the micro-level there are so-called *'niches'* of individual actors, alternative technologies and local practices that deviate from the status quo manifesting themselves as new ideas, initiatives or innovative techniques (Kemp *et al.*, 1998).

Complex adaptive systems

The typical division between (a) dominant structure at the middle level, (b) alternatives at smaller level and (c) long-term trends at the higher-level has proven to be a reasonable framework to describe transformation of the regime structure as a result of bottom-up innovations and changing landscape factors in retrospect. The regime can be seen as a complex adaptive system (Table 1), viewing it as an island of relative stability embedded in a changing landscape and not always capable of adapting due to its interdependencies between its actors and artifacts. Regime dynamics are very much dominated by strong internal relations and relatively weak external relations. Its internal dynamics are deeply embedded in social interactions

Table 1 Properties of complex adaptive systems (CAS), based on (Prigogine and Stengers, 1984)(Holland, 1995; Holling, 1987; Kauffman, 1995)

Properties of complex adaptive systems
– Many and diverse components and interactions
– Components are organized in a network configuration
– The system is open (exchange of matter, energy and information with external environment)
– Non-linearity
– Positive and negative feedback loops (reinforcing and dampening mechanisms)
– Nested organizational levels
– Multiple attractors (relative stable but dynamic equilibrium states) co-exist
– Attractors have stability domains, bounded by thresholds
– Components are able to learn and respond to the environment by changing behavior (interactions)
– Co-evolutionary interaction patterns may lead to irreversible pathways
– Higher level structures spring into being as result of lower level component interaction

These properties apply to social systems, leading to the conclusions that social systems are complex adaptive systems and that the behavior of this category of systems may have general features

and are consequently difficult to change. New agents trigger change by developing their own niche. Through increasing *network externalities* these niches become increasingly attractive to other agents (Arthur, 1988). Niches can break trough when macro socio-environmental conditions are favorable.

Typical of CAS is *pathdepedency*. When formed, the regime is able to maintain and reproduce its internal dynamics. Only when certain thresholds are exceeded, the equilibrium structures break and the adaptive system transforms. (Gersick, 1991) argues that disruption of the equilibrium has two sources: (1) internal changes that pull parts and actions out of alignment with each other or the environment, and (2) environmental changes that threaten the system's ability to obtain resources. Generally speaking, transitions are the result of two mutually reinforcing mechanisms: (1) destabilization of the regime on the one hand and (2) the emergence and up-scaling of innovations on the other. When both are absent the system is in equilibrium. In the acceleration phase they mutually reinforce each other strongly, leading to very rapid dynamics. Macro socio-environmental conditions can affect the mechanisms both ways. Describing transition of complex adaptive systems in terms of phases and multilevel interactions patterns yields the following dynamic pattern (Figure 3):

Phase 1. Predevelopment → approaching 'criticality': During the predevelopment phase, co-evolutionary regime dynamics increase regime interdependencies and as a result, the regime organization approaches criticality. Resilience decreases and the regime structure becomes increasingly vulnerable. Changing socio-environmental conditions impose stress on the regime structure, demanding efforts from actors in performing their functions in the supply-chain, policy or regulation domain. Innovations are still isolated and fragmented, improperly embedded and insufficiently developed enough to compete with the existing regime.

Phase 2. take off → triggering change and Build up of new regime: During the take off phase, the regime grows 'critical' and innovations start acting as perturbation of the status quo, triggering large scale change. As the system becomes increasingly 'critical', calamities affect all system domains due to high interconnectivity. In non-critical systems, calamities only have local effects. The same holds for innovations, which explains why is it so hard for them to break trough if the system is near criticality. Regime structures must first open up before innovations have a chance of penetrating the system. Regime dynamics collapse

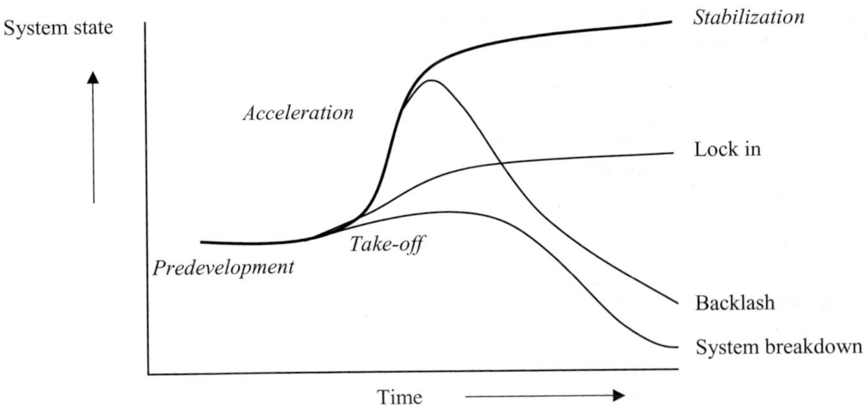

Fig. 3 Possible system pathways of a complex adaptive societal system. The transition is the desired pathway in achieving sustainable development. However, the complexity of the interaction processes limits control over societal developments which may lead to less desired pathways, such as the lock-in, the backlash or the system breakdown

when the systems key functions fall out. This may be the case when existing actors do not perform adequately. Simultaneously, there is build-up of innovation networks based on alternative ideas, concepts, theories and technology, which may lead to alternative behavior. Innovations must reinforce each other and align into innovation networks, which is a perquisite for behavioral change. Developing the network requires access to resources, like for example money and knowledge (e.g. the societal equivalents of nutrients and energy needed by biological systems to maintain homeostatic organization) in order to develop its assets. If the innovation network succeeds in becoming self-sustaining, in terms of maintaining and reproducing its own organization[1] it has survival advantages over innovations that do not cluster into self-sustaining networks.

Depending on co-evolutionary developments in the regime on the one hand, and the 'survival fitness' of the available attractors on the other, there can be three pathways (1) there remains a co-existence of more competing innovation networks (lock-in), (2) there is only one innovation network, which is reinforced by smaller innovation networks enabling it to grow (acceleration), or (3) there is a more or less chaotic world in which innovation networks are all insufficient in becoming self-sustaining and keep competing for the same resources. As a result, there is no adequate substitute for the destabilizing regime (system breakdown).

Phase 3. acceleration → Cascading effects: During the acceleration it becomes clear which innovation networks are 'fittest' and consequently are selected. Fitness is a relative term, referring to the survival chances of the innovation network in a co-evolving environment. Thus while the innovation networks compete for resources with each other, they simultaneously are influenced by destabilizing regime developments. Co-evolution in the end determines the outcome as a result of recursive developments between the macro developments, regime developments and innovations at the micro-level. During this phase,

[1] This capacity of biological systems to be able of reproduction their own organizations is referred to as *autopoiesis* (Varela *et al.*, 1974). H. Maturana later applied this to social systems as did sociologist Niklas Luhmann (Luhmann, 1984).

the 'old' regime transforms and consequently the 'selection environment' changes and is primarily based upon selection rules from the up-scaled innovation network.

Phase 4. → *Stabilization*: During the stabilization the new regime settles down. Transformation processes turn into optimization processes. Regime dynamics now are equilibrium dynamics in order to enhance efficiency. If this equilibrium is not reached, the new regime can still breakdown (backlash).

Transition dynamics in Dutch water management

The transition heuristic can be applied to the shift in Dutch water management from a scientific technocratic water management regime (Bosch and Ham, 1998; Ham, 1999; Lintsen, 2002) towards the integrated water management regime. Integrated water management (Saeijs, 1991) perceives water as part of a larger system, integrating engineering and ecological aspects into a broader societal system. This changing scope of water management is strongly reflected in subsequent National Policy Memoranda on Water Management (Rijkswaterstaat, 1964, 1982, 1989, 1998). Picking up the early signs, this change started around 1965 with the emergence of an environmental movement (Van der Brugge *et al.*, 2005). Since then, Dutch water management changed in terms of prioritizing water functions as well as in terms of safety strategies and since then, ecological functions of water became increasingly important, while the traditionally influential role of agriculture declined. Also, water managers now tend to broaden the riverbed and prohibit housing or economic activity in the river floodplains, instead of constructing dykes as a means to protect the people and investments. Slowly national water policy measures encouraged the integration of spatial planning and water management. Today, water is postulated a 'guiding principle for spatial planning'. As opposed to the 1960's, the properties of the water system are now assumed to be guiding socio-economic activity. Essentially, this means 180° turn in the relationship between water management and spatial panning. This clear fundamental break with tradition has, however, not yet reached a new equilibrium (Van der Brugge *et al.*, 2005). Applying the transition heuristic to the evolution of water management provides insight in which phase the transition is currently.

The predevelopment phase

Traditionally, water-related problems were being solved using technological means. Being an engineer, the core competences of the water manager were the construction of water infrastructure and calculating water flows. The management style was enabled through a scientific and technocratic *regime* that reigned throughout the 20th century (Bosch and Ham, 1998; Ham, 1999; Lintsen, 2002). This equilibrium can be overly simply sketched as the 'Water will follow' attractor.[2] The system dynamics were that of growing economic development, increasing population density and changing life-styles that were leading to increasing spatial claims of agriculture, industry, traffic, housing and infrastructure. In the attempt to meet these societal needs, water managers drained redundant water, canalized rivers and constructed dams, e.g. water 'followed' societal needs. The significant interventions in the water system could only be managed, maintained and controlled by continuing its technological

[2] Although the term 'attractor' has a specific meaning in systems theory, being the endpoint of a systems evolutionary trajectory (often illustrated in phase space), its meaning here is the range of management actions allowed by the regime. The water manager acts while it is constrained and at the same time enabled through its regime structures, leaving a limited set of choice options than are theoretically possible. In effect, locking the system in a subset of phase space.

mode of operating, manoeuvring itself onto this technological trajectory. As such, problems were being perceived as singular technological problems and technological solutions did not solve the problem on a more fundamental level, namely removing the pressures upon the system. Due to this path dependent management style, the financial costs of problems such as floods, droughts and diminishing water quality increased. As efficiency and cooperation initially countered these pressures and more complex governance structures came into existence trying to couple multiple scales, the water management system was approaching its spatial and economic thresholds.

Take-off → triggering change

After the 1993 and 1995 floods of the rivers Meuse and Rhine, an increasing number of people started to understand that the 'Water will follow' attractor had resulted in an unsustainable water system and that the problems were symptoms of the system exceeding its thresholds. Followed by high regional water levels in 1998 the perspective about water management started to change. The Committee 'Water management for the 21st century' claimed that 'Dutch water management was not sufficiently prepared to meet the challenges of climate change effects in the next century' (EC). The continuous subsidence of soil, the rising sea level and the decreasing capacity to retain water due to loss of nature would cause serious problems. The committee expected climate change to contribute to extreme events and higher discharge levels during winter while lower during summer. On the whole, strategists started to recognize the inherent problems of the current system and to acknowledge the spatial claim of water itself. They considered the way towards a more adaptive water system in order to anticipate future threats. The Committee proposed a new water management strategy based on two starting points: (1) water had to be guiding in spatial planning, and (2) water had to be retained, stored and drained when necessary.

Take-off → build up of the new regime

Many events, plans and developments have contributed to the visionary report of the Committee 'Water management for the 21st century'. Its foundations can be traced back to the late 1980's (Van der Brugge *et al.*, 2005). In a way, from there on, its history can be summarized as increasingly integrating different functions, with a strong emphasis on ecology and later on spatial planning. So, what is interesting here is the way in which these new discourses penetrated the water management practice.

The emergence of ecological orientation

In order to prevent a crisis such as the 1953 storm flood, Rijkswaterstaat (the ministry of water management) started an ambitious water defense program, the Delta Works. The Delta Works are a set of dams that have to protect the Dutch from the sea. The large dams had profound adverse effects on nearby ecosystems. Salt water based ecosystems had evolved into fresh water ecosystems, leading to dramatic consequences in local biodiversity (Bosch and Ham, 1998). Therefore the Delta Dienst, (the formal institute responsible for the construction of the Delta Works) founded an environmental department concerned with ecological research headed by H. L. F. Saeijs. A biologist himself, he brought over a hundred fellow biologists into the engineering world of water management. The research activities performed by the Delta Dienst led to a number of restoration projects. In 1985, the policy memorandum 'Dealing with Water' (RIZA, 1985) reported important elements of the ecological approach in water management. It reached a wide audience, partly due to the ecological calamities evoked by the Delta Works. The (eco)systems approach advocated in this document represented a new perception proposing water as an integral part of an ecosystem in relation with its community (Saeijs, 1991)

Important reasons explaining why the ecological perspective resonated in the water management regime was the involvement of a growing number of biologists. During the construction of the Eastern Scheldt storm surge barrier, one of the most prestigious dams, the Environmental Department of the Delta Dienst had grown into a group of over one hundred biologists and confronted the regime with the consequences of its practices (Bosch and Van der Ham, 1998). This in fact was a strong destabilizing factor in the regime as were the protests of the environmental movement. In the 1980's re-organization of the Ministry in order to integrate water quantity and water quality policies, many former Delta Dienst biologists were placed on strategic positions. Cross-fertilization between biologists and water engineers 'infected' Rijkswaterstaat with new ideas.

Another impetus towards the integration of ecological considerations and water management was provided by the award-winning plan 'Ooievaar' (Bruijn *et al.*, 1987) in a contest that was called 'Netherlands – Riverland' (organized by the E.O. Wijers Institute). The contest invited participants to come up with ideas about future water management. 'Ooievaars' vision departed from decoupling agriculture and nature preservation, claiming that agriculture was damaging and not preserving ecosystems. 'Ooievaar' in this sense broke with prevailing beliefs and questioned the (traditional) influence of agricultural demand in water management. In short, 'Ooievaar' planned the removal of agricultural exploitation in the river flood plains and instead created the original meandering riverbanks, rich of plants and animals. A number of 'Ooievaar' based experiments were started with success in different regions, e.g. the Duursche Waarden, in Rhenen and the Gelderse Poort (Bosch and Ham, 1998). The minister was quite fond of the plan, informing the media she had an alternative for expensive dikes along the river.

The emerging link with spatial planning

It was not before the end of the 80's that serious and conscious attempts were made to integrate the two policy fields of spatial planning and water management. Meaningful in this respect is the WWF-plan '*Levende Rivieren* (WWF, 1992) 'Levende Rivieren' ('Living Rivers') elaborated from 'Ooievaar' with stronger focus on the aquatic ecosystem and its flora and fauna. Wanting to restore broken food chains, 'Levende Rivieren' proposed the introduction of smaller channels in the river flood plains and by doing this it showed an alternative to planned dike enhancements. Small channels and excavation of clay-layers in the river floodplains would create more room for water and could thus present an alternative safety strategy to dike enhancements. Prior to that, smaller groups within Rijkswaterstaat had also explored the possibilities of integrating water policy with spatial planning, one of them resulting in the report '*Omgaan met de Omgeving*' (Rijkswaterstaat, 1992). 'Omgaan met de Omgeving' ('Dealing with the Surrounding Area') initiated a number of interdepartmental meetings of top officials discussing the future of this path of integration.

Co-evolving mechanisms

The above-described innovative plans can be interpreted as the innovation networks which laid down the foundations of a new water management regime, having a strong focus on spatial integration of different water functions. Three important developments at the macro-level that reinforced the innovation networks in destabilizing the techno-scientific regime were (1) the emergence of the environmental movement creating awareness about environmental pollution, which has led towards a more ecologically oriented approach; (2) the perception of flood risk had drastically changed after the floods and what initially were merely ambitions to integrate water practices and spatial planning, now had grown rapidly into a sense of urgency; (3) decentralization and privatization since the 1980's resulted in a shift of power that weakened the hierarchical structure in the water sector. Two important regime developments that have led to the institutionalization of the ecological approach have been the reorganization of Rijkswaterstaat, which offered ecologists positions within the ministry and the merger of the regional water management boards. Larger management units allowed for integration between quantity and quality measures. Both initiatives have had serious effects on the institutional arrangements in the sector.

In summary (Tables 2 and 3), the innovation networks came into existence in a changing landscape in which two new discourses emerged: (1) the inherent relation between ecology and water management (water quality policy) and (2) between spatial planning and water management (water quantity policy), and at the same time enabling this landscape by transferring it into new approaches. Both discourses have been reinforced by crises, such as the ecosystem damage induced by the Delta Works, and the floods of 1993 and 1995. The current transformation dynamics in Dutch water management are foremost the attempts of innovating issues concerning water quantity.

Acceleration?

We demarcate the acceleration phase as the point in time when the selection rules for policy and implementation change according to the wishes of the emerging regime. Thus, there is a transfer of selection power from the traditional regime towards the new regime. Although this appears to be the case with the report 'Water management for the 21st century', in practice it is not. The logic and coherence of the concepts at the (strategic) abstract level is the strength of the report, however, in the real world there are many practical hurdles. The selections rules are still very much unclear. For instance, there is need of institutional renewal in terms of actors, responsibilities and tasks; there is need of technology and scientific knowledge, possibly new regulations regarding land ownership and insurance etc. Illustrative in this respect are the 'water test' which secures that water is taken into account in spatial planning policy, debates about designating flood areas and debates about merging regional water management with provincial layers of government. Debates are ongoing and although the outcome is not crystal clear, its direction might be. The strategy seems to be widely acknowledged, but the actual implementation is difficult and consequently the actual physical and institutional changes are issues of debate. The regime is in the middle of a process of re-configuration, which requires alternative modes of operating, cooperation and regulation. Hence the transition is still in the middle of the transformation process between the two regime attractors. The choices that are now being made are very much decisive for the trajectory of the transition.

Water transition management

Taking the complexity and uncertainty as a starting points, transitions require innovative approaches in order to manage them (Loorbach and Rotmans, In press; Rotmans *et al.*, 2001). Essentially, transition management stimulates transition processes by organizing the build-up of the societal structures needed to realize the new regime. Because the road is unclear, experimentation is essential in order to learn how. Results feedback into the vision, which may then

Table 2 Scheme of developments at three levels of scale (macro, meso, micro) that have influenced the system state of water management in the Netherlands over a time period (1975–2004)

System state 1975	– Events	System state 1985	– Events	System state 1995	– Events	System state '05
			Micro			
Supranational	– Growing environmental awareness – Economic growth – Limits to growth		– Rio Summit ('92)		– Johannesburg summit ('02) – Climate change – Sea level rise – EU Water framework Directive	
National	– Delta Works calamities (ecological impact)		– 1st National environmental policy plan – 1st Nature policy plan – Floods ('93, '95)		– National environmental policy plan 2, 3, 4 (NMP2-4) – Environmental management act	
			Meso			
– Technocratic water management	– Delta Works – 2nd National policy memorandum water management	– *Water system management*	– 3rd National Policy Memorandum water management ('89) – Re-organization Rijkswaterstaat	– *Integral water management*	– WB21 ('99) – 4th National policy memorandum water management ('98)	– *Integrated water resource management*
– Engineering approach	– Protests against water management approach	– Engineering approach	– Re-organization regional water boards	– Room for water	– Delta Plan Rivers	– Adaptation and retention
– Hierarchical organization (top-down)		– Hierarchical organization (top-down)	– Decentralization	– Democratic organization (Stakeholder participation)		– Participatory Policy process
Priorities: – *Safety* – *Agriculture*		Priorities: – Safety – Agriculture – Ecology		Priorities: – Safety – Nature development – Agriculture – Spatial planning		Priorities: –Safety – Spatial planning – Nature development – Agriculture
			Macro			
	– Environment dept. in Delta dienst				– Room for rivers ('95)	
	– Dealing with water ('85) – Restoration projects		Plan ooievaar ('87) Living rivers ('92) Dealing with the surrounding area ('92)		– Tackling flooding ('98)	

System states are described in terms of management concept, approach and priorities (1975,1985, 1995, 2004). Based on (Loorbach, 2004)

Table 3 Key aspects and differences between the water management style of the 21st century and the water management style that was dominant throughout the 20th century

Aspects of water management	1970's	2000
Problem perception	Singular	Interrelated
Management perspective	Problem solving	Anticipation
Scale	Local water problem	Water system structure,
Management style	Technological solutions	Spatial solutions
Strategy	Pumping, drainage, dikes	Retention, natural storage
Approach	Planning	Process
Competences	Disciplinary	Interdisciplinary
Staff	Engineers	Engineers, biologists, public managers, spatial planners
Institutional organization	Hierarchical, top down	Networks, participation

Also in this table, the differences are dichotomized for the sake of clarity. A more appropriate conceptualization would be continuous scales

have to be adapted. Experimentation is done in niches that are embedded in larger innovation networks. This coordinated search process must eventually lead to the up-scaling of the innovation networks. Applied to water management this would mean developing strategies that for instance enable the cooperation between spatial planners and water managers. Transition management is based on the following underlying management principles (Rotmans *et al.*, 2005)

1. The phase of the transition is guiding for the employing management strategies and instruments.
2. A mix of top-down steering, network steering and self-steering instruments should be used, depending on the transition dynamics at hand.
3. Multi-level governance is required in which the objectives and instruments vary at the different levels but have to be attuned to reinforce each other.
4. Stakeholders have to participate and to be aligned
5. Long-term goals must be adaptive to emergent innovations and macro-developments.
6. Timing and type of intervention is crucial. Non-equilibrium dynamics should be used to innovate.

A transition manager tries to combine content with process. He therefore must understand which structures the innovation networks have to be building up and how they can achieve this. It is therefore crucial that he understands transition patterns, e.g. which processes play at each level in each phase. Hence, transitions management is much more than implementing a strategic vision, it is a joint search and learning process directed at developing innovations and new arrangements that will start to reinforce each other. The transition manager thus coordinates multi actor process at strategic, tactical and operational levels. The operational model developed by Loorbach and Rotmans, (In press) has four activity clusters (Figure 4): (1) the establishment and development of a transition arena; (2) the creating of long-term integrated visions, transition pathways and agendas; (3) mobilizing actors and knowledge development through experimenting and (4) monitoring and evaluating the transition process.

The transition management approach starts with creating a so-called transition arena of 10–15 people (Loorbach and Rotmans, In press; Rotmans *et al.*, 2000). These participants are selected based upon their skills as being able to think across domains, being creative, having relevant knowledge of the field and having networking capabilities. Important is that

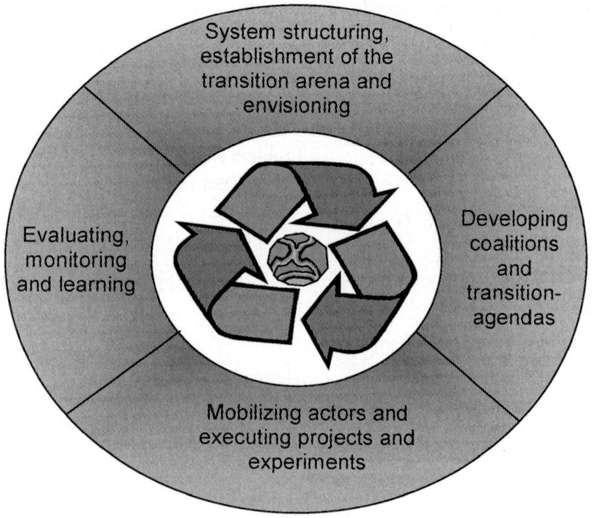

Fig. 4 Transition management is a cyclical coordinated multi-actor process at strategic, tactical and operational levels and is organized around four co-evolving activity clusters (1) the establishment and development of a transition arena and envisioning process; (2) developing coalitions, transition pathways and agendas, (3) mobilizing actors and knowledge development through experimenting and (4) monitoring and evaluating the transition process (Loorbach and Rotmans, In press)

the transition arena has to be set up outside the traditional institutional settings to avoid early media and political attention that could possibly drain the process. The selected participants should join on personal account rather than representing their home organization or institution, in order to avoid a rather narrow focus on the short-term stakes and vested interests of their occupational background. However, the transition arena has to be of a trans-disciplinary nature representing different but existing perspectives on the problem.

A transition arena concerning the transition in water management for example might involve policy makers, water engineers, ecologists, spatial planners, landscape architects, farmers, but also experts from related sectors such as the construction sector (housing), urban planners etc. Confrontation between the different perspectives would enrich the problem definition and might lead to alternative 'out of the box' solutions. It is helpful to structure the discussion by using a multi-phase and multi-level system approach in order to give meaning to these developments and understand the direction of its trajectory. Based on the joint representation of the water system sustainability visions for the water system have to be formulated through back – and forecasting methods. The vision consists of a set of qualitative images that illustrate and visualize a future sustainable water system (Dirven *et al.*, 2002) These images should contain physical and spatial elements as well as elements of the new water management style, such as risk management in terms of anticipative and adaptive water management strategies, 'openness' towards other policy domains and institutional organization with regard to participation from stakeholders. Subsequently, transition pathways towards the envisioned system have to be developed. Next, experiments are being set up in order to learn (learning-by-doing and doing-by-learning). Evaluation leads to a next round of experimenting and if necessary the adjustment of the vision.

In the Dutch case, focus now should be on translating the strategy into new institutional regime arrangements and practical instruments for local water managers. Both the institutional arrangements and the instruments have to be developed in co-evolution, so that the institutional arrangements enable the instruments. This is a process that has to be coordinated and is one of the things that is currently lacking in the Netherlands. There is much experimentation going on, but there is a lack of coordination that brings together these learning experiences. Developments that are interesting from this point of view are the water-test that forces water management to be involved in spatial planning and the debate around a merger of regional water management boards with provincial government that is responsible for regional spatial planning. Both the test and the merger influence the transition trajectory but no one seems to be concerned with how they are related, or should be. Yet, the up-scaling of the regional water management boards implies that water managers will increasingly be confronted with existing sectoral and regional policies.

As long as such 'gaps' remain between the strategic level and the operational level, the transition remains in the take-off phase. The institutional structure of the regime is herein a key lever that is decisive in the progress and future direction of the transition. Such institutional changes have serious reinforcing power for the transition when performed well. On the other hand, if ill-performed it can either seriously slow down the transition, or block the desired direction. It therefore is very important to act upon a shared vision of future water management.

The transition of European water management

As we have argued, the problems Dutch water management is being confronted with, are more or less comparable across all European Member States because the problems arise as a result of pressures that are deeply rooted in our modern Western society and a transformation process in which the institutional and technological change hampers. The Water Framework Directive operates against this background. The WFD was an attempt to rethink European water management. On the one hand, the WFD tries to increase the coherence of the fragmented objectives; on the other the European Commission came to recognize that European water policy has to address the increasing awareness of citizens and stakeholders about water. The implementation process of the WFD is a rather complex process in which the institutional arrangements have to change from national geo-political entities into (cross-boundary) river basin management regimes. Its focus is clearly on the up-scaling of the ecological approach and to a large extent neglects the spatial dimension of water management. Therefore, it can be argued that the WFD is an important step in the transition towards more adaptive water systems, but still has to undertake the even larger steps in order to involve water quantity issues and the spatial and institutional consequences. To this end it is worthwhile to analyze to what extent the WFD and transition management are based on similar principles.

One of the remarkable things about the WFD is that it is a legislative framework and the transpositions of the EU directives to the member states leaves room to manoeuvre with regard to the form and means of the implementation. At the same time there is the Common Implementation Strategy (CIS), which to some extent seems to be contradictory with this. The results from pilots in nine river basins should lead to a general implementation strategy, which then can be applied to river basins in all Member States. However, there are always differences between river basin management regimes in terms of institutional structures, problems or societal functions. River basins are heterogeneous and may be in very different transitional phases. Particular instruments therefore could be very effective in one of the pilots, but suboptimal in other basins due to local institutional arrangements for instance. The timed usage of various types of instruments may be much more effective and too stringent use of the implementation manual could even hold back desired ongoing developments.

The WFD tries to account for this through public participation. Preamble 14 of the directive states that the overall success of the directive relies on public participation. Subsequently, preamble 46 emphasizes the importance of informing the general public in order to ensure participation in the planning process. The most important article with regard to public participation is article 14. It prescribes three main forms of participation (2000/60/EC):

– Active involvement in all aspects of the implementation of the Directive, especially - but not limited – to the planning process.
– Consultation in three steps of the planning process
– Access to background information

Active involvement means that stakeholders actively participate in the planning process by discussing issues and contributing to the solutions. Consultation means that the public can react to plans and proposals and this should be done at least three times. Access to background information should be given at all times. WFD instructs the Member-States to encourage the first, but ensure the latter two. According to the 'Guidance document on Public Participation' the main reason for public participation is to ensure compliance with the directive: 'Public participation improves decision-making by ensuring that decisions are soundly based on shared knowledge, experiences and scientific evidence, that decisions are influenced by the views and experience of those affected by them, that innovative and creative options are considered and that new arrangements are workable and acceptable to the public'. Transition management recognizes the need for sharing knowledge, however, this does not necessarily guarantee innovation. One of the reasons explaining why the water-related problems are persistent, is the strong interconnectivity and the internal logic of the existing institutions. Changing one piece of the puzzle requires many more pieces to change in order to make the puzzle fit again. It can be argued that the participatory process prescribed by the WFD does not stimulate innovation, since predominantly traditional stakeholders are involved, e.g. stakeholders with particular interests in the current regime. Stakeholder evaluation could block radical innovative options because they do not 'fit' their vested interests. Stakeholder participation is absolutely necessary but can also block the path towards fundamental institutional change. In order to overcome the stakeholder dilemma, a transition arena should not involve stakeholders with vested interests, but innovative niche players. Allowing them to carry out experiments, they are able to develop knowledge and experience, to create networks, to communicate and to improve skills, e.g. build up an infrastructure. Developing these innovations in terms of network externalities and learning experiences makes it less costly and more attractive for actors to eventually join the innovation network.

Table 4 shows some differences between the WFD and transition management applied to the field of water management. Both in terms of the content as in terms of the nature of the process the overall objective is different. The WFD is concerned with good quality status for all waters, meaning general protection of the aquatic ecosystem, specific protection of valuable habitats, protection of drinking water resources and bathing water. The objective of transition Management is developing sustainable and adaptive water systems. Although both consider the river basin the optimal unit for water management, the implications for the process are profound. Transition management tries to develop strategies to integrate water quality, water quantity and the societal functions the water provides the region. From a transition management perspective, this requires a transformation of two co-evolving systems, namely the societal system and the aquatic ecosystem. The WFD mainly considers the process to be a one-directional implementation process of water quality standards (although this may have consequences for water functions). Participation is needed in order to facilitate and legitimize the process. As argued, this could reinforce existing power configurations, instead of transforming them.

Another difference is the time horizon used in the transition management and the WFD. The WFD has to be implemented in 2015. From then on, every six years the river basin

Table 4 Key aspects and differences between the Water Framework Directive and transition management also in this table, the differences are dichotomized for the sake of clarity

	WFD	Transition management
Objectives	Good quality for all waters	Adaptive water systems
Means	Integrating water quality directives	Integrating ecological, economic and socio-cultural water functions
Management unit	River basin management	Integrated water system approach
Problem perception	Implementation problem	Persistent problem
Process	Decision making process and implementation of river basin management plan	Transition processes
Approach	Management boards, advisory boards and public participation	Transition arena, coalitions, innovation networks and development rounds
Participation	Balancing interests and increasing public support	Dealing with complexity uncertainty and pluralism
Participatory process	Consulting parties based on draft river basin management plan	Co-production of visions, strategies and experimenting (learning-by-doing)
Time horizon	Implementation of WFD by 2015. River basin management plans every six years.	Long term perspective(30 years) and time-varying development rounds

A more appropriate conceptualization would be continuous scales

management plan has to be revised. Although transition management can also be divided in development rounds of approximately five years, these rounds are more or less ongoing and cyclic. There is a constant iteration process between learning experiences and the long-term vision. The long terms goals are guiding for the short term, but also adaptive to new developments and experiences. Using only the tight time-schedule planning and implementation every six years may again excludes innovations that take more time to develop.

Conclusions

Due to the increasing complexity of our modern society, water functions are becoming more and more interrelated. European water management is being confronted with complex, uncertain and multi-facetted problems when solving water related problems. The persistence of these problems is caused by the strong interwovenness of water institutions, management structures and dominant practices throughout the water system. This results in a tight and well-organized water management regime of policy departments, regional water management boards, engineering offices, which is hard to change. Due to this so-called regime structure radical innovations have trouble to break through and change the dominant way of doing things.

The particular case of the Netherlands generates insights into the general nature of the transition process of European water management and showed that persistent problems require innovative approaches, such as transition management. The transition analysis shows that the Dutch transition is in the take-off phase and near the acceleration phase. The new water management strategy, such as the retention-store-drain strategy, broadening of riverbeds, the designation of flood-areas and co-operation between water managers and spatial planners seem to be accepted. Nonetheless, the change of actual practices remains difficult because there are considerable gaps between abstract strategies, the enabling institutions and practice, hampering the shift towards the acceleration phase.

There are differences and similarities between transition management and the European Framework Directive. Although the WFD does stimulate forms of integration and participation, the current format might not create the innovative climate necessary for transition processes. Enriching the WFD with transition management principles may help to improve that.

Acknowledgements The authors would like to thank the reviewers for their useful and constructive comments. The authors also thank Flor Avelino and Derk Loorbach.

References

Arthur BW (1988) Competing technologies. In: Dosi G (ed) Technological change and economic development. Printer, London/New York

Berkhout F, Smith A, en Stirling A (2003) Socio-technical regimes and transitions contexts. SPRU Electronic Working Paper Series

Bosch A, Ham Wvd (1998) Twee eeuwen Rijkswaterstaat, 1798–1990. (Two centuries Rijkswaterstaat, 1798–1990). European Library, Zaltbommel

Bruijn DD, Hamhuis D, Nieuwenhuijze Lv, Overmars W, Sijmons D, Vera F (1987) Ooievaar: De toekomst van het Rivierengebied. (Stork: the future of the riverland). Gelderse Milieufederatie, Arnhem, The Netherlands

Dirven J, Rotmans J, Verkaik A (2002) Samenleving in transitie: Een vernieuwend gezichtspunt (Essay). Innovatienetwerk Agrocluster en Groene Ruimte, Den Haag

EC (2000) EU Water Framework Directive (No. 2000/60/EC). Brussels

Geels FW, Kemp R (2000) Transities vanuit sociotechnisch perspectief (Essay). MERIT, Maastricht

Gersick CJG (1991) Revolutionary change theories: a multi-level exploration of the punctuated equilibrium paradigm. Acad Manage Rev 16(1):10–36

Ham Wvd (1999) Heersen en Beheersen, Rijkswaterstaat in de twintigste eeuw. (Rule and Control, Rijkswaterstaat in the 20th century). European Library, Zaltbommel, The Netherlands

Holland JH (1995) Hidden order: how adaptation builds complexity. Helix books/Perseus Books, Cambridge, Massachusetts

Kauffman S (1995) At home in the universe: the search for laws of complexity. Oxford University Press, Oxford

Kemp R, Schot J, Hoogma R (1998) Regime shifts to sustainablilty through processes of niche formation: the approach of strategic niche management. Technol Anal Strateg Manage 10:175–196

Lintsen H (2002) Two centuries of central water management in the Netherlands. Technol Cult 43:549–568

Loorbach D, Rotmans J (In press) Managing transitions for Sustainable Development. In: Wieczorek AJ, Olsthoorn X (eds) Industrial Transformation - Disciplinary approaches towards transformation research. Kluwer

Luhmann N (1984) Soziale systemen. Suhrkampf, Frankfurt

Nelson RR, Winter SG (1982) An evolutionary theory of economic change. Belknap Press of Harvard University Press, Cambridge, Massachusetts

Prigogine I, Stengers I (1984) Order out of chaos: man's new dialogue with nature. C.O. New Science Library, Boulder

Ravetz JR (1999) What is post-normal science. Futures 31(7):647–653

Rijkswaterstaat (1964) Eerste nota Waterhuishouding (1st national policy memorandum on water management). Ministry of Transport, Public Works and Water Management, The Hague

Rijkswaterstaat (1982) Tweede nota Waterhuishouding (2nd national policy memorandum on water management). Ministry of Transport, Public Works and Water Management, The Hague

Rijkswaterstaat (1989) Derde nota Waterhuishouding (3rd national policy memorandum on water management). Ministry of Transport, Public Works and Water Management, The Hague

Rijkswaterstaat (1992) Omgaan met de Omgeving (Dealing with the surrounding area). Ministry of Transport, Public Works and water management, The Hague, The Netherlands

Rijkswaterstaat (1998) Vierde nota Waterhuishouding (4th national policy memorandum on water management). Ministry of Transport, Public Works and Water Management, The Hague

RIZA (1985) Omgaan met water (Dealing with water). Rijkswaterstaat, The Hague, The Netherlands

Rotmans J (1994) Transitions on the move. Global dynamics and sustainable development. Rijksinstituut voor Volksgezondheid en Milieu, Bilthoven, The Netherlands

Rotmans J, Kemp R, van Asselt M (2001) More evolution than revolution: Transition management in public policy. Foresight 03(01):17

Rotmans J, Kemp R, van Asselt M, Geels F, Verbong G, Molendijk K (2000) Transitions & transition management: the case of a low emission energy supply. ICIS/MERIT, Maastricht

Rotmans J, Loorbach D, van der Brugge R (2005) Transitiemanagement en duurzame ontwikkeling: Co-evolutionaire sturing in het licht van complexiteit. Beleidswetenschap

Sabatier PA, Jenkins-Smith HCJ (1999) The advocacy coalition framework, an assessment. In: Sabatier PA (ed) Theories of the policy process. Westview Press, Oxford

Saeijs HLF (1991) Integrated water management. A new concept: from treating of symptoms towards a controlled ecosystem management in the Dutch delta. Lands Urban Plan 20:245–255

Smits R, Kuhlmann S (2004) The rise of systemic instruments in innovation policy. Int J Foresight Innov Policy 1(1/2):4–32

Van der Brugge R, Rotmans J, Loorbach D (2005) The transition in Dutch water management. Regional Environmental Change, Online first(1)

Varela F, Maturana H, Uribe R (1974) Autopoiesis: The organization of living systems, its characterization and a model. Biosystems 5:187–196

WWF (1992) Levende Rivieren (Living rivers). World Wildlife Fund, Zeist, The Netherlands

Water Resour Manage (2007) 21:269–294
DOI 10.1007/s11269-006-9053-z

ORIGINAL ARTICLE

Some foci of integrated water resources management in the "South" which are oft-forgotten by the "North": A perspective from southern Africa

Roland E. Schulze

Received: 20 April 2005

Abstract Following some definitions of IWRM within a context of integrated catchment management, and a summary of the major goals and strategies as well as scale considerations in IWRM, this paper highlights some differences between IWRM in Lesser Developed Countries (LDCs), i.e. the so-called "South", and Developed Countries (DCs), i.e. the so-called "North", by outlining characteristics of DCs and LDCs which shape their respective needs in IWRM. Thereafter inherent problems in regard to IWRM in LDCs are identified. This is followed by examples from four case studies in southern African catchments which focus on some of the uniquenesses of IWRM issues in LDCs which, in the author's experiences, are often forgotten by theorists and practitioners from the "North", *viz.* that

- while catchment studies tend to emphasise mainstem river discharge characteristics, these are not the sources of rural water supply problems in LDCs (a case study from the Thukela HELP catchment in South Africa);
- water poverty is acute in many meso-scale catchments and is likely to be exacerbated by global warming (again, a case study from the Thukela catchment);
- water quality problems for the rural poor, who are still without potable water supplies, frequently revolve around the biological health of rivers, rather than those related to chemical or physical water quality (a case study from the Mgeni catchment in South Africa); and
- climate change may have severe impacts on both within-country reservoir management and out-of-country outflow obligations to downstream countries on already stressed catchments dominated by high water demanding irrigated crops (a case study from the Mbuluzi catchment in Swaziland).

In each case study simulation modelling has been used as a tool in IWRM. A concluding section therefore focuses on some selected problems which have been identified by the author in regard to hydrological modelling in LDCs. These revolve around issues of governance, human resources and practicalities.

R. E. Schulze
School of Bioresources Engineering and Environmental Hydrology, University of KwaZulu-Natal, Pietermaritzburg, South Africa
e-mail: schulzer@ukzn.ac.za

Keywords Integrated water resources management · Lesser developed countries · Water poverty · Climate change · Biological river health · Downstream water obligations · Hydrological modelling

Introduction

The "South" and "North" have become loosely entrenched terms to distinguish not only between the economies of lesser developed countries (LDCs) vs developed countries (DCs), but also between their differences with respect to water management problems and their overall and integrated management. This paper commences with a perspective on Integrated Water Resources Management (IWRM) which encompasses a discussion on definitions, goals/strategies and scale issues before some differences between IWRM in LDCs and DCs are highlighted, as viewed from a southern Africa perspective. Four hydrological modelling-based case studies which focus on aspects of IWRM in southern Africa, and which the author contends are often forgotten or neglected by colleagues from the "North" when they consider problems of IWRM, are then presented. They cover issues of mainstem vs tributary hydrological characteristics, water poverty, biological water quality and upstream vs downstream problems. All four case studies relating to IWRM in southern Africa having been hydrological model-based, a discussion then follows on problems frequently encountered in simulation modelling in developing countries, before the paper is concluded.

A Perspective on Integrated Water Resources Management

It may be argued that there is no one unified view of Integrated Water Resources Management, or IWRM, for it depends on the geography, history, the cultural values, level of development and planning objectives found in a catchment. This introduction, therefore, presents one perspective on IWRM, influenced to a large extent by personal experiences.

Defining IWRM and its relationship with integrated catchment management

From the plethora of definitions which abound in the literature, IWRM may be conceptualised as a framework for the co-ordinated planning and management of land, water and other environmental resources for their equitable, efficient and sustainable use (from the UK; Calder, 1998) according to the balanced views and goals of relevant stakeholders (added from Grigg, 1999 from the USA) in order to maximise the resultant social welfare in an equitable manner without compromising the sustainability of vital ecosystems (from Global Water Partnership, 2000). These concepts are imbedded in the South African Department of Water Affairs and Forestry's definition of IWRM as

> "a philosophy, a process and a management strategy to achieve sustainable use of resources by all stakeholders at catchment, regional, national and international levels, while maintaining the characteristics and integrity of water resources at the catchment scale within agreed limits" (DWAF, 1998).

Implicitly or explicitly these definitions place IWRM beyond being simply the management of water quantity and quality, or a catchment manager's "wish list", while simultaneously it is not as overarching and broad a socio-economic nor politico-institutional concept as Integrated Catchment Management, ICM, which UNESCO (1993) defines as

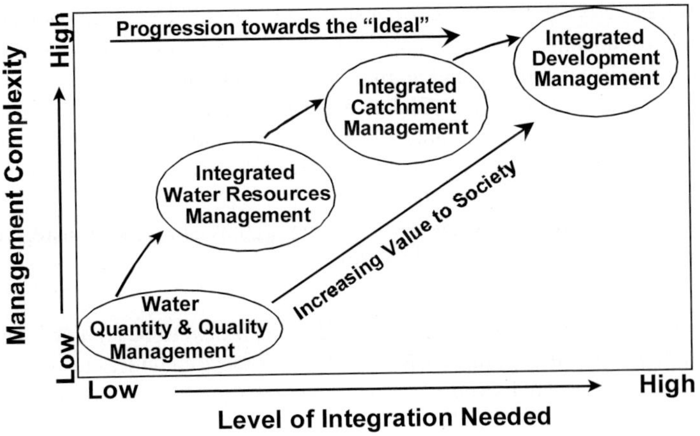

Fig. 1 The relationship of IWRM and ICM and their value to society to the level of management complexity and level of integration needed (after Ashton, 2000)

> "the process of formulating and implementing a course of action involving natural and human resources in a catchment, taking into account social, economic, political and institutional factors operating within the catchment and the surrounding river basins to achieve specific social objectives".

IWRM may thus be viewed as a vital, albeit incomplete, subset of ICM particularly if socio-political aspects do not receive the same emphasis as biophysical factors in management scenarios (Schulze, 1999). Diagrammatically, IWRM may be placed into a context of varying levels of management complexity and levels of integration needed, as conceptualised by Ashton (2000) in Figure 1.

Approaches which embody some of the major goals and strategies of IWRM

Of the many approaches that have been proposed, six are viewed by this author as embodying some of the major goals and strategies of IWRM (DWAF, 1996, 1998; Calder, 1999; Schulze, 1999, 2004; Ashton, 2000; Frost, 2001). While there is a degree of overlap between some of these six proposed approaches, they are nevertheless presented separately. They are as follows:

A Systems Approach, i.e. recognising individual components, as well as linkages between them, and addressing the needs of both human and natural systems (DWAF, 1996); recognising water and land management at *local* catchment level to be mutually dependent, then enabling and ensuring an *upward integration* of strategic water management at scales beyond those of the local catchments, i.e. scaling up from local to larger catchments; seeking solutions by an incrementally evolving and iterative process rather than by attaining one optimal solution; using a blend of "soft system" tools focusing on the human dimension, together with "hard system" methodologies such as mathematical simulation models and their decision support systems (Calder, 1999; Schulze, 1999); and recognising that solutions should focus on underlying causes and not merely their symptoms.

An Integrated Approach, where Integration implies "joined-up", "together", "holistic" and "with integrity" (Schulze, 1999) and, in IWRM, beyond being only "comprehensive" (DWAF, 1998). Integration includes co-ordinated development and management of

- land and water
- surface and groundwater
- catchment and adjacent coastal and marine environments and
- upstream vs downstream interests (Global Water Partnership, 2004). It has to occur across
- the socio-political system of equitable allocation needs and institutional management, the
- anthropogenic system of land use (and mis-use) with its characteristics of hydrological intensification and extensification, the
- engineered system of dams, sewage works or inter-catchment water transfers, and the
- aquatic system with its variable instream flow requirements,

all of which are considered by the stakeholders in a catchment to be significant and relevant key issues of concern to the set of IWRM objectives of the area under consideration.

While integration is not achieved easily where social, political, administrative and natural boundaries do not coincide (which is usually the case), two types of integration are required (Jewitt and Görgens, 2000; Frost, 2001). The first is *horizontal integration*, which takes place within the same hierarchical level, where that level can be at macro-scale or micro-scale, and where integration could be

- either between nations sharing a river, or
- between different water use sectors within the same river basin, such as domestic vs industrial vs agricultural vs environmental, or
- between upstream vs downstream users, or
- between activities of adjacent land uses/users within a catchment, while the second is
- *vertical integration*, where collaboration/co-ordination crosses a range of political, legislative or management sectors; alternatively of modelling systems or components of a natural system (such as river basins or aquatic ecology) which function at different vertical scales within the same sector (Figure 2).

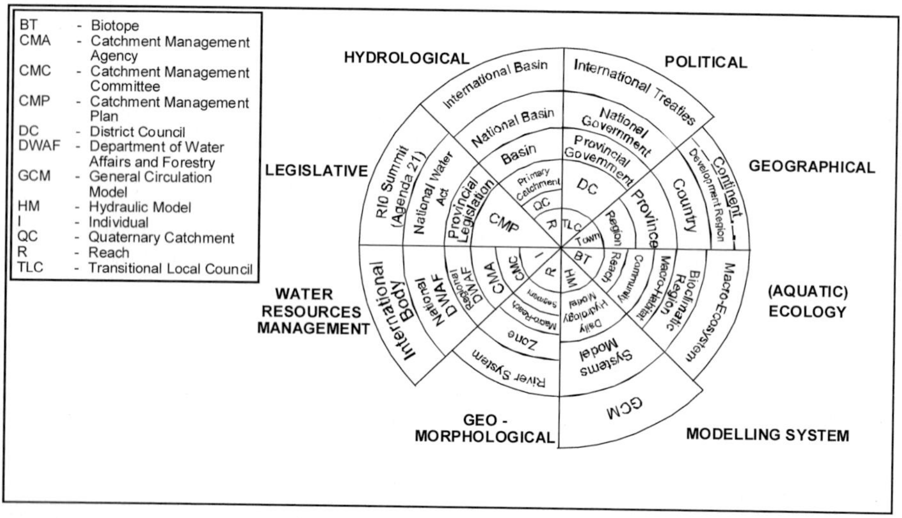

Fig. 2 Examples of vertical and horizontal integration, taken from an IWRM approach in the Kruger National Park, South Africa (after Jewitt and Görgens, 2000; modified by Schulze, 2004)

A Management Approach, in more generic IWRM terms, is perceived to imply a maximisation of the use of resources; minimising consequences of the above over the long term; even reversing the consequences of previously damaged systems by catchment rehabilitation and re-naturalisation of stream/riparian zone systems; seeking the well-being and enhancing the quality of life of the inhabitants within the area of study and seeking equitable solutions which must be fair and just to all concerned (Schulze, 1999).
Management must be seen to recognise the intrinsic value and importance of water and not merely its availability to satisfy economic needs. In more pragmatic terms, management in IWRM includes land and water to be managed conjunctively, because every land use decision becomes a water resources decision (Falkenmark *et al.*, 1999); water to be managed at the *lowest* appropriate level using a bottom-up rather than a top-down, approach; water allocation to take account of all affected stakeholders, including the often non-vocal poor and the environment; that water should be recognised as an economic good and that the principles of demand management be applied, with appropriate pricing policies, to encourage efficient usage of water between competing sectors such as domestic, agriculture, industry and the environment.

A Stakeholder Approach is perceived by the author as recognising the importance of the involvement of individuals, landowners and government agencies, in a participatory process where all decisions around the management and sustainable use of land and water resources are made.

A Partnership Approach, thereagainst, emphasises common objectives as well as defining the collective rules, responsibilities and accountabilities of every individual who, and every water use and administrative agency which, participates in the process of decision making on use and management of land and water resources at all levels, down to that of the village and even the individual household (DWAF, 1996; Calder, 1999). It thus reflects a commitment to the principle of stewardship at all levels of management (Ashton, 2000).

A Balanced, Sustainable Approach requires close attention needing to be given to decisions designed to achieve a sustainable blend, often compromise, between long term and viable economic development for all the catchments' dependants (local, national and international), equitable access of water resources to them and protection of resource integrity (DWAF, 1996; Calder, 1999). As such, a balanced approach attempts to optimise the relationship between the capacity of available resources to provide sustainable services (e.g. water of a given quantity and quality – a basic human need) and utilisation of the resource, including consumptive water uses such as agricultural or industrial, and non-consumptive water uses such as environmental requirements, as well as waste disposal (Ashton, 2000). Sustainability, it should be noted from the above, is not to be confused with zero growth (Ashton, 2000).

IWRM and scale issues

General

Catchment processes take place across a range of spatial and temporal scales, as do land use practices, socio-economic processes and levels of governance. The scale at which water-related integrated management should take place is, therefore, not one with a straightforward answer (Frost, 2001). As a general statement, however, experience in southern Africa shows that the appropriate temporal and spatial scales of operation in IWRM are those scales at which the policy makers, catchment managers and stakeholders of an IWRM plan believe that they can achieve their set(s) of objectives, depending on the problem(s) at hand (Schulze, 1999; Frost, 2001).

This is likely to depend on the life expectancy of a planning or management option, or the time it takes for such a plan to become operational. In turn, this will be defined, *inter alia*, by how effectively an area can be managed; what level of development previously had been attained in the area of interest, the homogeneity of the catchment with respect to bio-physical resources (water, agriculture), human resources, wealth and ease of communication with stakeholders; all of which will be influenced by constraints of politics, finances and bureaucracy (Schulze, 1999). Within an overarching "scale of operation" of IWRM plan will, therefore, be imbedded a hierarchy of intermediate and internal smaller space and shorter time scales, to define interim stages of implementation, or goals or milestones (Schulze, 1999).

Each catchment manager should thus work within a logical framework of scales, which need to range up and down, because the physical and social systems dealt with in IWRM are not bounded. One cannot, therefore, in IWRM be too prescriptive in terms of ideal scale(s), minimum scale(s) or maximum scale(s) (Schulze, 1999).

Spatial scale considerations in IWRM

IWRM has to take cognisance of all, or some, of global scale issues (e.g. water conventions, climate change, El Niño-Southern Oscillation or ENSO scale events); international scale problems (e.g. international river agreements); national issues (e.g. national water management agendas); catchment scale issues; local government scale initiatives; community scale issues; and in poorer countries also household scale problems (which there may include household food security and/or household water poverty).

Spatial scale issues in IWRM often reflect the level of development of a country, e.g. in poorer countries or poorer regions within a country, the space scale tends to be much smaller, determined by factors such as the distance range at which one can mobilise communities, or land availability around a village or access to local water sources (Schulze, 1999; Frost, 2001). The larger the spatial scale the more difficult management becomes, according to Frost's (2001) observations in rural Africa, in regard to

- the range of resources available; and
- the number and diversity of stakeholders (who have different skills, different interests, different resource endowments, as well as different capacities for management), implying that agreement/consensus is not easy and that plans of action become more complex and time-consuming.

When focusing at too *broad a scale*, it is often impossible to keep in view the "fine grained variation" embodied in all the various processes and there is a risk of overlooking local features, needs, circumstances, and/or aspirations, especially of the poor within the catchment (again from Frost, 2001; working in southern Africa). On the other hand, when focussing on too *fine* a *scale*, there is a danger of losing sight of the wider context of IWRM and losing sight of the overall governing processes of IWRM (Frost, 2001).

Temporal scale considerations in IWRM

From the above discussions on spatial scale considerations in IWRM it is clear that time scales in IWRM should not be viewed as static, but rather as a hierarchy of overlapping scales (Schulze, 1999). A number of types of time scales are identified and need to be considered in juxtaposition with one another in IWRM (Schulze, 1999). These include

- *climate scales* at intra-seasonal, inter-seasonal and decadal (re. climate change) time frames, which 'drive'
- *river flow scales*, which for surface water issues range from high flow/drought "cycles" related to ENSO at multiple year scales; and the inter-seasonal variability associated with that; the seasonality and concentration of streamflows within a year; intra-annual variability; the forecastability of river flows from the near real time, through a lead time of days and up to a season ahead; and studies on extremes such as floods; on the other hand, for groundwater the temporal recharge patterns and water table fluctuations are of importance;
- *aquatic habitat time scales*, which are determined by magnitudes, variabilities, frequencies and durations of low and high flows as biological triggers, and which are highly influenced by upstream land use management as well as by in-stream reservoir management;
- *agricultural time scales*, where for crops the intra- and inter-seasonal timeframes are important whereas for forestry, inter-seasonal to decadal timeframes are of greater significance;
- *economic time scales*, ranging from longer term international to national, to regional, local and to shorter term individual rural subsistence household time scales;
- *political time scales*, which need to distinguish between essentially stable government structures vs potentially unstable government structures and inter-election time scales for national to local governance structures;
- *management and planning time scales*, often of the order of 10–20 years; and
- *wealth/development level time scales*, where wealthy countries tend to have longer term planning horizons while for poorer countries they tend to be shorter (Schulze, 1999).

In summary, it needs re-emphasising that the scale at which IWRM is best initiated is the scale at which people are impacting on land and water resources and land/water resources are impacting on people. Thus, in Europe when the Rhine is impacted, large spatial scale IWRM is the order of the day while in southern Africa, for example, the mainstems of many larger rivers' waters *per se* may have been barely impacted by anthropogenic activities, but some of its individual tributaries' flows may be severely affected by upstream land and channel management, and hence effective IWRM generally takes place there at smaller spatial scales with shorter time horizons.

Some differences between IWRM in lesser developed countries (LDCs) and developed countries (DCs), viewed from a southern African perspective

Characteristics of DCs and LDCs which shape their needs in IWRM

Characteristics influencing IWRM in developed vs lesser developed countries, viewed from a southern African perspective and experiences, are summarised in Table 1.

Because of the high levels of expectation of IWRM in developed countries, a pro-active perspective and a generally non-life-threatening environment and infrastructure, IWRM there can focus more on quality of life and environment as well as long-term issues, which include preservation of the environment, with a focus on aquatic ecosystems, the re-naturalisation and rehabilitation of the catchment and its receiving streams, water quality related matters, demand management of water, and potential impacts of climate change on water resources (Schulze, 1999).

As a consequence of poorer infrastructure in lesser developed countries, higher vulnerability to natural events and often being in survival mode, IWRM there frequently has to address more immediate issues (Schulze, 1999) which in southern Africa, for example, would include

Table 1 Characteristics influencing IWRM in more developed vs lesser developed countries (after Schulze, 1999)

Developed countries	Lesser developed countries
Infrastructure	
• High level of infrastructural development, with infrastructure generally improving	Infrastructure often fragile and frequently in a state of retrogression
• Infrastructure decreases vulnerability to natural disasters (e.g. floods, drought)	High vulnerability to natural disasters; heavy damage and high death toll
• High ethos of infrastructure maintenance	Low ethos of infrastructure maintenance
• High quality data and information bases available, well co-ordinated	Data and information bases not always readily available
Capacity	
• Scientific and administrative skills abundantly available	Limited scientific and administrative skills available
• Expertise developed to local levels	Expertise highly centralised
• Flexibility to adapt to technological advances	Often in survival mode; technological advances may pass by
Economy	
• Mixed, service driven economics buffered by diversity, highly complex interactions	High dependence on land, i.e. agricultural production; at mercy of vagaries of climate
• Economically independent and sustainable	High dependence on donor aid, NGOs
• Multiple planning options available	Fewer options available in planning
• Take a long term planning perspective	Take a shorter term planning perspective
• Countries wealthy, money available for planning and IWRM	Wealth of countries limited, less scope for planning and IWRM
Socio-political	
• Population growth low or even negative	High population growth rates and demographic pressures on land
• Generally well informed public with good appreciation of planning	Poorer informed public, less appreciation of science/planning
• High political empowerment of stakeholders	Stakeholders often not empowered, afraid to act or to exert pressure
• Decision making decentralized	Decision making centralised
Environmental awareness and management	
• High level of expectation of planning and IWRM	Lower level of expectation and attainment of goals
• Desire for aesthetic conservation	Need for basics for living

providing basic potable water supplies to households (vs providing water of the highest quality), managing the water supply (vs demand management), poverty alleviation (vs quality of life enhancement), "harnessing" the local environment (vs sustaining it), seeking short term needs of the weeks and season ahead (vs long term perspectives), overcoming the vagaries of climate variability, both intra- and inter-seasonal (vs being concerned with climate change), or creating a basic infrastructure (vs maintaining, improving an existing one).

Problems regarding IWRM which can thus be identified for LDCs

With the tendency for many concepts on IWRM and ICM to emanate largely from the developed world, a focus is necessary on problems of IWRM in LDCs. In this paper the focus will be on problems experienced in southern Africa. First, certain generalities (many of which would also apply outside southern Africa) are stated. These include, for example, that decisions on water management are often made "from a distance" in a far-away capital city; that poor peoples' water needs are frequently overlooked or underestimated in broader scale IWRM; that amongst stakeholders there are major disparities in wealth, influence with government, opportunity, skills, resource endowments and capacity for management as well as for economic performance (Frost, 2001); that government project failures abound because funds have run out, or they are behind schedule, or operation and maintenance are inadequate; that the main need is for basic infrastructural development to provide for water security; and that priorities pertaining to environmental issues are frequently lowered and, where considered, sometimes need to focus on economic benefits such as erosion and river control.

If pre-conditions for successful implementation of IWRM are considered, the following of Farrington's and Lobo's (1997) points pertain specifically to the LDC context, *viz*. application of local catchment planning methodologies that are both technically sound and participatory; building on local peoples' (vernacular, indigenous) knowledge, experience and practice; planning initiatives that are accessible to, and involve, local community organisations and which include appropriate capacity building and technical support; and development of a framework of local-level collaboration amongst NGOs, CBOs (community-based organisations) and government departments with relevant government agencies. Both Farrington and Lobo (1997) and Frost (2001) lament that government-led initiatives which emphasise physical planning at the broader scale will often not be sustainable, particularly in southern Africa, because of the lack of necessary local 'ownership' and 'buy-in' of the stakeholders on the ground, as well as lack of cohesiveness in purpose amongst land and water users.

A further set of IWRM problems in LDCs relate to donor community involvement (Howe and Dixon, 1993). They cite, for example, lack of co-ordination between donors from different countries in the same development area; lack of consideration by donor/lenders of host country driven national programmes, strategies and priorities regarding land and water issues; leading (at times) to situations whereby developing countries often cannot formulate and implement their own water-related strategies/priorities owing to financial dependence on the international donor community; donor countries "selling" their own modelling or dam building technologies (whether or not they are wholly appropriate) because foreign aid is often tied to the use of donor country consultants and their expertise; and with subsequent difficulties arising in regard to project maintenance, back-up or post-audit. Similar problems arise in the lack of capacity building and institutional development in the field of IWRM to render LDCs technically self-sustaining, with the continuous failure of appropriate capacity building leading to continued dependence on external assistance (cf. also Table 3).

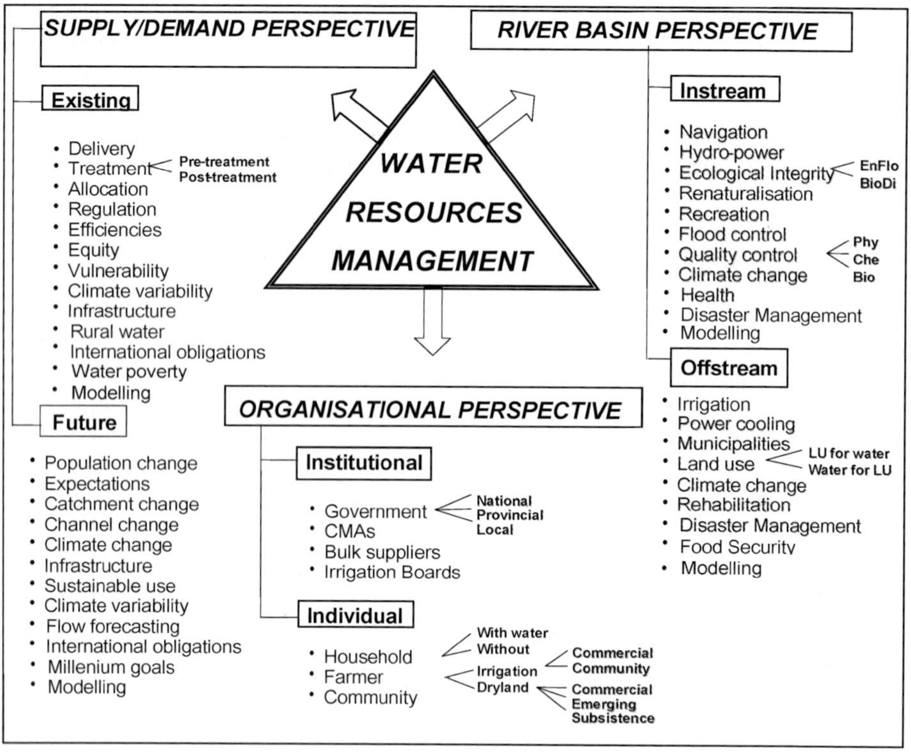

Fig. 3 Perspectives of water resources management (Schulze, 2005b)

Hydrological modelling based case studies on foci of IWRM in southern Africa which are oft-forgotten by the "North"

Background

When it comes to the nitty-gritty of effecting IWRM on an actual catchment, individuals usually focus on one of three closely interlinked perspectives illustrated in Figure 3, *viz.*

- the more engineering-oriented *supply/demand perspective* of the catchment manager on the ground, who has to balance existing biophysical and socio-econo-political pressures as well as planning for the future;
- the more ecologically-oriented *river basin perspective* with both instream and offstream considerations; and
- the *organisational perspective*, which has to find the balance between a more top-down institutional with the more bottom-up individual component.

Irrespective of the thrust of ones individual perspective of those in Figure 3 (and any one perspective without the other two does not hold in IWRM), there are issues which clearly stand out in the figure as foci more relevant to the underdeveloped "South", certainly to southern Africa, than to the developed "North". These include, for example, issues on equity, water poverty, rural water supply, water issues at household level, water and health, rehabilitation (of overgrazed lands) or food security. Four hydrological model based case studies have been selected from the author's research experience in southern Africa which emphasise some of the issues of LDCs which are oft-forgotten by practitioners and academics alike from the "North".

Focus 1: Catchment studies tend to emphasise mainstem river discharge characteristics; however, these are seldom the sources of rural water supplies in LCDs

Streamflow studies on major African catchments are frequently undertaken by scientists from the "North", often with substantial funding through institutions from the "North", and (inadvertently perhaps) the perception seems to be held that the catchments' inhabitants have access to those major rivers' waters. This, however, is seldom the case for the indigenous rural populations of South Africa's large catchments, as will be illustrated below.

The Thukela, on the east coast of KwaZulu-Natal province in South Africa and one of the UNESCO-HELP catchments, covers 29 036 km^2 with the main river rising in the Drakensberg mountain range at >3 000 m altitude in the west and flowing eastwards for ~180 km before entering the Indian Ocean (Figure 4). It is a highly diverse catchment, physiographically as well as climatically and socio-economically (Dlamini and Schulze, 2004a).

Hydrologically the Thukela has been delineated into 113 interlinked, cascading, meso-scale operational subcatchments (SCs). The human footprint in the Thukela consists of a complex (largely historical) juxtapositioning of developed and underdeveloped areas, with large tracts of former segregated "Homelands" mainly in the north and east suffering from severe water poverty at meso-scale levels (Figure 5, Water Poverty Index <60 and even <40).

Using the widely verified daily time step, conceptual-physical *ACRU* agrohydrological model (Schulze, 1995) with each of the 113 SCs "driven" by a 45-year daily dataset of rainfall and temperature together with relevant soil and baseline land cover information (Dlamini and

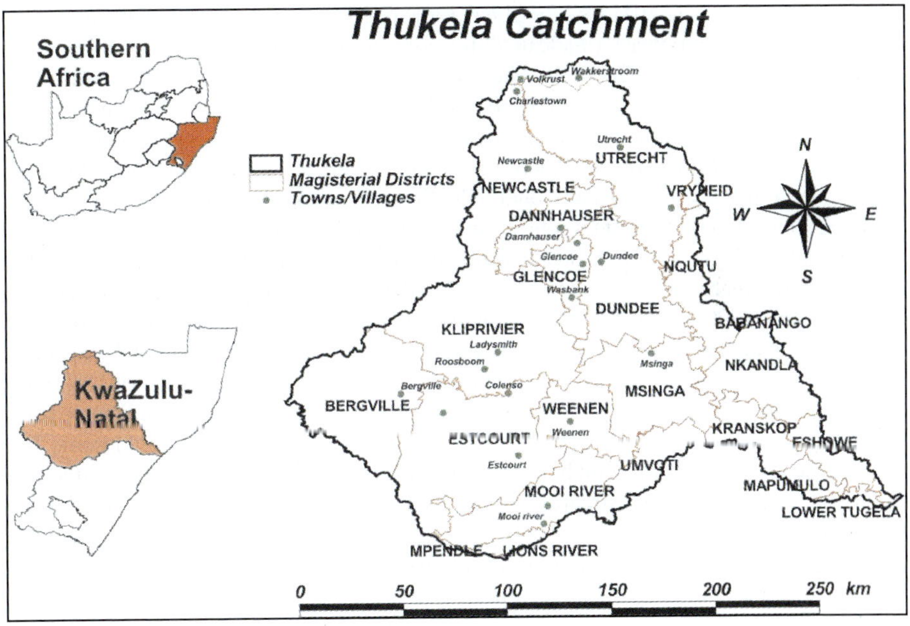

Fig. 4 Location of the Thukela catchment within South Africa

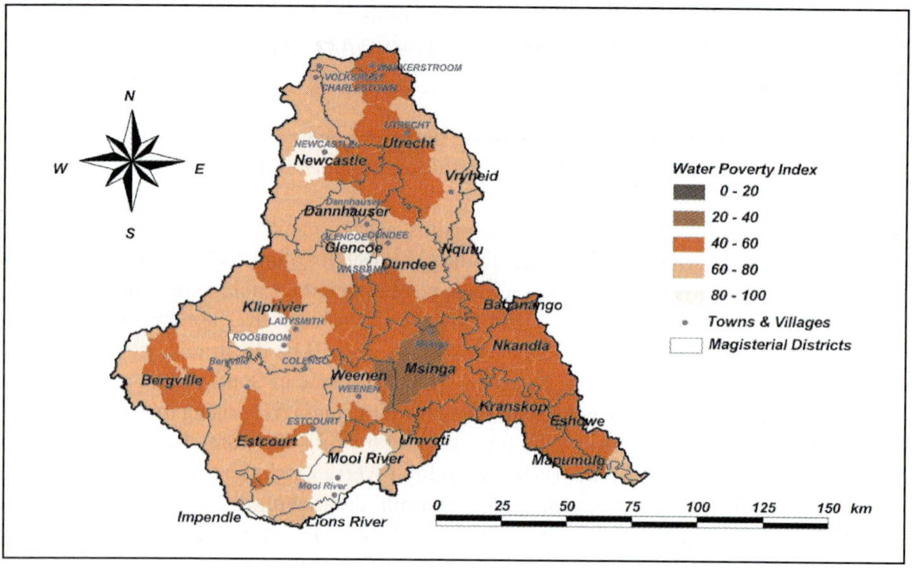

Fig. 5 Spatial patterns of water poverty in the Thukela catchment (Dlamini and Schulze, 2005)

Schulze, 2004a), mean annual runoff, MAR, and the inter-annual coefficient of variation of annual runoff, CV_{an} (%), have been generated and mapped, both for individual SC generated flows and for accumulated flows, i.e. including all upstream flows at any individual SC outlet (Figures 6 and 7).

Individual subcatchment MARs (Figure 6, top) are generally low in the north and central-east, there where water poverty (made up of five indicators of which water availability is only one) tends to be worst (cf. Figure 5). These are also the areas where flow variability tends to be highest (Figure 7, top). Accumulated MARs, on the other hand, are discernibly higher and more sustained along the mainstem Thukela and its major tributaries (Figure 6, bottom), and the accumulated flows of the SCs of these major rivers also display considerably lower CV_{an} (%) than their equivalent individual SCs (Figure 7, bottom vs top).

These patterns have important implications on water resources development for indigenous rural inhabitants. First, generally lower flows are generated where these communities live. Secondly, while there appears to be abundant available water along the mainstem Thukela, rural poor communities in South Africa (and in much of the rest of sub-Saharan Africa) do NOT live along mainstems of river systems with their more sustained perennial flow patterns. They rather reside, usually as individual households, in subcatchment tributaries. A further constraint is that *within* a subcatchment the communities/households tend to be located close to the watershed boundaries where their local supply of surface water would come from first order headwater streams which, in the Thukela, are often ephemeral in flow, rather than from the second or third order streams within a meso-catchment which have higher and more consistent flows.

This is an example where perceptions of relatively abundantly available water, often held by colleagues from the "North" do not, in this author's experiences, match realities of the "South", certainly not in rural South Africa, particularly as the scale of managing ones water resources in many parts of LDCs is at meso-catchment rather than at whole-catchment level.

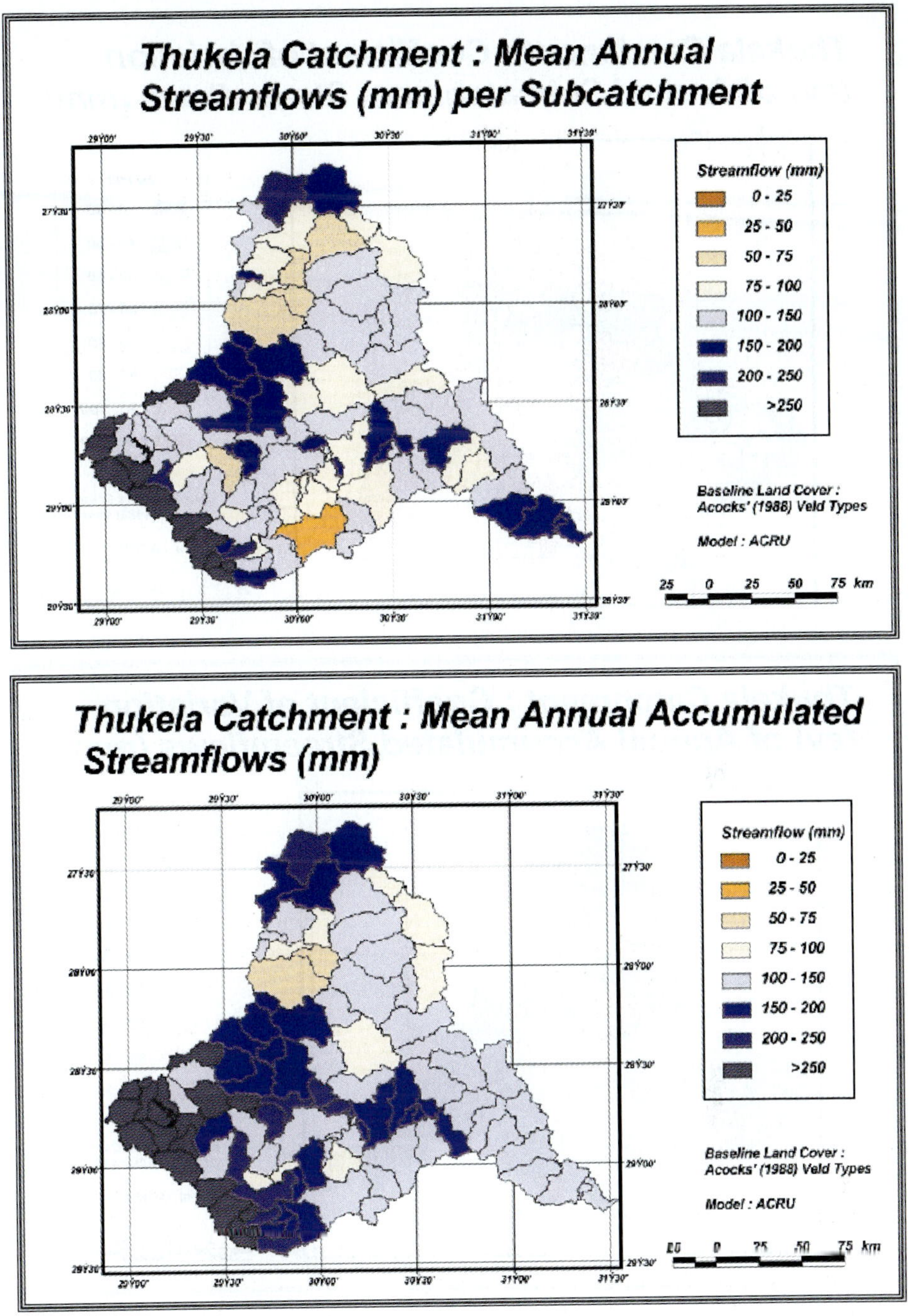

Fig. 6 Mean annual individual subcatchment (top) and accumulated subcatchments' streamflows (bottom) in the Thukela catchment, South Africa (after Dlamini and Schulze, 2004a)

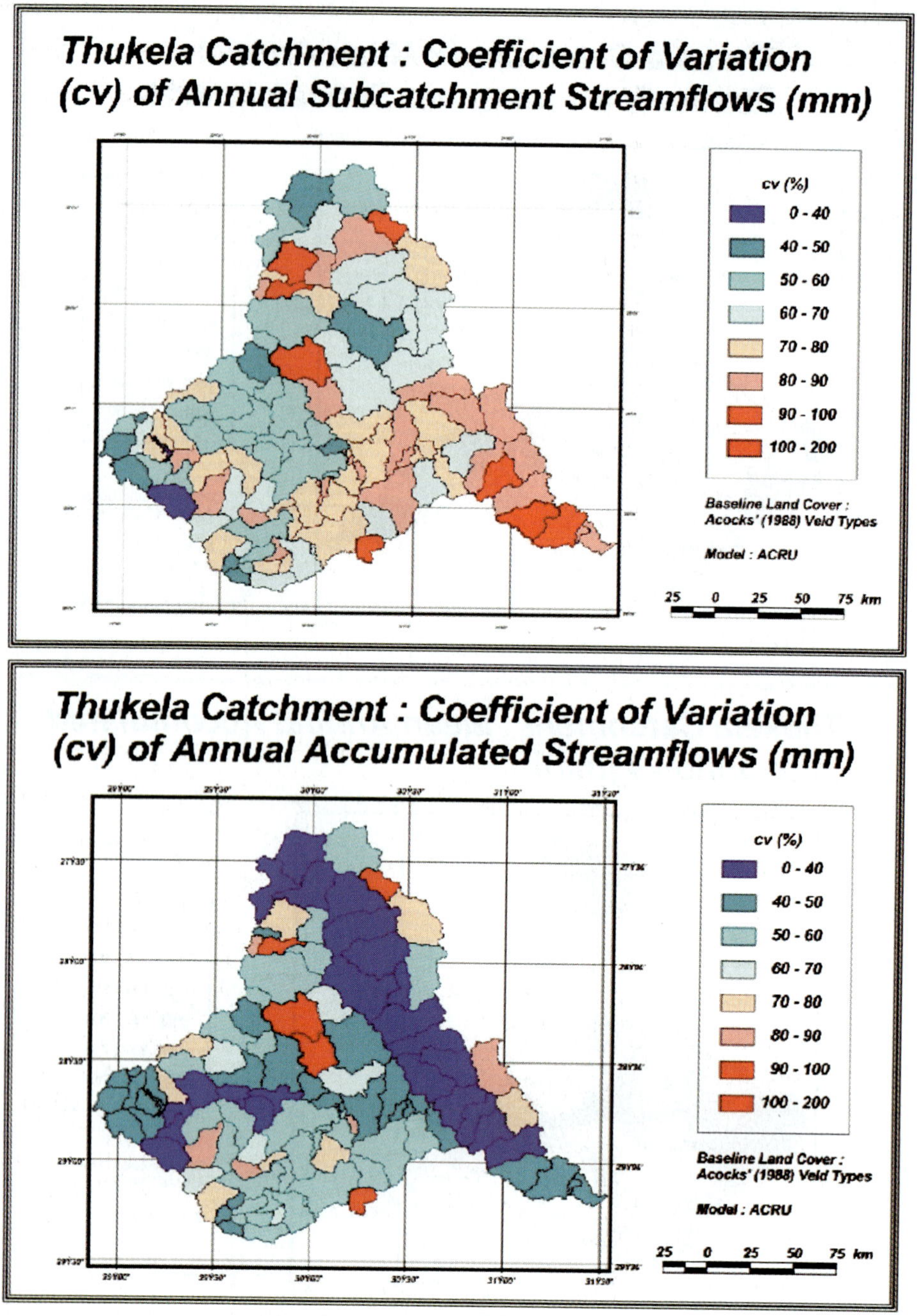

Fig. 7 Inter-annual coefficients of variation of streamflows for individual subcatchments (top) and accumulated upstream subcatchments' streamflows in the Thukela catchment, South Africa (after Dlamini and Schulze, 2004a)

Table 2 The components of the WPI and variables used in computing the components at meso-catchment scale in the Thukela catchment, South Africa (after Dlamini and Schulze, 2005)

Component	Description	Variables used in computations
Resource	Surface and groundwater; quality and reliability; soil moisture (absolute water availability)	Annual per capita water availability
Access	Access to domestic water supply and sanitation; access to water for productive use (including rights) and difficulty in accessing water (equity)	% households using unprotected water sources; average time taken to collect water; land under irrigation as a % of total irrigable land
Capacity	Level of human and financial capacity to manage water system (sustainability)	% of poor households with income <X; % completed high schooling or tertiary education; mortality per 1 000 of children under 5
Use	Level of water use by different sectors of economy (productivity, efficiency, sustainability)	Domestic water use; industrial water use; agricultural water use
Environment	Level of environmental integrity; water needs for the environment (sustainability)	% degraded land

Focus 2: Water poverty, already acute in many meso-scale catchments, could be intensified by global warming if other socio-economic upliftment were to remain unchanged in future

Since the 1980s, many water scarcity/stress/poverty assessment models have been developed worldwide. One of the most comprehensive is the Water Poverty Index, WPI, by Sullivan *et al.* (2002). The WPI is a sophisticated multi-disciplinary, multi-level, framework-based and composite method for investigating water scarcity and its relation with human welfare. The WPI consists of five components which deal with the water resource, access to it, capacity to manage it, use it and environmental factors. The component descriptions are summarised in Table 2, and the variables which are used in computing each equally weighted component in a meso-scale catchment study of the WPI in the Thukela catchment (described in the previous section) are, likewise, summarised in the table (Dlamini and Schulze, 2005). Data for the computation of the WPI were derived from the 1996 and 2001 Population Censuses at enumerator district level (of which there are >2 000 in the Thukela catchment) and from hydrological modelling in 113 subcatchments of the Thukela catchment, some results of which were shown previously in Figures 6 and 7.

Water poverty in many parts of the Thukela catchment is acute, as illustrated already in Figure 5. The contention in the context of this paper is, however, that if the last four components in Table 2 were to remain unchanged, then global warming would add an additional stress to the "resource" component of the WPI in the Thukela catchment. To test this hypothesis, a hypothetical, but plausible, climate change scenario for 2 $\times$ CO_2 conditions was derived for the Thukela catchment from four GCM scenarios for South Africa used in a previous study (Perks *et al.*, 2000; Schulze and Perks, 2000) and substantiated in a recent study by Engelbrecht (2005). The plausible scenario consisted of perturbing present climate input files used in each of the 113 SCs of the Thukela by 2°C for daily maximum and minimum temperatures (which would enhance potential and actual evapotranspiration, but reduce soil moisture and hence runoff generation), while simultaneously decreasing daily rainfalls by

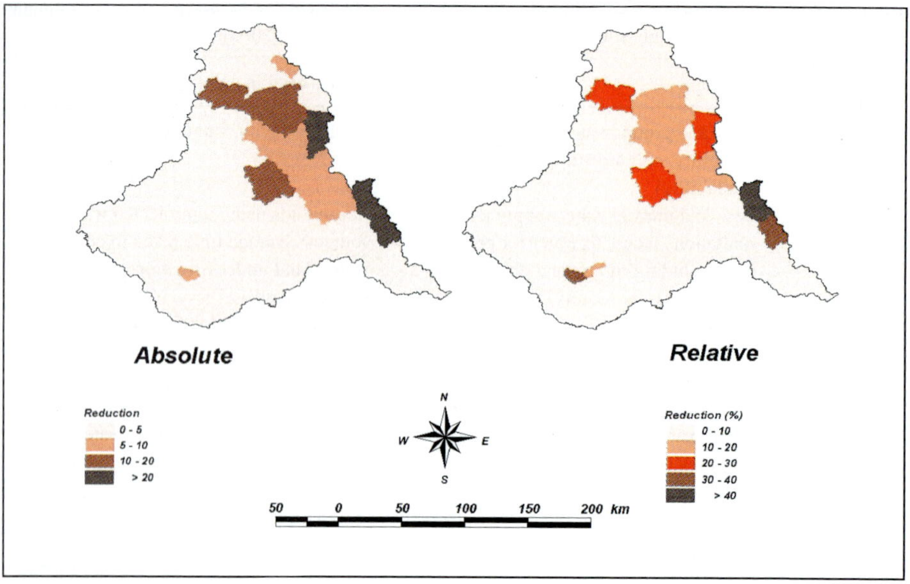

Fig. 8 Absolute (left) and relative (right) reductions in the resources component of the Water Poverty Index in the Thukela catchment for a plausible 2 × CO_2 climate change scenario (Dlamini and Schulze, 2005)

10% (which would not only reduce overall streamflow, but would alter the partitioning of the stormflow and baseflow components of total streamflow).

It should be recalled from the previous case study that *reducing* the composite WPI, or any one of its five constituent variables, implies greater water stress and hence water poverty. Figure 8 (top) shows that for the "resources" component in the Thukela catchment *absolute reductions* up to 20 points and more out of 100, which is equivalent to *relative* (percentage) *decreases* up to 40%, of the resources component of the WPI are displayed in parts of the Thukela. The net effect on the composite WPI is, similarly, that overall water poverty would get worse, again in certain areas more so than in others.

Should, therefore, Millenium Development Goals regarding delivery of potable water to households NOT be met everywhere, and climate change with its anticipated further increases in temperature and streamflow variability occur as predicted under most current plausible scenarios, water poverty could, indeed worsen. With that, potentially also incidences of malaria, cholera and/or bilharzia could rise. Such additional potential stresses are very much on the minds of water resources planners of southern Africa – often (it is perceived) with more consciousness than with their colleagues in the "North".

Focus 3 : Water quality problems for the poor in the "South" frequently revolve around biological water quality, rather than physical or chemical water quality

Key concerns with respect to river water quality in the "North" often revolve around non-point source chemical pollution from nitrate leaching and phosphate wash-off from intensive, high fertilizer input agricultural practices. Many developing areas of the "South" and certainly in southern Africa, however, derive their non-point source pollution in receiving streams from, first, relatively uncoordinated rapid growth of large informal urban or peri-urban shack-type settlements, often with no direct household water supplies or water-borne sewerage and,

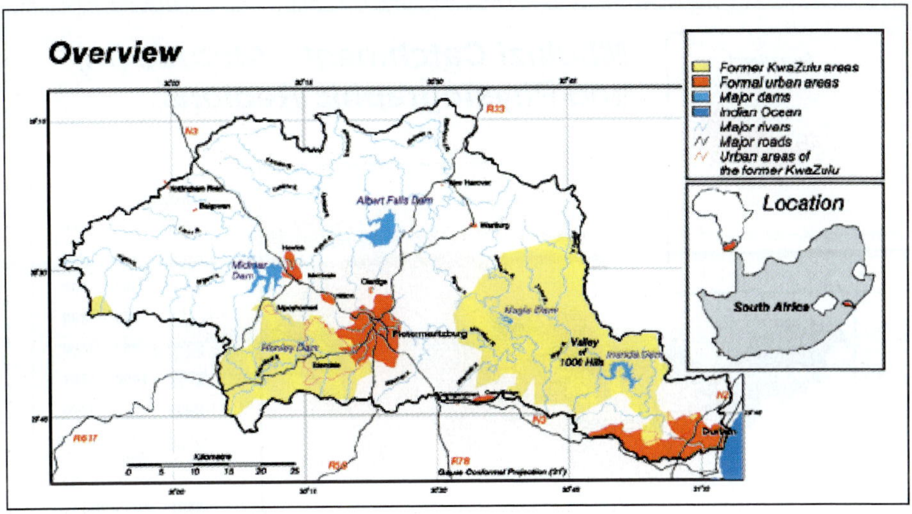

Fig. 9 Location of the Mgeni catchment in South Africa (after Kienzle *et al.*, 1997)

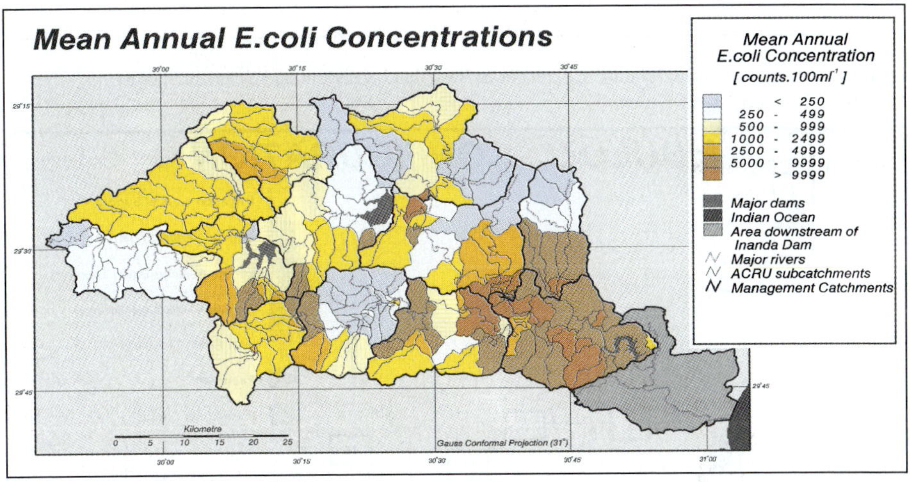

Fig. 10 Mean annual *E. coli* concentrations per subcatchment in the Mgeni catchment, South Africa (after Kienzle *et al.*, 1997)

secondly, from a livestock-based economy in which stocking densities frequently exceed the rangelands' carrying capacities, with consequent severe overgrazing and resultant high surface wash-off.

In sub-tropical climates characterised by high intensity convective showers this results in the receiving streams transporting considerable loads of solid particles to which appreciable concentrations of human and livestock derived pathogens are absorbed, indicated by the presence of *Escherichia coli*. These pathogens in the streams can result in serious biological water quality problems, compounded by the fact that the untreated river water is still being used widely for domestic and recreational purposes, which can then result in severe human health problems (i.e. diarrhoea) when *E. coli* concentrations exceed critical limits.

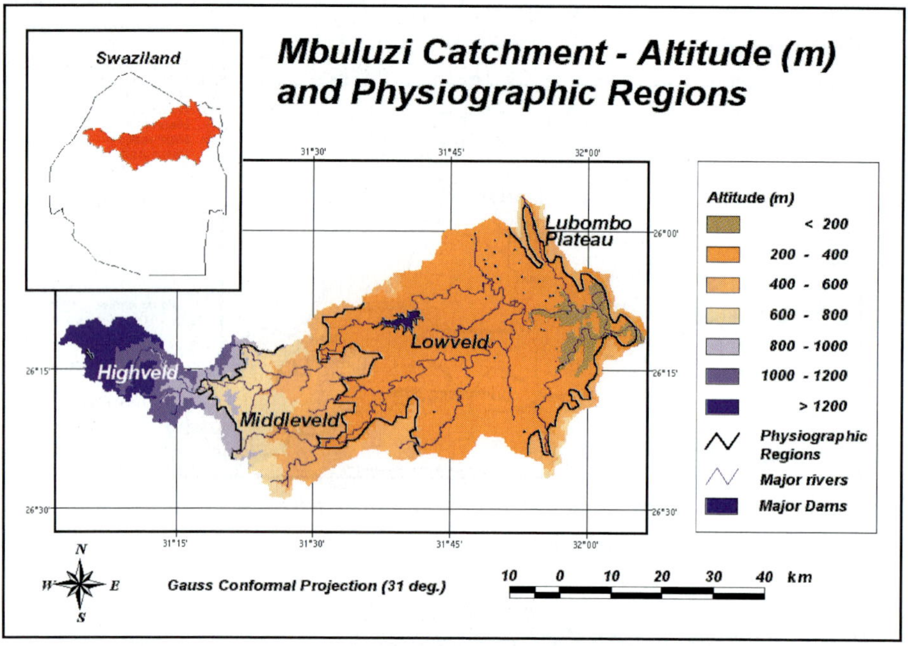

Fig. 11 Location map of the Mbuluzi catchment in Swaziland (Dlamini and Schulze, 2004b)

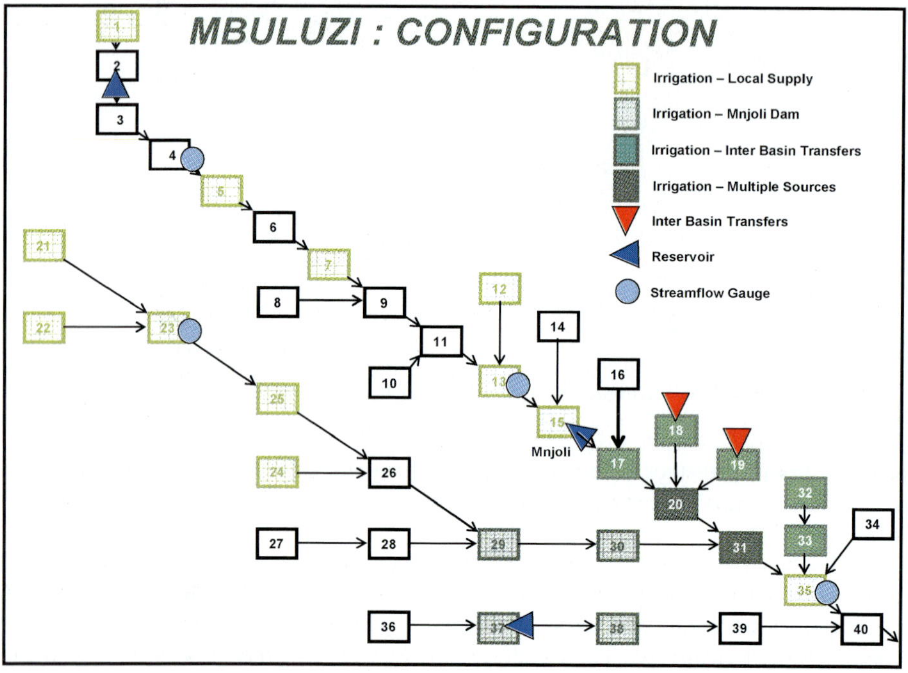

Fig. 12 Configuration of the Mbuluzi catchment into subcatchments for purposes of daily agrohydrological modelling (Schulze and Dlamini, 2005)

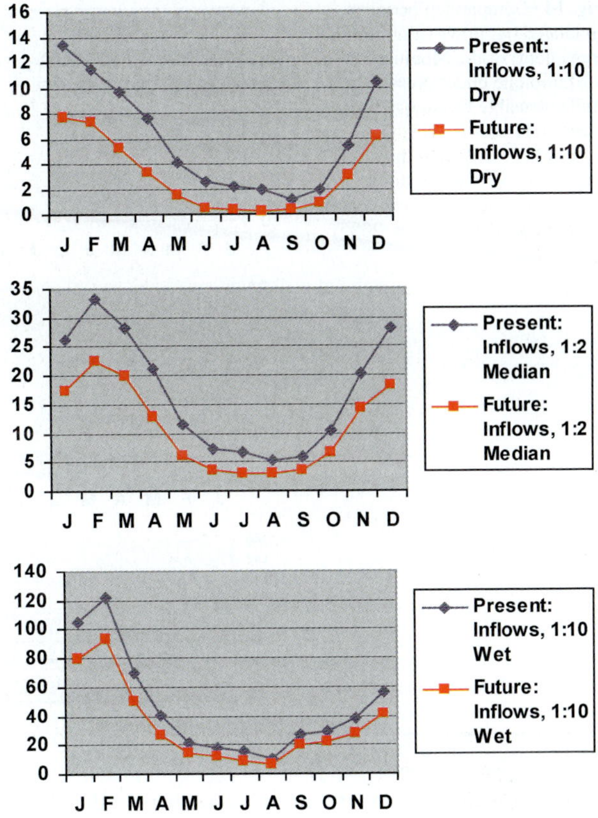

Fig. 13 Comparison between simulated inflows (mm/month equivalent) into Mnjoli Dam under present (p) and a plausible future (f) climate scenario [$T_f = T_p + 2\,^{\circ}\mathrm{C}$; $P_f = (P_p - 10\%)$] for the driest year in 10, for median flow conditions and the wettest year in 10 (Schulze and Dlamini, 2005)

The problems of biological river health as an IWRM challenge is illustrated from a case study in the 4 079 km^2 Mgeni catchment in South Africa (Figure 9), which partially borders the southern watershed boundary of the Thukela catchment. This important catchment, while constituting a mere 0.33% of South Africa's area, supplies water to ~15% of the country's 43 million population and produces ~20% of the country's Gross National Product (Kienzle *et al.*, 1997). Configured into 137 subcatchments over which was superimposed a 250 m digital elevation model for application with the *ACRU* agrohydrological simulation model (Schulze, 1995), daily concentrations of *E. coli* were generated from *ACRU* model algorithms which take into account daily sediment yields and sources of *E. coli*. These sources of *E. coli* are derived from subcatchment livestock densities and humans, with emphasis on that portion of the population residing within 250 m of receiving streams and using pit latrines from which raw sewage discharges into the natural water courses following runoff-producing rainfalls (Kienzle *et al.*, 1997).

Figure 10 identifies source areas of *E. coli* within the Mgeni system. Mean annual concentrations range from 30–18 200 counts/100 ml, with high concentration areas closely associated with former KwaZulu homeland areas (Figure 9) which are characterised by degraded landscapes as a result of overgrazing, as well as with dense informal peri-urban settlements where poor sanitation conditions prevail, and with pockets of high intensity cattle feedlots.

This case study on the biological river status illustrates another dimension which practitioners of IWRM in lesser developed countries frequently have to contend with, and which is

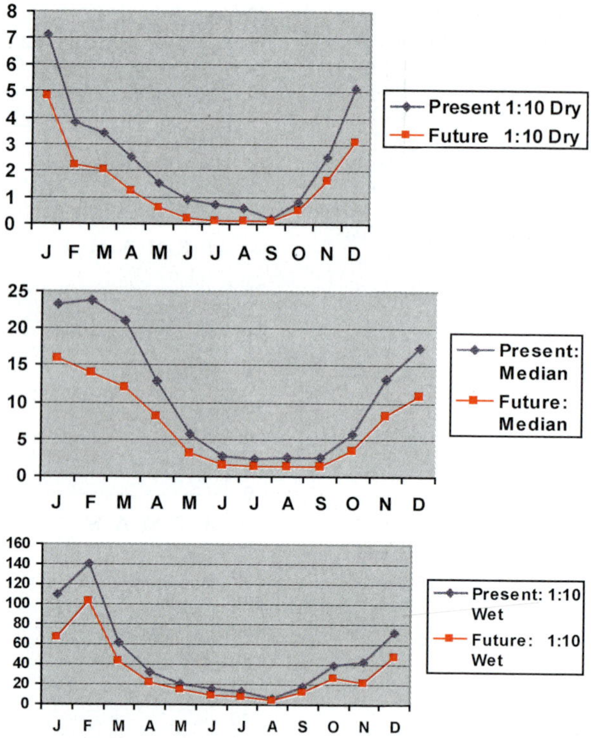

Fig. 14 Comparison between simulated out flows (mm/month equivalent) of the Mbuluzi into Mozambique under present (*p*) and a plausible future (*f*) climate scenario [$T_f = T_p + 2\,^\circ\mathrm{C}$; $P_f = (P_p - 10\%)$] in the driest year in 10, for median flow conditions and the wettest year in 10 (Schulze and Dlamini, 2005)

often forgotten by academics and practitioners of the "North". Not only are water treatment costs to kill off the *E. coli* pathogens substantial, but such high concentrations also raise questions on the urgent necessity for water supplies to be piped into rural areas as well as for land use control with respect to overgrazing, although the latter could cause potential cultural conflict with peoples to whom livestock numbers may mean more than livestock quality.

Focus 4: Climate change is likely to have severe impacts on both within-country reservoir management and out-of-country outflow obligations to downstream countries on already stressed catchments dominated by high water demanding irrigated crops

At first perusal the title of Focus 4 appears as much a problem of the "North" as it would be a problem of lesser developed countries, especially as climate change impacts on water resources will be a world-wide phenomenon. Likewise, water obligations by upstream countries to downstream countries are not a "South"-restricted problem. As a general observation, however, catchment land uses in DCs are often more diversified than those of the semi-arid LDCs, while the monoculture of a single cash crop under irrigation, can dominate a LDC's GDP more so than that of DCs. Furthermore, even under current climatic conditions already most semi-arid regions of the LDCs tend to experience a higher inter- and intra-annual climatic and, consequently, streamflow variability than do most DCs with their more temperate, sub-humid hydroclimates.

A case study from landlocked Swaziland's 2 959 km^2 Mbuluzi catchment (Figure 11) illustrates the severe impacts which anticipated climate change could have, first on the upstream inflows into the dominant storage reservoir in the catchment, *viz*. the Mnjoli Dam; secondly,

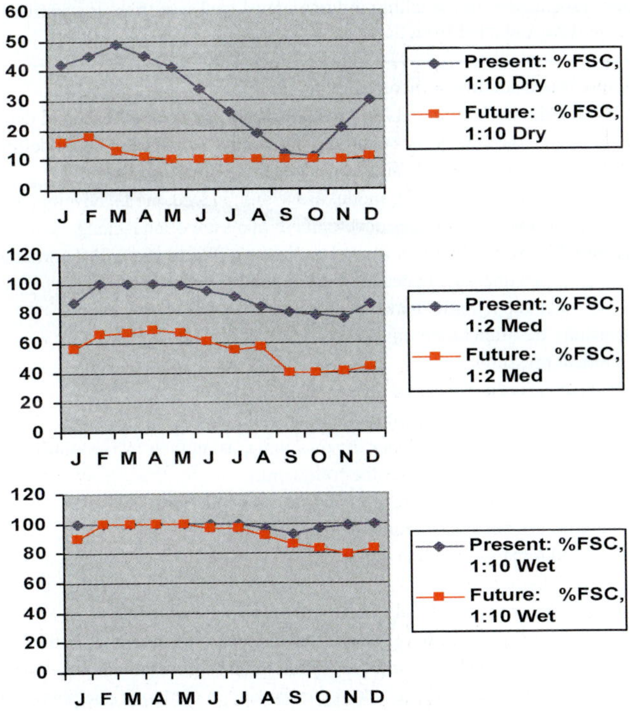

Fig. 15 Percentages of full supply capacity (FSC) of Mnjoli Dam under present (*p*) and a plausible future (*f*) climate [$T_f = T_p + 2\,^\circ\mathrm{C}$; $P_f = (P_p - 10\%)$] in the driest year in 10, for median conditions and the wettest year in 10 (Schulze and Dlamini, 2005)

the reservoir's performance when it is a major downstream supplier to >24 000 ha of all year round high water demanding irrigated sugarcane in a semi-arid climate; and thirdly, the potential consequences on the heavily impacted Mbuluzi's outflows into downstream Mozambique, which itself places heavy demands on the waters of the Mbuluzi (Schulze and Dlamini, 2005). Figure 12 illustrates schematically the configuration of the Mbuluzi catchment into 40 subcatchments, each with distinctive soils, land uses, water uses and daily climate data files for purposes of agrohydrological impacts modelling with the daily timestep, multi-purpose *ACRU* modelling system (Schulze, 1995).

In order to assess impacts of climate change, present climate conditions were perturbed for a hypothetical, but plausible, $2 \times CO_2$ scenario for Swaziland (Perks *et al.*, 2000; Schulze and Perks, 2000; Engelbrecht, 2005) by enhancing maximum and minimum temperatures by 2°C while reducing daily rainfalls by 10%.

Figures 13 and 14 illustrate, first, the wide ranges both of the Mbuluzi river's inflows into Mnjoli Dam and of the outflows from Swaziland into Mozambique for 1:10 dry hydrological conditions, median flow conditions and 1:10 wet year flows, as well as the very marked seasonal cyclic intra-annual flow patterns between summer (December–March) and winter (May–October). More important in the context of this paper, however, are the severe flow reductions under conditions of climate change. Together with a 2 °C temperature increase, the 10% reduction in precipitation is simulated to reduce reservoir inflows in median years by 34% and in 1:10 dry years by 47%, while the outflows into Mozambique are reduced by similar amounts (34% and 44% respectively). Patterns of the monthly percentages of full

Table 3 Problems associated with modelling in lesser developed countries, as identified by practitioners and post-graduate students (adapted from Schulze, 2005)

1. Government, Governance and Infrastructure Related Problems
- Concepts, and the "legitimacy", of models and modelling are not yet established in departments of state
- It remains difficult to convince the water resources "hierarchy" of the value of modelling
- There is still a strong belief of "measure and analyse" among older managers
- There is insufficient long-term support for model usage and, indeed, model development from the top, in part because long term water resources development plans are often lacking
- "Directional pressure" is exerted by international bodies (e.g. World Bank, IMF) what to pursue with their money (e.g. sustainability instead of building infrastructure)
- Moves to prioritise water supply will "leave locals out of the loop"
- Hydrological decisions are often based on
  - politics and political pressures
  - personal (but subjective) experiences of influential people
- Model results therefore often have little influence over policy formulations
- Modelling is frequently undertaken by donor organisations, using their own countries' models and staff
- The legacy of donors in modelling includes the following:
  - donors often leave behind impressive reports and recommendations, but
  - the recommendations are not always implemented;
  - expertise resides, and remains, with the donors;
  - there is a feeling of "donor dominance";
  - they leave behind little, or no, real local capacity to operate their models
- Governments often do not support local expertise; they believe "foreigners are better"!
- There is often no leadership in technical and conceptual skills in government departments
- Frequent changes in government lead to a lack of consistency in planning; goalposts change and new paradigms are introduced
- Bureaucracy prevails; long times are taken for authorisation of water related projects

2 Human Resources Related Problems
- There is little interest in modelling
- Local engineers/hydrologists are in a "comfort zone" with existing techniques
- Only few engineers/hydrologists have intimate modelling experience/expertise because
  - there is insufficient training in modelling;
  - there is little back-up for modelling; or
  - modellers tend to be academics rather than practitioners
- There is little teamwork in modelling; it tends to be an "individual pursuit"
- There is confusion among model users when choices have to be made which model to apply
- A "rugged personality" syndrome often exists with the individual "bullying" personal views on models/modelling

3. Practical Problems
- Data problems exist, for example:
  - lengths of records for model input and/or verification are short and/or poorly quality controlled
  - input data are often housed in different institutions
  - data are difficult to access
  - data are often collected for data's sake, and not for modelling purposes
  - each donor brings their own countries' monitoring equipment
- Models developed in developed/donor countries may be
  - too data demanding
  - focussing on processes that are not appropriate to the hydroclimates of LDCs;
  - too complex for users to interpret results effectively;
  - not answering the questions on the real on-the-ground problems

(*Continued on next page*)

Table 3 (*Continued*)

• Problems of power politics are at play as to who should disseminate model output/information to potential users: – politicians? -water resources planners? – educators? -regional planners?
• The use of "old" models prevails because of institutional inertia
• There is a lack of facilities (hardware, software, laboratories, libraries)

supply capacity (FSC) of Mnjoli Dam are illustrated in Figure 15. Already under present climatic conditions, irrigation demands in the still hot, but very dry winter months May to October reduce the Dam to around only 12% of its FSC in the driest year in 10; under a plausible future climate, Figure 15 shows that in such dry years the Dam would be at its dead storage level of 10% of FSC in 7 months of the year. Even in median years reservoir levels are 30–40% below present levels throughout the year with anticipated climate change.

What are potential implications of this simple, but plausible, climate change scenario to the water managers of the Mbuluzi catchment? With reductions in both water yield and assurance of supply, as well as a likely deterioration in water quality under warmer future climates, international water obligations to downstream Mozambique are likely to have to be re-negotiated. Curtailments of water releases from Mnjoli Dam are one further option to be considered, but that could have implications on the national economy, for sugar is Swaziland's primary export crop. Certainly irrigation scheduling will have to be re-visited to optimise water use efficiencies, and deficit irrigation is a likely route to go. The challenges to IWRM of such a scenario are, thus, immense.

Problems of simulation modelling in lesser developed countries

There are many "hard" and "soft" tools available to water managers to identify problem areas, to act as decision aids and to provide solutions in IWRM, of which simulation modelling is but one such tool. In each of the four case studies presented above, a daily timestep hydrological model has been applied as a tool to identify a specific problem at hand. All the modelling was, however, undertaken at an institute (University of KwaZulu-Natal) with considerable hydrological modelling expertise. Many problems are generally experienced in the application of hydrological models in LDCs. Some of these are listed in Table 3 under three broad headings, *viz.* those which are government, governance and infrastructure related, those which are human resources related and practical problems. Factors mentioned under specific headings frequently overlap with those under the other two headings. The list in Table 3 is not exclusive, for it originates from problems identified to the author by practitioners and post-graduate students from LDCs during more than a decade of lecturing on hydrological models at the UNESCO-IHE Institute for Water Education in the Netherlands.

The problems identified in Table 3, of course, may not occur at all, or may occur to a greater or lesser extent, in individual countries of the "South" and many problems listed extend well beyond modelling *per se* and into the realm of governance in general. Equally, however, some of the problems identified in Table 3 may apply in countries of the "North"! Nevertheless, much still needs to be achieved in many countries of the "South" to give modelling the status that this author believes it should have.

Discussion and conclusions

IWRM is an ideal that is difficult enough to realise and effect in developed countries with their high levels of existing infrastructural development and maintenance, high quality data, abundant levels of scientific and administrative skills down to local levels, strong mixed economics, long term planning perspectives, relatively high stakeholder involvement and their desire to pursue issues surrounding quality of life and of the environment (cf. Table 1). In many countries of the "South", which do not have the economic foundation and human capacity of many of the above-named attributes of the "North", the realisation of IWRM which embodies systems, integration, management, stakeholder, participatory and sustainability approaches often appears a far-off dream, despite frameworks and ideals enunciated in the Plan of Implementation of the 2002 World Summit on Sustainable Development or the Partnership of Africa's Water Development. By necessity in LDCs, space scales of IWRM as well as time scales, are smaller and shorter than in the "North". All the above are issues oft-forgotten by theorists and practitioners of IWRM from the "North".

Many decisions in IWRM are heavily dependent on hydrological simulation modelling. Again, the "South" is typically beset with many problems in regard to modelling, be they related to governance, infrastructure, human resources, human nature or of a more practical nature. These, once more, are realities of IWRM which are frequently not appreciated by academics, governments, consultants or funders from the "North".

Each of the four case studies presented has illustrated that much of IWRM in southern Africa deals with managing water scarcity in some form or other, be it

- reduced availability of water in the headwaters and external subcatchments where the indigenous rural poor tend to live, away from the mainstem rivers with their more sustained streamflows; or
- the added stress of climate change on rural water resources in areas now already characterised by water poverty; or
- the high pathogen counts in receiving streams, on which many people rely as their source of domestic water supply, but which is really unfit for human consumption; or
- having to cope with reduced reservoir inflows as a result of climate change, coupled with higher demands for irrigation water in a warmer climate and reduced outflows to a downstream riparian zone, with possible international repercussions.

While these case studies were specific to southern Africa, their context is nevertheless considered to mirror much of what is experienced generally by countries of the "South".

Acknowledgements Findings reported in this paper emanate from projects funded by the Water Research Commission of South Africa and the European Commission. Their support is acknowledged gratefully, as is the contribution by PhD student Dennis Dlamini.

References

Ashton P (2000) Integrated catchment management: balancing resource utilisation and conservation. CSIR, Pretoria, RSA, Aquatic Biomonitoring Course Notes, pp 1–11

Calder IR (1998) Water resource and land use issues. SWIM Paper 3, International Water Management Institute, Colombo, Sri Lanka

Calder IR (1999) The blue revolution: land use and integrated water resources management. Earthscan, London, UK, p 192

Dlamini D, Schulze RE (2004a) Meso-scale indicators of water poverty in the Thukela catchment, South Africa, under baseline land cover conditions. In: Schulze RE, Pike A (eds) Development and evaluation of an installed hydrological modelling system. Water Research Commission, Pretoria, RSA, WRC Report No 1155/1/04, pp 150–179

Dlamini D, Schulze RE (2004b) How do simulated sediment yields vary spatially and temporally under different land uses and states of degradation or rehabilitation : a case study from the Mbuluzi catchment in Swaziland. In: Schulze RE (ed) Modelling as a tool in integrated water resources management: conceptual issues and case study applications. Water Research Commission, Pretoria, RSA, WRC Report 749/1/04, Chapter 12, pp 222–232

Dlamini D, Schulze RE (2005) Water poverty within a context of climate change: a case study at meso-scale in the Thukela catchment. In: Schulze RE (ed) Climate change and water resources in Southern Africa: studies on scenarios, impacts, vulnerabilities and adaptation. Water Research Commission, Pretoria, RSA, WRC Report 1430/1/05, Chapter 24, pp 387–394

DWAF (1996) The philosophy and practice of integrated catchment management: implications for water resource management in South Africa. Water Research Commission, Pretoria, RSA, Report TT81/96, p 140

DWAF (1998) Water law implementation process: a strategic plan for the Department of Water Affairs and Forestry to facilitate in the implementation of catchment management in South Africa. Water Research Commission, Pretoria, RSA, Report, KV107/98, p 66

Engelbrecht F (2005) Simulations of climate and climate change over southern and tropical Africa with the conformal-cubic atmospheric model. In: Schulze RE (ed) Climate change and water resources in Southern Africa: studies on scenarios, impacts, vulnerabilities and adaptation. Water Research Commission, Pretoria, RSA, WRC Report 1430/1/05, Chapter 4, 57–74

Falkenmark M, Andersson L, Carstensson R, Sundblad K (1999) Water: a reflection of land use. Swedish Natural Science Research Council, Stockholm, Sweden, p 128

Farrington J, Lobo C (1997) Scaling up participatory watershed development in India: lessons from the Indo-German watershed development programme Overseas Development Institute, London, UK. Natural Resource Perspectives 17

Frost PGH (2001) Reflections on integrated land and water management. In: Gash JHC, Odada EO, Oyebande L, Schulze RE (eds) Freshwater resources in Africa. BAHC International Project Office, PIK, Potsdam, Germany, pp 49–56

Global Water Partnership (2000) Integrated water resources management. Global Water Partnership Technical Advisory Committee, TEC Background Paper No. 4, GWP, Stockholm, Sweden

Global Water Partnership (2004) Catalysing change: a handbook for developing integrated water resources management (IWRM) and water efficiency strategies, Global Water Partnership Technical Committee, TEC, GWP, Stockholm, Sweden, p 48

Grigg NS (1999) Integrated water resources management: who should lead, who should pay? J Am Water Resour Assoc 35:527–534

Howe CW, Dixon JA (1993) Inefficiencies in water project design in an operation in the third world: an economic perspective. Water Resour Res 29:1889–1894

Jewitt GPW, Görgens AHM (2000) Scale and model interfaces in the context of integrated water resources management for rivers of the Kruger National Park. Water Research Commission, Pretoria, RSA, Report 627/1/00, p 184 plus Appendices

Kienzle SW, Lorentz SA, Schulze RE (1997) Hydrology and water quality of the Mgeni catchment. Water Research Commission, Pretoria, RSA, p 88

NWA (1998) National Water Act of South Africa Act 36 (1998) Government printer, Pretoria, RSA, p 200

Perks LA, Schulze RE, Kiker GA, Horan MJC, Maharaj M (2000) Preparation of climate data and information for application in impact studies of climate change over South Africa. University of Natal, Pietermaritzburg, School of Bioresources Engineering and Environmental Hydrology. ACRUcons Report 32, p 72

Schulze RE (1995) Hydrology and agrohydrology. Water Research Commission, Pretoria, RSA, p 552

Schulze RE (1999) Integrated catchment management: summary of a CEH workshop on ICM. Centre for Ecology and Hydrology, Wallingford, UK, p 11 (Available from author)

Schulze RE (2000) Transcending scales of space and time on impact studies of climate and climate change on agrohydrological responses. Agric, Ecosyst Environ 82:185–212

Schulze RE (2004) River basin responses to global change and anthropogenic impacts. In: Kabat P *et al* (ed) Vegetation, water, humans and the climate. Springer Verlag, Berlin, Germany, pp 339–374

Schulze RE (2005) Hydrological modelling: concepts and practice. UNESCO-IHE Lecture Series, Delft, Netherlands, p 132

Schulze RE, Dlamini DJM (2005) Case study 3: potential impacts of a hypothetical, but plausible, climate change scenario on within-country reservoir management for irrigation, and out-of-country flow obligations in the Mbuluzi catchment, Swaziland. In: Schulze RE (ed) Climate change and water resources in Southern Africa: studies on scenarios, impacts, vulnerabilities and adaptation. Water Research Commission, Pretoria, RSA, WRC Report 1430/1/05, Chapter 14, pp 249–254.

Schulze RE, Perks LA (2000) Assessment of the impact of climate change on hydrology and water resources in South Africa. University of Natal, Pietermaritzburg, School of Bioresources Engineering and Environmental Hydrology, ACRUcons Report 33, p 118

Sullivan CA, Meigh JR, O'Regan D (2002) Evaluating your water: a management primer for the water poverty index. Centre for Ecology and Hydrology Wallingford, UK, p 94

UNESCO (1993) Integrated water resource management: meeting the sustainability challenge. IHP Humid Tropics Programme, Series 5, UNESCO Press, Paris, France

Water Resour Manage (2007) 21:295–313
DOI 10.1007/s11269-006-9054-y

ORIGINAL ARTICLE

The GLOWA Volta Project: A framework for water resources decision-making and scientific capacity building in a transnational West African basin

Charles Rodgers · Nick van de Giesen · Wolfram Laube · Paul L. G. Vlek · Eva Youkhana

Received: 20 April 2006

1. Introduction

The Volta River Basin occupies over 400,000 km^2 within the sub-humid to semi-arid West African savanna zone. The basin is shared by six riparian nations, among which Ghana (40% of basin area) and Burkina Faso (43%) are the most important in terms of population, water use and economic activity. Basin precipitation averages around 1,000 mm per year, with a steep south to north gradient, and less than 10% becomes usable as runoff due to high evaporation rates. Historically, rainfall is erratic and unreliable, a situation likely to be exacerbated as a consequence of global climate change. Basin inhabitants are largely rural and poor, with per capita incomes falling well below Sub-Saharan African standards, and only 37% (Burkina Faso) to 62% (Ghana) have access to improved sources of drinking water. Basin population is expanding by over 2.5% annually, effectively doubling every 28 years. Irrigation, the dominant consumptive use of water in the northern and central basin, competes directly with hydro-power generation in the south for available water resources, and the demand for water to serve these and other uses is projected to increase dramatically over the next two decades.

The GLOWA Volta Project (GVP), initiated in 2000 and funded by the German Government, is designed to provide a comprehensive, integrated analysis of the physical and socioeconomic determinants of the hydrologic cycle within the Volta Basin, with a specific focus on the impacts of global environmental change. GLOWA Volta is an interdisciplinary project involving climatologists, hydrologists, geographers and other physical scientists working in coordination with agricultural economists, sociologists and anthropologists. The overall

Integrated Assessment of Water Resources and Global Change: A North-South Analysis, 23–25 February 2005, Bonn

C. Rodgers · W. Laube · P. L. G. Vlek · E. Youkhana
Center for Development Research (ZEF), University of Bonn

N. van de Giesen
Department of Water Management, Delft University of Technology

project objective is to design and to implement a scientifically sound Decision Support System (DSS) for the sustainable development of water resources in the Volta Basin under changing conditions.

The objective of this paper is to provide a description of the water resources situation, and of the community of stakeholders, within the Volta Basin. In particular, it proposes to examine the structure and activities of the GLOWA Volta Project, and the ongoing development of the DSS as a mechanism for improved stakeholder participation in the increasingly challenging task of basin water resources management. In the following sections, descriptions of the Volta Basin climate, physical and human geography and a brief synopsis of the status of water resources allocation are provided. In Section 3, important Volta water sector actors and institutions are identified. In Section 4, the GLOWA Volta Project is described, and significant activities and accomplishments are discussed. In the final section, the development of a prototype Decision Support System for the White Volta Basin in collaboration with the Ghanaian Water Resources Commission (WRC) is described. The GLOWA Volta Project approach will be shown to emphasize interdisciplinary research, participation in research networks, consultation with basin stakeholders at all levels of involvement, and investment in scientific capacity building.

2. The Volta Basin

2.1. Physical features

The Volta Basin is located in West Africa between 5°.30 N–14° 30 N and between 2°.00 E and 5°.30 W (refer to Figure 1). The basin encompasses the majority of Ghana (70% of land area) and Burkina Faso (63%) and lesser proportions of Togo, Benin, Mali and Cote d'Ivoire, respectively (Table 1). It is in general a low relief basin, with elevations ranging from sea level to 920 m, a mean elevation of 257 m and correspondingly low channel grades. The lower Volta is fed by three major tributaries. To the west, the Black Volta (147,000 km^2) drains western Burkina Faso and small areas within Mali and Cote d'Ivoire; the White Volta (106,000 km^2) drains much of northern and central Ghana and Burkina Faso, and to the east, the Oti (72,000 km^2) drains the northwestern regions of Benin and Togo. The three tributaries join in Northern Ghana to form Volta Lake, impounded behind the Akosombo Dam. This dam and reservoir, completed in 1964, stores roughly 150 billion cubic meters (BCM) and has an installed hydropower generation capacity in excess of 900 MW (Sutherland *et al.*, 2004). With a surface area of 8,500 km^2, Lake Volta is among the world's largest artificial lakes.

Table 1 Spatial distribution of Volta Basin between Riparian Nations

Country	Country area of Volta Basin (km^2)	% of Basin in country	% of country in Basin
Burkina Faso	178,000	42.65	63.0
Ghana	167,692	40.18	70.0
Togo	26,700	6.40	47.3
Benin	17,098	4.10	15.2
Mali	15,392	3.69	1.2
Côte d'Ivoire	12,500	2.99	3.9
Total	417,382*	100%	

Source. Andah and Gichuki, 2003

Fig. 1 The Volta Basin, showing political boundaries and important tributaries, *source*. Andah and Gichuki, 2003

The Volta encompasses at least four climatic zones, from lowland rainforest in the southern portions of the basin, where annual precipitation can exceed 2,000 mm, to Sahel-Sudan in the north, where average rainfall is well below 1,000 mm per year and potential evaporation is considerably higher. Basin-wide rainfall averages 1,025 mm per year (1936–1963), of which roughly 9% becomes river discharge, as measured by Akosombo Dam outflows. Climatic patterns are strongly influenced by the movement of the Inter-Tropical Convergence Zone (ITCZ), which generates unimodal as well as bimodal rainy seasons. In the South, there are two rainy seasons with peaks in June/July and September/October. The North has only one wet season, from May through November, with peak rainfall occurring in September. Annual rainfall exhibits extensive spatial and temporal variability, and unreliable precipitation patterns make rainfed agriculture a risky undertaking throughout much of the basin. Mean annual temperatures approach 30 °C and humidity varies between 90% in coastal areas to below 20% in the North during the harmattan[1] and the dry season. Estimates of annual potential open water evaporation range between 1350 mm in the South to 2000 mm in the Northern parts of the country (MoWH, 1998; van Edig *et al.*, 2002).

Basin runoff exhibits higher temporal variability than the rainfall input due to nonlinear response and threshold effects, making the extent of utilizable water resources within the basin highly sensitive to variation in precipitation, and to alterations in land use and cover. Andreini *et al.* (2000) estimated an empirical annual rainfall-runoff relationship of the form:

$$Q = 0.529 \cdot (P - 343) \quad (Q, P \, \text{in km}^3/\text{year})$$

This implies that roughly half of precipitation volume becomes discharge, but only after 340 km^3, or 85% of the average annual rainfall, is received. Mean basin yield was around 35 km^3/year prior to the closing of Akosombo Dam (1936–1963), and 31 km^3/year over the period 1967–1998. Time series analysis demonstrates that small changes in precipitation lead to proportionately larger changes in runoff, which is an issue of great concern given the likelihood of alteration in rainfall patterns as a consequence of changes in global circulation. Preliminary research by Kunstmann and Jung (2004) appears to identify numerous statistically significant climatic trends, notably decreasing precipitation and increasing temperature. GVP climatologists in Ghana and at IMK-IFU (Garmisch-Partenkirchen) are currently investigating probable changes (delays) in the date of onset of the rainy season over the last several decades, a phenomenon widely reported by farmers in the Basin (Laube, personal communication).

2.2. Human geography

The demographic, socioeconomic and health status of Volta Basin inhabitants can be summarized via selected indicators from the World Bank-World Development Indicators database (2004) and Human Development Report (2000), presented in Table 2. These aggregate statistics are national in scope, but should be representative of conditions within the basin in most cases, particularly for Ghana and Burkina Faso. By African standards, the basin is densely settled, with Ghana possessing roughly three times the mean population density of Sub-Saharan Africa (SSA). Per capita income in Volta Basin countries tends to be lower than the SSA average, although Ghana appears somewhat more prosperous when income is evaluated in Purchasing Power Parity (PPP) terms. Much of Ghana's affluence is located in urbanized regions to the south, however, outside basin boundaries. Foreign assistance is correspondingly greater than the regional average, particularly to Ghana and Burkina Faso. Life expectancy at birth is broadly consistent with SSA as a whole, which is to say well below potential. Among riparian residents, Ghanaians enjoy the longest lives, due in part to a low incidence of HIV/AIDS relative to the SSA region, although malaria persists as a serious problem. However, considerable progress has been made toward eradicating onchocerciasis

[1] The Harmattan, typically occurring in December – February, is characterized by hot, rainless conditions; and by the presence of atmospheric haze originating in the Sahara Desert.

Table 2 Selected Volta Basin socio-economic indicators

Development indicator	Units	Year	Burkina Faso	Ghana	Togo	Benin	Mali	Cote d'Ivoire	Sub-Saharan Africa
Population density	Per km^2	2003	44.3	89.8	89.4	60.7	9.5	52.9	29.8
GDP per ca	1995 USD	2003	295	447	323	456	338	732	585
GDP per ca PPP	1995 ID	2003	1,023	1,943	1,328	978	864	1,277	1,618
Foreign aid	% of GNI	2002	14.8	10.8	3.8	8.3	15.2	9.6	6.3
Life expectancy female	Years	2002	43.5	55.6	50.7	54.5	41.9	45.7	46.6
Life expectancy male	Years	2002	42.3	54.3	48.6	51.0	39.9	44.8	45.1
Adults 14–49 with HIV/AIDS (a)	%	1997	7.17	2.33	8.52	2.06	1.67	10.06	–
Malaria incidence (a)	Per 100	1997	–	11.94	–	11.92	3.69	6.99	–
Rural population	%	2003	82.2	54.6	64.8	55.4	67.7	55.1	63.9
Improved water source, urban	%	2000	66.0	91.0	85.0	74.0	74.0	92.0	83.2
Improved water source, rural	%	2000	37.0	62.0	38.0	55.0	61.0	72.0	44.8
Improved sanitation, urban	%	2000	39.0	74.0	69.0	46.0	93.0	71.0	74.7
Improved sanitation, rural	%	2000	27.0	70.0	17.0	6.0	58.0	35.0	42.4
Labor force in agriculture	%	1990	92.4	62.2	65.5	63.5	83	54.2	–
Cereal yields	Kg/Ha	2003	994	1,039	1,003	1,216	895	1,134	1,050
Fertilizer consumption	100 g/ha	2001	82.3	27.6	76.5	155.5	90.1	201.6	125.9
Irrigated cropland share	%	2001	0.63	0.19	0.68	0.53	2.94	0.97	4.23

Sources. World Bank World Development Indicators (2004), Andah and Gichuki, 2003 (river blindness), which was until recently endemic to the Northern regions of the basin (Andah and Gichuki, 2003).

Basin inhabitants are overwhelmingly rural, as throughout much of SSA. In Ghana, the percentage of Basin residents living in rural areas is greater than the national average suggests, since the densely settled southern and coastal areas, including Kumasi and Accra, lie outside basin boundaries. Rural residents in general are less likely to have access to improved water supply and sanitation, particularly in Burkina Faso. Roughly two in three Basin inhabitants, and nearly 9 out of 10 in Burkina Faso, are employed in agriculture. Agricultural productivity is low, at roughly 1 metric ton of cereal yield per hectare. Although this is consistent with agricultural performance throughout SSA, these yields are well below the world average of 2.2 MT per hectare, and compare even less favorably with East Asian yields, which average over 3.2 MT per hectare. Erratic and unreliable precipitation contributes significantly to the yield gap, although fertilizer consumption, at less than 3 kg per ha in Ghana, is only around 20% of SSA average (12.6 Kg/Ha) and far below the world average of nearly 100 Kg/Ha. The factor of greatest significance from the perspective of water management is the extremely low percentage of cropland within the basin and surrounding areas that is currently under irrigation – less than 1% in Ghana and Burkina Faso, as against 4.23% throughout SSA and 20% globally (WDI, 2004). Paradoxically, efforts to mitigate poverty and low agricultural productivity through irrigation investment will position the agricultural sector increasingly as a competitor to the power generation sector, arguably no less critical to overall economic development.

2.3. The structure of demand and competition for water

The downstream locations of Akosombo and Kpong dams, near the basin outlet, frame the primary conflict over Volta water resources: upstream consumptive use against downstream hydropower generation. The dam at Akosombo, originally proposed in the 1920's, was constructed by an international consortium in 1961 largely to provide (highly subsidized) hydropower to the Volta Aluminum Company (Valco) (Andreini *et al.*, 2000). The availability of inexpensive hydropower was, and remains an important engine of economic growth for Ghana, critical to the viability of the mining and industrial sectors, and for surrounding countries which purchase Volta hydropower.

Ghana is but one of six riparian states sharing Volta waters, however. A significant proportion of Volta Lake inflow originates in the upstream countries of Burkina Faso and Togo. Efforts to develop water resources for hydropower production, irrigation and other purposes in these nations negatively impact storage in Volta Lake, although the impacts of climatic variability currently exceed the impacts of upstream abstraction.[2] Two major storage reservoirs have already been built in Burkina Faso: Bagre (1.7 BCM live storage) on the White Volta and Kompienga (2.05 BCM) on the Oti (Obeng-Asiedu, 2004). Three additional dams are planned. If the proposed hydropower dam in Bui Gorge on the Black Volta is built (6 BCM), it may further complicate efforts to maintain storage in Volta Lake at optimal levels for hydropower production. The assertion that hydropower generation is a "non-consumptive" use of water means little when the point of generation lies downstream from competing, consumptive uses.

[2] For example, the low levels of Lake Volta in 1998 reflected low rainfall in the preceding rainy season (Van de Giesen *et al.*, 2001).

Table 3 Current and projected water demand

Country	Sector	Millions cubic meters per year					Change relative to 2000%		
		1990	2000	2010	2020	2025	2010	2020	2025
Benin	Domestic	–	56	196	336	448	250	500	700
	Irrigation	–	152	548	1,225	1,600	261	706	953
	Livestock	–	40	94	133	175	135	233	338
	Total	–	248	838	1,694	2,223	238	583	796
Burkina Faso	Domestic	67	85	106	132	149	25	55	75
	Irrigation	43	203	384	554	639	89	173	215
	Livestock	37	46	61	78	88	33	70	91
	Total	147	334	551	764	876	65	129	162
Côte d'Ivoire	Domestic	–	4	5	12	14	25	200	250
	Irrigation	–	19	57	166	276	200	774	1,353
	Livestock	–	1	2	3	3	100	200	200
	Total	–	24	64	181	293	167	654	1,121
Ghana	Domestic	82	138	192	272	284	39	97	106
	Irrigation	75	565	1,871	3,605	3,733	231	538	561
	Livestock	18	26	41	63	67	58	142	158
	Total	175	729	2,104	3,940	4,084	189	440	460
Mali	Domestic	5	9	13	16	18	44	78	100
	Irrigation	126	180	219	291	311	22	62	73
	Livestock	4	34	74	123	142	118	262	318
	Total	135	223	306	430	471	37	93	111
Togo	Domestic	51	68	92	123	145	35	81	113
	Irrigation	43	50	91	133	171	82	166	242
	Livestock	15	19	22	30	36	16	58	89
	Total	109	137	205	286	352	50	109	157
Riparian states	Domestic	–	360	604	891	1,058	68	148	194
	Irrigation	–	1,169	3,170	5,974	6,730	171	411	476
	Livestock	–	166	294	430	511	77	159	208
	Total	–	1,695	4,068	7,295	8,299	140	330	390

Sources. MWH 1998, Andah and Gichuki, 2003; Laube and van de Giesen, 2005

Consumptive water use itself is anticipated to expand dramatically over the coming 20 years, driven primarily by population growth and expansion of surface irrigation (Table 3). There is at present little irrigated area within the basin – official estimates are roughly 8,000 out of a potential 1.2 million hectares in Ghana, and 20,000 out of a potential 165,000 Ha. in Burkina Faso (Obeng-Asiedu, 2004), although this figure almost certainly excludes much of the farmer-developed small scale irrigation. In Burkina Faso, actual irrigated area may be closer to 50,000 (van de Giesen, personal communication). Important irrigated crops within the basin include rice, which receives supplemental irrigation during the rainy season, and tomatoes, onions and maize during the dry season (MoFA) Much of the small scale irrigation is provided by small earthen dams which impound intermittent flow in small catchments in the North of Ghana and throughout Burkina Faso. These small reservoirs, while perhaps of appropriate scale for decentralized management, have high surface area to volume ratios, and thus may evaporate as much or more water than they make available for irrigation (Liebe, 2002). The number of such dams and associated irrigated commands is expanding rapidly, particularly in Burkina Faso.

As Table 3 indicates, Ghanaian irrigation water consumptive demand is projected by MoWH (1998) to expand by a factor of six between 2000 and 2025; and in Burkina Faso, by a factor of three over the same period (Andah and Gichuki, 2003). Overall, domestic water supply (provided largely by wells in rural regions with water of suitable quality) is projected to increase from 360 to 1,058 MCM; livestock use from 166 to 511 MCM, and irrigation consumption from 1,169 to 6,730 MCM. Overall abstractive use will increase from roughly 1.7 km^3 in 2000 to 8.3 km^3 in 2025, nearly all of which will be withdrawn at points upstream of the Akosombo hydropower generating facilities.

As growth in agricultural output and productivity, adequate domestic water supply and sanitation and power generation are all essential at this stage in the Basin region's economic development, the competition between consumptive uses and hydropower generation presents hard choices. Exacerbating this situation is the absence of an Inter-state Basin Compact, Authority or similar legal framework to regulate and allocate flows across international borders and to enable coordinated development of the Basin's water resources. Encouraging progress is being made toward such a governing structure, however, through the joint IUCN – Global Water Partnership project "Improving Water Governance in the Volta River Basin," commencing in 2004 (Odame-Ababio, 2004). Another development with the potential to alter the economic calculus underlying Volta water allocation is the recent approval by the World Bank of funding for the West African Gas Pipeline Project (World Bank, 2004). This US$ 590 Million project will bring Nigerian natural gas to coastal locations in Benin, Togo and Ghana. *Ex ante* economic analysis (World Bank, 2002) suggests that electric power can be generated thermally using natural gas at 0.042 USD per KWH, as compared with 0.072 for hydropower. These and other evolving economic and political developments highlight the need for an integrated water resources DSS for the Volta Basin.

3. Stakeholders and water management institutions

As each of the riparian nations has a unique set of laws and institutions regarding the management of water resources, the following narrative will focus on Ghana, for which the GVP has the most complete information. Prior to the enactment of the 1992 Constitution, no single institution had authority over the water sector, although a (non-statutory) riparian doctrine was widely respected. Following the intent of Article 269 of the new Constitution, in 1996 the Ghanaian Parliament passed the Water Resource Commission Act (Act 522, 1996). This Act empowers the Water Resource Commission (WRC) to manage the nation's water resources, effectively acting on behalf of the President in whom all water rights are vested under a doctrine of public trust. Under the Act, no person or organization has the authority to divert, dam, store, abstract or use water, or to develop or maintain physical structures for the use of water resources apart from WRC's consent. The WRC was created in part to remedy the problems of unclear, conflicting or redundant mandates that characterized the sectoral approach to water resources management that prevailed prior to passage of Act 522. Primary responsibilities of the new WRC, consistent with the philosophy of IWRM, include the following (Mensah (1999), in Laube and van de Giesen, 2005):

- to propose, co-ordinate and monitor plans and activities related to the development, improvement utilisation and conservation of water resources
- to grant water rights
- to collect, collate, store and disseminate data or information on water resources
- to require water user agencies to undertake research and experiments into water resources

Table 4 Composition of the Ghanaian Water Resources Committee (WRC)

	Institution
Water User Agencies	Volta River Authority (VRA)
	Ghana Water Company Limited (GWCL)
	Irrigation Development Authority (IDA)
	Forest Commission (FC)
	Minerals Commission (MC)
Research Organisations	Water Research Institute (WRI)
	Hydrological Services Department (HSD)
	Meteorological Services Division (MSD)
Regulatory Institution	Environmental Protection Agency (EPA)
Others	Independent Chairman
	Executive Secretary of the WRC
	Chief's Representative
	NGO's Representative
	Women's Representative

Source. Laube and van de Giesen, 2005

– to monitor and evaluate programmes for the operation and maintenance of water resources
– to advise the government on water resources
– to advise pollution control agencies on the management and control of the pollution of water resources

The WRC consists of an executive secretariat (currently an acting secretariat), and representatives of the primary agencies, institutions and scientific organisations involved in water resources management, as well as representatives of a range of stakeholder communities (Table 4). Although all agencies represented on the WRC have ceded some degree of power and autonomy over water resources decision-making, most seem content to voice their concerns and forward their agendas within the commission (van Edig *et al.*, 2002b: 35–36).

The WRC has drafted a Water Policy, yet to be adopted by the Ghanaian Parliament, and a range of water use regulations and tariffs for the abstraction of raw water, approved in 2002. The WRC has also initiated two Basin Pilot Projects, in the Densu and White Volta Basins, where Integrated Water Resource Management (IWRM) strategies will be introduced and tested at the local level. Efforts to build awareness and solicit stakeholder participation in the two pilot basins have been made via public stakeholder forums at which the Basin Commission concept was introduced. Attempts have also been made to identify major water users, in order to bill their abstractions, in three out of the eight Ghanaian regions. Furthermore, a Water Resource Information System (WRIS) will be set up to collect, collate and disseminate relevant information and to fill existing data gaps. Concrete steps to establish international water sharing and management mechanisms within the Volta-Basin are also under way. (MoWH, 2002; Odame-Ababio, 2004).

Underlying statutory law and legislation, however, are longstanding regional traditions, practices and institutions regulating water use. Rivers, creeks, lakes, ponds and water holes are locally important sources of drinking, livestock and irrigation water in many rural areas, and in some deprived urban areas. Such waters are typically perceived as communal property, often held in trusteeship by chiefs or priests and priestesses. Chiefs and spiritual leaders effectively control these water resources, and enforce the local rules governing their use by threatening violators with spiritual or social sanctions. Sometimes these traditional authorities may act in co-operation with local administrative or political bodies. Enforcement of traditional rules sometimes proves to be difficult – traditional and spiritual leaders do not always abide by the trusteeship principle and may compromise the interests of those they represent for economic or political advantage. Furthermore, some of the traditional rules are losing influence as members of communities undergo Christianisation, or the secularizing effects of modernity.

Some traditional norms and values carry only ceremonial meaning whereas others, although spiritually anchored, have practical implications for resource allocation and use. For example, prohibitions on fishing during certain times of the year, or on livestock watering in the vicinity of fetching points, have important environmental, sanitary and distributional outcomes. One important traditional norm is the 'riparian doctrine'. Most Ghanaians believe that those who own land adjacent to or containing water bodies have the right to use this water as they see fit, but *not* the right to deny others access to the resource. This can even apply to privately dug wells, since in many Ghanaian societies it is morally unacceptable to deny others access to drinking water, and denial might in any case lead to severe conflicts. Although the 'riparian doctrine' and other traditional water rights are not recognized formally by the WRC, it rhetorically admits traditional knowledge and practice into the water resources management process. In general, however, because local institutions and norms regarding water resources vary considerably throughout the ethnically and culturally heterogeneous social landscape of Ghana, such traditions cannot be fully incorporated at the level of national or regional regulation, but may well influence local or district-level laws and practices. (Laube and van de Giesen, 2005).

Many of the most important actors in the Ghanaian water resources sector are not formally affiliated with the Government or WRC. Multilateral lending institutions (e.g., World Bank), international and faith-based NGOs (Water Aid, Catholic Relief Services, World Vision, . . .) and bilateral assistance agencies (GTZ, KFW, CIDA, DANIDA, EU Development Fund) play important roles in the development of water resources at scales from basin to local, including the financing and construction of wells, small reservoirs, water supply and sanitation and irrigation systems in small towns and rural areas. Their actions are not in every case fully coordinated with Government water resources development policies, however (Laube, personal communication.)

4. The GLOWA Volta Project: A decision support system for the Volta Basin

The GLOWA Volta Project (http://www.glowa-volta.de), initiated in 2000, is one of five major scientific projects funded by the German Federal Ministry of Education and Research (BMBF) and designed to examine the problems of global climate change and water resources availability, management and use at the scale of major river basins. GLOWA research sites include the Danube and Elbe Basins in Western Europe, the Jordan in the Southern Mediterranean, and the Draa (Morocco), Oueme (Benin) and Volta River Basins in North and Western Africa. GLOWA projects aim to develop and to utilize a range of innovative instrumentation, simulation tools and system integration methods to enable the rational development, management and allocation of increasingly uncertain water resources in each setting (http://www.glowa.org).

Core scientific objectives of the GLOWA Volta Project (GVP) are the following:

1. Analysis of the physical and socioeconomic determinants of the hydrologic cycle in the face of projected global climate change, emphasizing (a) natural variability of precipitation, variations caused by human activities and their effect on the hydrologic cycle; (b) interactions between the hydrologic cycle, the biosphere and land use, (c) water availability and conflicting water uses, and (d) institutional and legal aspects of water resources management; and

2. Development of a scientifically sound Decision Support System (DSS) for the assessment, sustainable use and development of water resources in the Volta Basin.

The GVP is an interdisciplinary research project, with respect both to the network of partner research organizations involved and to the structure of research clusters constituting the Project. The research network in the "North" includes the Center for Development Research (ZEF) at Bonn; the Institute for Climatic Research (IMK-IFU) at Garmisch-Partenkirchen, DLR (German Space Agency) at the University of Wurzburg, and the University of Wageningen, The Netherlands. Important research partners in the Volta Basin include the Water Research Institute (WRI), Savannah Agricultural Research Institute (SARI), Soil Research Institute (SRI) and the Institute of Statistical, Social and Economic Research (ISSER), respectively, in Ghana; and the Institut de l'Environment et de Recherches Agricoles (INERA) in Burkina Faso. Interdisciplinary research themes (clusters) include the following (ZEF, 2002):

- *Atmosphere Cluster*: Projects A1:Regional Climate Simulations and Evapotranspiration Tagging; A2: Hydro-Meteorological Monitoring System; and A3: Onset of the Rainy Season
- *Land Use Cluster:* Projects L1: Land Use Change and Detection; L2: Soil Characterization; L3: Vegetation Dynamics; L4: Modeling Spatial and Temporal Upscaling of Erosion and Hydrological Processes; and L5: Land Use Change Prediction Model (LUCC)
- *Water Use Cluster:* Projects W1: Runoff and Hydraulic Routing; W2: Water and Livelihood; W3: Institutional Analysis
- *Technical Integration and Decision Support Cluster:* Projects D1: Technical Integration of Socio-Economic and Environmental Modeling Subsystems; D2: Household Decision-making and Policy Response; D3: Experimental Application of Scientific Knowledge (Policy Pilot Study); and D4: Initiation of Policy Dialog at Basin Level.

The GVP was designed as a 9-year project, and is currently midway through Phase II of three planned phases. Phase I (2000–2003) consisted largely of establishing the GVP "infrastructure" – the network of biophysical instrumentation sites, the first rounds of socioeconomic surveys and fieldwork, and the establishment and calibration of working relationships with partner institutions within Ghana. The first cadre of 13 M.S. and Ph.D. students from Volta Basin countries completed their academic training and conducted scientific fieldwork in Phase I. During Phase II (2003–2006), research activities have expanded to Burkina Faso. Mesoscale climate modeling conducted at IMK-IFU is increasingly successful in simulating the West African climate, and short-term forecasts for the region are now available online (http://www.glowa-volta.de/atm/forecast.htm). Phase II research efforts are focused on the functional integration of mathematical simulation and modeling tools which will serve as the backbone of the Volta Basin Decision Support System (DSS). A prototype DSS is under development as a component of the WRC White Volta Pilot Project, and should be completed and tested within Phase II. During Phase III (2006–2009), the scope of the DSS will be expanded to encompass the entire Volta Basin, inclusive of transboundary flows, and the system will be transferred to research partners, including GVP-trained scientists, within the Basin.

Although GVP is a scientific project, the research agenda is ultimately guided by the requirements and interests of Basin stakeholders, broadly construed. GLOWA research in West Africa faces a number of exceptional challenges not typically encountered in the "North". Among the most important are the large scale and geographic heterogeneity of the Basin combined with scarcity of existing hydrologic, meteorological and land use data. Reliable socioeconomic data on Basin inhabitants, their livelihood activities, the nature and extent of their uses of water, and the water-related health risks they face are likewise limited in extent. In addition, the Volta is an international basin, encompassing portions of six West African countries and lacking effective transboundary water management protocols, thereby greatly expanding the requirements for institutional analysis. Finally, appropriate scientific expertise and supporting knowledge infrastructure are in relatively short supply within the region, although these skills and resources are essential for the successful implementation of integrated water resources management (IWRM) in the Basin. The following paragraphs describe the methodological approaches the GVP adopted to address these and other challenges. They will be seen to have several factors in common, including interdisciplinary structure, utilization of research networks, frequent consultation with basin stakeholders, and investment in scientific capacity building.

4.1. Climatic observatory and bio-climatic research network for West Africa

Volta Basin historical hydrologic and climatic data are in many areas inadequate for supporting informed water resources decision-making. The GVP strategy to address this data gap involves the assimilation of complementary data from direct measurement and remote sensing within state-of-the-art simulation models. High quality bio-climatic data is acquired through the establishment of a network of six heavily instrumented "super sites" arrayed along North-South (Ejura – Tamale – Navrongo - Boudtenga) and East-West (Kompienga – Boudtenga – Dano) transects in Ghana and Burkina Faso, respectively, and representative of each important climatic-ecological zone. At each site, water and energy budgets are calculated on the basis of direct measurements. In addition, nutrient fluxes, soil erosion and transport and other important biophysical phenomena are measured. GVP staff scientists work alongside and train Ph.D. candidates from Europe and Africa as well as locally recruited technicians, and site instrumentation is transferred to host country institutions over the course of the project. Through these activities, local technical capacity is enhanced in the process of generating the required project data.

A wide range of remotely sensed environmental data is acquired and interpreted by DLR (German Space Agency) scientists and GVP doctoral students, and by scientists at collaborating institutions including the Savannah Agricultural Research Institute (SARI) in Tamale, Ghana. Image classification is facilitated by the availability of ground truth collected at the "super sites," and at other GVP research locations throughout the basin. The integrated analysis of measured and remotely sensed data permits the identification of scaling rules characterizing the spatial behavior of important climatic, hydrologic and land cover variables. To date, Intsiful (2004) successfully developed upscaling laws for roughness length, surface albedo, surface emissivity, insolation factor and several other parameters using 1D and 3D SVAT models (LSM, MM5) linked to an advanced nonlinear parameter estimation tool (PEST). These data and rules provide the basis for calibration and validation of a suite of simulation models, including the mesoscale climate model MM5 (Grell *et al.*, 1994), the physical hydrology model WaSIM (Schulla and Jasper 1998), and a prototype land conversion model, GV-LUDAS (Berger *et al.*, 2004). These models are used in complementary modes, e.g., with MM5 generating boundary conditions for WaSIM, and both providing boundary conditions for land conversion simulations.

GLOWA Volta data acquisition activities are further augmented through the integration of GVP field research into a larger regional *Biophysical Observation Network (BON)* operated in conjunction with BIOTA West Africa (http://www.biota-w02.de), and INERA (Burkina Faso). The creation of a West African biophysical observation network not only increases the spatial coverage and extent of data available to support water resources decision-making within the Basin, it also expands the effective resources of all participating institutes through the sharing of site maintenance costs. GVP and partners are currently negotiating to expand this network further through integration with the African Monsoon Multidisciplinary Analysis (AMMA) flux monitoring project, designed to make continuous observations on a selection of typical land covers under a broad range of climatic conditions over 2 years (Lloyd and Taylor, 2005).

4.2. Socioeconomic research

Socio-economic data on Volta Basin inhabitants, their economic activities and their water requirements and uses are also deficient, yet indispensable to the design of sound water management protocols. This is an area in which mathematical models are largely incapable of supplying otherwise missing data. To remedy this deficiency, GVP designed and is implementing structured household surveys in Ghana (2002, 2005) and Burkina Faso (2005). Six broad research issues are investigated: (1) access of households to safe water supply, (2) determinants of household water demand, (3) household expenditures on water, (4) water-related health indices, (5) factors motivating migration and (6) factors driving land use change. In order to integrate the socioeconomic surveys with parallel GVP investigations in the physical and natural sciences, a *common sampling framework* was developed. Reported data on water use, livelihood and health are linked to the physical environment in which respondents live, through geo-referencing of interview locations and coordinated sampling and testing of soil and water by trained technicians. Water quality samples are collected from the sources actually used by survey respondents, and soil samples collected from the plots they cultivate. This permits the cross-referencing and cross-validating of physical and socioeconomic data sets, and allows GVP scientists to evaluate the participants' actions, choices and responses in the objective context of their physical surroundings (Berger *et al.*, 2002).

Additional insight is acquired through anthropological and sociological studies within the basin. These take place on two primary scales: local studies, in which GVP social scientists integrate themselves into community activities to observe firsthand and to participate in water-related activities. Through such case studies, GVP scientists have acquired a more precise picture regarding the structure of local authority with respect to land and water use decisions (Laube, 2004). At the national and regional levels, GVP researchers are identifying and evaluating the respective roles of institutions and actors within the Ghanaian water sector.

4.3. A scientifically sound decision support system for Volta Basin stakeholders

The primary objective of the GVP is the development of a scientifically sound Decision Support System (DSS) to assist decision-makers within the Volta region (a) in *anticipating* the potential consequences of global climate change as manifested within the Basin; (b) in identifying and, where appropriate, in implementing *avoidance strategies* to pre-empt such consequences; finally (c) in identifying and preparing to implement *adaptation strategies*, where avoidance strategies are unavailable or unlikely to succeed. The analysis must include endogenous trends within the basin, such as deforestation, nutrient depletion, population increase and migration, which interact with global trends to define the state of water resources.

It must be stressed that the DSS is not a "software and data" entity as such, although integrated mathematical simulation models linked to databases are central and indispensable components of the DSS. The decision support process involves the active participation of decision-makers and executive bodies, such as the WRC in Ghana, as well as those affected

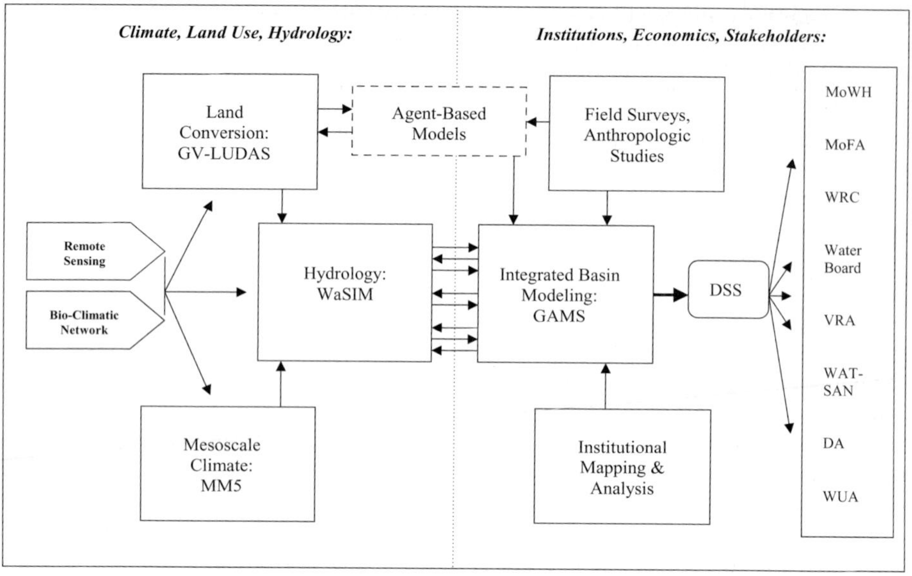

Fig. 2 Schematic representation of decision support system for the Volta Basin

by the decisions, which include such stakeholders as the VRA (hydropower generation), MoFA (irrigated agriculture), and the power consumers and farmers who make up their constituencies. Equally indispensable components of the DSS are mechanisms by which the views and interest of each stakeholder community are represented in the policy simulation process. An illustration will be provided with reference to the WRC/GVP White Volta Pilot Project.

The DSS is represented schematically in Figure 2, which is conceptually partitioned into physical and social science components. The physical science domain is indeed dominated by integrated simulation models: climate (MM5), land conversion (GV-LUDAS) and hydrology (WaSIM), linked via software applications currently under development in collaboration with the Department of Computer Science and Informatics, University of Bonn. Individual components can also serve as stand-alone decision support tools: MM5, for example, will provide short term- to seasonal forecasts which will assist farmers in determining safe planting dates.

The primary DSS integration task, however, involves combining output from physical simulation models with socioeconomic data and institutional rules and constraints in order to test the likely consequences of a set of policy decisions, such as investment strategies, in the face of simulated climate and land use changes. Fundamental to this integration is the interface between the linked atmosphere – land surface – flow system models and the social/policy domain. The policy simulation model under development for the Volta is the integration of a network flow hydrologic model with physical process simulation models of irrigated agricultural production, hydropower generation, municipal (domestic) and industrial demand; and with economic relationships characterized by the relative prices and values of water in its respective uses. The linked climate-hydrologic models provide boundary conditions, and institutional rules, including water rights and international water sharing protocols, provide constraints. The prototype Volta Basin model (Obeng-Asiedu, 2004) has evolved from the simulation-optimization model of the Aral Sea Basin, described in Cai *et al.*

Table 5 GLOWA Volta Advanced Degrees completed or in progress, December 2004

Research area	Europe	Ghana	Burkina Faso	Other Africa
Climatology	1	3		
Hydrology, Water Engineering	1	4	1	4
Geography, Remote Sensing	4	1	1	2
Agriculture, Forestry	1	2	1	
Soils, other Natural Sciences	1	2		
Mathematics, Computer Science	1	1		
Economics, Ag. Economics		4		2
Anthropology, Sociology	4		1	

(2003) and subsequently developed for the Maipo Basin, Chile (Rosegrant *et al.*, 2000), the Brantas Basin, Indonesia (Rodgers *et al.*, 2003) and other locations. The integrated economic-hydrologic model is coded in GAMS (Brooke *et al.*, 1998) and linked to large-scale non-linear optimization solvers (CONOPT, MINOS). An important component of the policy simulation process is the development of *use cases*, which begin with the research or policy question, for example the effectiveness of a particular type of adaptive intervention or investment, and then identify the simulation tools, institutional rules and assumptions and other required inputs, and the flow of data between model components that will be required to generate meaningful outputs. This process begins and ends with stakeholders, who provide the initial questions and rules, and who make use of the outputs.

4.4. Scientific capacity building

The most tangible measure of the GVP's progress to date is the number of scientific staff who have received advanced academic training and credentials through their participation in the project. Scientific capacity building is an explicit project goal, recognized from the outset as essential for success. Capacity building is, moreover, a bidirectional process. Students from the Volta Basin and surrounding countries receive advanced scientific training in Europe, but they also bring with them indispensable knowledge about the physical and human geography, politics and institutional settings of their home countries. This knowledge is particularly useful in the context of field research, which must be conducted in a manner sensitive to local custom and practice.

As of January 2004, 42 individuals have enrolled in GVP-supported graduate degree programs, the great majority in Doctoral programs (Table 5). To date, 13 have completed their degrees. Many have returned to affiliated institutions in Ghana, while others have been retained as postdoctoral scholars, continuing their research within the GVP. As Table 5 indicates, the distribution of graduate degrees reflects the interdisciplinary character of the project: the majority are awarded in natural science, but with significant numbers in the decision and social sciences as well.

5. White Volta Pilot Project

In 2000, the Ghanaian WRC conducted a Water Resources Management Problems Identification, Analysis and Prioritization Study in order to target specific river basins for intervention. In the White Volta Basin, encompassing much of northern and eastern Ghana, the WRC identified the following water sector problems, in order of priority: (a) flooding, (b) water shortage, (c) lack of comprehensive institutional and legal framework, and inadequate management information and data for water management, (d) high fluoride concentrations in groundwater, (e) water pollution and improper land use, and (f) high salinity of groundwater. In some cases, these problems were of national concern. The White Volta was selected as a pilot basin in which IWRM policies and practices could be tested and evaluated. An additional rationale for selecting the White Volta was that it is an international basin, shared with Burkina Faso, where some technical cooperation on water resources management had already been initiated (Amphoma, 2004).

The specific objectives of the White Volta pilot program are:

– To identify and motivate stakeholders for collaboration and participation in IWRM;
– To establish the institutional framework for IWRM;
– To develop a Water Action Plan; and
– To monitor and evaluate the IWRM process, all with a view toward wider application of the lessons learned in the White Volta setting.

The White Volta Pilot Project (WVPP) provides an illustration of the collaborative, decentralized and stakeholder-driven philosophy of the WRC and GVP. Initial WVPP activities included a stakeholder identification process, followed by a consultative stakeholder workshop held in November 2002; and the subsequent constitution of the Basin Board (Advisory Committee). The Board membership consists, *inter alia*, of riparian District Assemblies, research and data collection institutions, water and sanitation NGO's, religious bodies and traditional authorities (Amphoma, 2004). The WRC is currently staffing the WVPP office in Bolgatanga, shared by researchers from GVP and the CGIAR Challenge Program on Water and Food-funded Governance and Modeling project (G&M). Wider co-operation will involve the IUCN, GWP and GEF through their transboundary Land and Water Management Program.

The structure of this collaborative project is summarized in Figure 3, defining the roles and responsibilities of the representatives of the WRC (Basin Officer), GVP (Scientific Coordinator) and G&M (Postdoctoral Scientist). The WRC Basin Officer has overall managerial authority, serves as Basin Board Executive Secretary, and co-ordinates stakeholder activities. He is responsible for developing and implementing the Basin Action Plan, and for reviewing progress toward project objectives. Among his important responsibilities are registering of water use rights, collecting water fees, and generating public awareness of IWRM. He is also liaison with WRC headquarters. The primary task of the GVP Scientific Coordinator is to supervise and to participate in the development of the White Volta DSS. He provides scientific support to the Basin Officer by addressing WVPP data requirements, and by making the scientific process transparent to stakeholders. He will also be involved in scientific capacity building activities, and produce scientific documentation and scholarly publications. The Food and Water Challenge Program – Governance and Modeling Project is a collaboration between the International Food Policy Research Institute (IFPRI), ZEF, and partners in Ghana. The primary responsibility if the G&M resident scientist is to interact with the community of stakeholders in order to develop an understanding of how water use decisions are made at household and community level. These processes will be modeled subsequently using agent-based techniques (Berger *et al.*, 2004), and will in addition provide an important interface with the DSS.

Early evidence of effective engagement by stakeholders was observed at the White Volta Pilot Project Inception Workshop and Stakeholder Consultation held in Bolgatonga, Upper East Region in July 2005. Working groups focused on (1) Institutional Capacity Building for IWRM, (2) existing framework and history of collaboration in the White Volta Basin and (3)

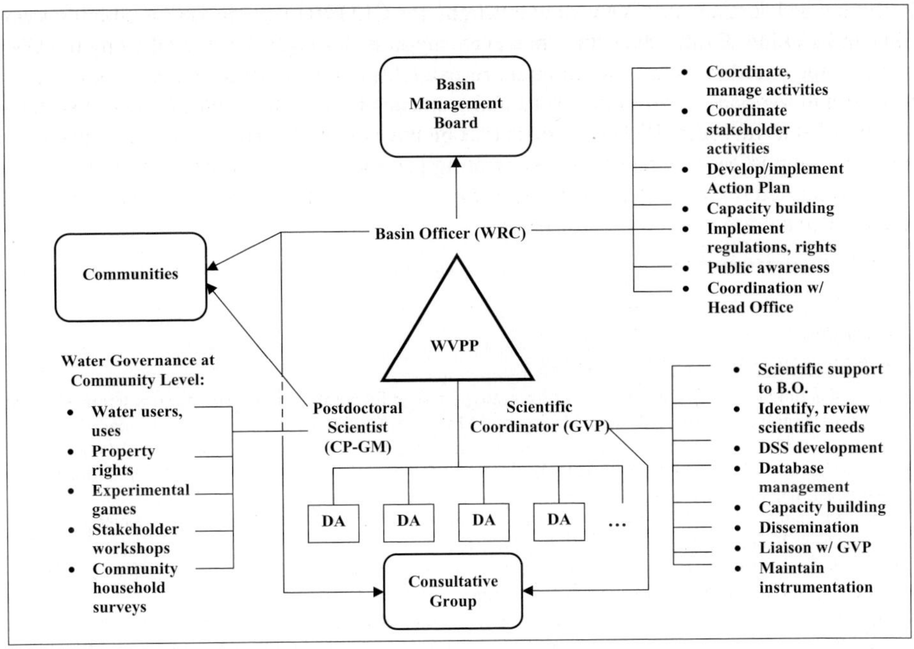

Fig. 3 Structure of White Volta Pilot Project

priority water management concerns within the sub-basin. Participants identified the need for an umbrella group representing stakeholders at all levels, the need for improved groundwater data collection, assessment and management and the need to establish research initiatives on flooding, water quality, waterbourne disease; and on water rights and the funding of water resources activities as high priorities within the basin (WRC, final communiqué July 15, 2005). The White Volta Pilot Project prototype DSS is scheduled for testing in 2006. The lessons learned will be incorporated into the Volta Basin DSS, which is the primary output of Phase III of the GLOWA Volta Project.

6. Summary and conclusions

The problems facing the water sector in the Volta Basin are, in most respects, generic to an increasing number of basins worldwide. These include increasing demand overlaid on static or possibly declining resources, increasing intersectoral competition, deteriorating water quality, weak, ineffective or conflicting rights and governance structures, and limited financial resources available for investment. The situation in the Volta Basin is distinguished largely as a matter of *degree* – the extremity of the Basin's vulnerability to the likely impacts of global climate change, the disproportionate dependency of Basin inhabitants on increasingly risky rainfed agriculture exacerbated by the exceptionally low levels of irrigation investment and irrigated area, the overall fragility of the West African economy and its sensitivity to movements in commodity prices, and the susceptibility of inhabitants to a wide range of waterborne and water-related illnesses. The list could be expanded.

Effective integrated management of water resources at the scale of large basins is difficult under the best of circumstances. Where scientific and financial resources and data are limited, the challenge is greater still. Viewed in this light, the GLOWA Volta Project is a highly ambitious undertaking. Evidence to date, however, suggests that the approach taken by the GVP, emphasizing coordinated interdisciplinary research bridging natural and social sciences, participation in regional research networks and emphasis on scientific capacity building, holds the potential to realize the project's primary objectives in the form of a scientifically sound Decision Support System capable of generating genuinely useful output. However, success in generating credible outputs from the White Volta Pilot Project will represent a critical milestone toward meeting these objectives.

References

Andah W, Gichuki F (2003) Volta basin profile: strategic research for enhancing agricultural water productivity (Draft). Accra: International Water Management Institute, Challenge Program on Water and Food

Amphoma B (2004) The enabling environment for Integrated Water Resources Management (IWRM) in the Volta Basin: the Case of the White Volta Basin. Workshop Paper, Promoting Bilateral Cooperation through Informed Dialog in the Volta Basin, Ho, June 8–10 2004

Andreini M, van de Giesen N, van Edig A, Fosu M, Andah W (March 2000) Volta Basin Water Balance. ZEF Discussion Papers on Development Policy No 21, Bonn

Berger T, Asante F, Osei-Akoto I (2002) GLOWA-Volta common sampling frame – selection of survey sites. Bonn, ZEF unpublished Research Documentation

Berger T, Park SJ, van de Giesen N, Vlek Paul LG (2004) Agent-based integration of disciplinary knowledge of resource use dynamics. J River Basin Manage, Special Issue on the GLOWA Volta Project. Forthcoming

Brooke A, Kendrick D, Meeraus A, Raman R (1998) GAMS: a user's guide. GAMS Development Corp, Washington, DC

Cai X, McKinney DC, Lasdon L (2003) An integrated hydrologic-agronomic-economic model for River Basin management. J Water Resour Plann Manage 121(9):4–17

Center for Development Research (ZEF) (2002) GLOWA Volta phase II – From concepts to application. Center for Development Research (ZEF), Bonn

Grell G, Dudhia J, Stauffer D (1994) A description of the fifth-generation penn state/ncar meso-scale model (MM5). NCAR Technical Note, NCAR/TN-398+STR

Intsiful J (2004) Upscaling of land surface parameters through inverse-svat modeling. Ecology and Development Series No 20. Center for Development Research, Bonn

Kunstmann H, Jung G (2004) Impact of regional climate change on water availability in the Volta Basin of West Africa. Submitted to IAHS, October 2004

Laube W, van de Giesen N (2005) Ghanaian water law and policy: institutional and hydrological perspectives. In: Wallace JS, Wouters P, Pazvakavambwa S (eds) hydrological information in water law and policy: current practices and future potential. Kluwar, Forthcoming

Laube W (2004) Promise and perils of water reform: perspectives from Northern Ghana. Journal of River Basin management, special issue on the GLOWA Volta Project. Forthcoming

Liebe J (2002) Estimation of water storage capacity and evaporation losses of small reservoirs in the upper East Region of Ghana. Diploma Thesis, Geographische Institute der Rheinischen Friedrich-Wilhalms-Universitat, Bonn

Lloyd C, Taylor C (2005) AMMA task team 2 reference document

Mensah KB (1999) Water law, water rights and water supply (Africa). Ghana – study country report. Silsoe, Department for International Development: 28

Ministry of Works and Housing (MoWH) (1998) Ghana's water resources: management challenges and opportunities. Accra, Ministry of Works and Housing

Ministry of Works and Housing (MoWH) (2002) Background Paper on Water resources management and ghana poverty reduction strategy. Final Report. Accra, Ministry of Works and Housing:70

Obeng-Asiedu P (2004) Allocating water resources for agricultural and economic development in the Volta Basin. European University Studies, Series V, Economics and Management, Vol 3096

Odame-Ababio K (2004) Water politics in the Volta Basin – defusing conflict, risks. Workshop Paper, Promoting Bilateral Cooperation through Informed Dialog in the Volta Basin. Ho, June 8–10 2004

Rodgers C, Siregar M, Wahida Sumaryanto, Hendradja B, Suprapto S, Zaafrano R (2001) Integrated economic-hydrologic modeling of the Brantas Basin, East Java, Indonesia: issues and challenges. In: integrated water resources management in a river basin context: institutional strategies for improving agricultural water management, Proceedings of an International Workshop held in Malang, Indonesia January 15–19, 2001. International Water Management Institute, Colombo, Sri Lanka

Rosegrant M, Ringler C, McKinney DC, Cai X, Keller A, Donoso G (2000) Integrated economic-hydrologic water modeling at the Basin Scale: The Maipo River Basin. J Agric Econ (IAEA) 24(1):33–46

Schulla J, Jasper K (1998) Model description WASIM-ETH (Water Balance Simulation Model ETH). ETH-Zurich, Zurich

Sutherland JW, Ben S, Kwame A, Osafu-Kissi A (2004) Innovative approaches to sustainable hydro power production in the volta basin–The VRA Initiative. Workshop Paper, Promoting Bilateral Cooperation through Informed Dialog in the Volta Basin, Ho, June 8–10 2004

Van de Giesen M, Andreini N, van Edig A, Vlek P (2001) Competition for Water Resources of the Volta Basin. In: Regional Management of Water Resources (Proceedings of a symposium held during the sixth IAHS Scientific Assembly at Maastricht, The Netherlands, July 2001.) IAHS Publication no. 268, 2001

van Edig A, Laube W, van de Giesen N (2002) Internationale und nationale Wasserkonflikte: Institutionelle und rechtliche Hintergründe der Wassernutzung des Volta-Flusses am Beispiel Ghanas. Wasserkonflikte in der Dritten Welt. G. M. *et. al.* Mainz, Johannes Gutenberg-Universität Mainz 15:75–94

van Edig A, Engel S, Laube W (2002) Ghana's Water Institutions in The Process of Reform. Reforming Institutions for Sustainable Water Management. SN Bonn *et al.*, German Development Institute

World Bank, MIGA (2004) Project Appraisal Document on A Proposed IDA Partial Risk Guarantee in the Amount of US$ 50 Million for Ghana and a Proposed MIGA Guarantee in the Amount of US$ 75 Million for Sponsors' Equity to the West African Gas Pipeline Company Limited for the West African Gas Pipeline Project. Report No 30335–AFR

World Bank (November 11, 2004) Updated Project Information Document (PID). Africa – 3A – West African Gas Pipeline (IDA S/UP). Report No.: AB493

World Bank (2004) World Development Indicators on-line database (http://publications.worldbank.org/WDI/)

Water Resour Manage (2007) 21:315–329
DOI 10.1007/s11269-006-9055-x

ORIGINAL ARTICLE

Integrating a climate change assessment tool into stakeholder-driven water management decision-making processes in California

David R. Purkey · Annette Huber-Lee · David N. Yates · Michael Hanemann · Susan Herrod-Julius

Received: 20 April 2006

Abstract There is an emerging consensus in the scientific community that climate change has the potential to significantly alter prevailing hydrologic patterns in California over the course of the 21st Century. This is of profound importance for a system where large investments have been made in hydraulic infrastructure that has been designed and is operated to harmonize dramatic temporal and spatial water supply and water demand variability. Recent work by the authors led to the creation of an integrated hydrology/water management climate change impact assessment framework that can be used to identify tradeoffs between important ecosystem services provided by the California water system associated with future climate change and to evaluate possible adaptation strategies. In spite of the potential impact of climate change, and the availability of a tool for investigating its dimensions, actual water management decision-making processes in California have yet to fully integrate climate change analysis into their planning dialogues. This paper presents an overview of decision-making processes ranked based on the application of a 3S: Sensitivity, Significance, and Stakeholder support, standard, which demonstrates that while climate change is a crucial factor in virtually all water-related decision making in California, it has not typically been considered, at least in any analytical sense. The three highest ranked processes are described in more detail, in particular the role that the new analytical framework could play in arriving at more resilient water management decisions. The authors will engage with stakeholders in

D. R. Purkey
Stockholm Environment Institute

A. Huber-Lee
International Food Policy Research Institute

D. N. Yates
National Center for Atmospheric Research

M. Hanemann
University of California, Berkeley

S. Herrod-Julius
U.S. Environmental Protection Agency

these three processes, in hope of moving climate change research from the academic to the policy making arena.

Keywords Climate change · Hydrologic models · Stakeholders · Water planning · Water resource planning models

Background

The two central challenges in California water management are: (1) to overcome the spatial and temporal mismatch between where and when the precipitation occurs and where and when needs arise to use water, and (2) to balance the competing needs for water for off-stream uses in agriculture and urban areas, and for in-stream use for aquatic ecosystems. In California, the mismatch of demand with supply reaches dramatic proportions: two thirds of the state's precipitation occurs north of Sacramento, while over two thirds of the state's water use occurs south of Sacramento; in addition, over 80% of the total precipitation occurs between October and March, while about 75% of all water use in California occurs between April and September. The challenge is, thus, to ensure that water is available in the right place and at the right time for both humans and ecosystems.

Since work by Gleick (1987), Lettenmaier *et al.* (1991) and others, it has been recognized that climate change in California could exacerbate both of these problems. Climate change is likely to cause more winter precipitation to fall as rain than snow and to lower the water content of the snow on the ground, leading to an increase in winter runoff as a fraction of total runoff, an increase in the frequency of winter floods, and an earlier start of spring snowmelt. In the absence of additional storage, these changes would have the effect of reducing the state's effective water supply. A large volume of subsequent research has generally reinforced these conclusions, including recent work by Dettinger and Cayan (1995), Glieck and Chalecki (1999), Miller *et al.* (2003), Lund *et al.* (2003), Brekke *et al.* (2004), Dettinger *et al.* (2004), Stewart *et al.* (2004), and VanRheenen *et al.* (2004). Moreover, there is evidence that some of these changes are already under way, with clear signs of a warming trend in California over the past two decades (Dettinger *et al.*, 2004) and the peak snowmelt runoff now occurring one to three weeks earlier in various watersheds of the Sierra Nevada.

Most analyses of the effects of climate change on the California water system are based on simulations of global climate change prepared for use in the 2001 report of the Intergovernmental Panel on Climate Change (IPCC). The newest results from two of the major General Circulation Models, the Hadley model and PCM, became available at the end of 2003, and these were downscaled to California and analyzed (Hayhoe *et al.*, 2004). Of the two models, Hadley is considered a medium climate sensitivity model and PCM a low climate sensitivity model. Both models show sharply different climate impacts for California in more recent versions, relative to previous work.

While the details differ among the models (and depend on the specific emissions scenario being considered), both models now suggest that a sharp increase in summer temperatures in California is a plausible future scenario. Previous versions of these models had shown a warming of about 1–4 °C in the winter by the end of the century, and a similar degree of warming in the summer. The new versions suggest a slightly higher level of warming in the winter (about 2–4.5 °C), but a substantially warming in the summer, amounting to 2.5–4.5 °C under a low emissions scenario (B1) and a dramatic increase of 4.5–9.5 °C under a high emissions scenario (A1). A consequence of the temperature increase is a sharp decline in the

Sierra Nevada snowpack. All of these future changes are subject to the level of uncertainty associated with the CGMs and emissions scenarios used to produce them.

By the beginning of April in a "good" water year, the total amount of water stored in the Sierra snowpack roughly equals the total amount stored in major reservoirs; thus, the snowpack effectively doubles the ability to store water for warm-season uses. By mid-century, the snowpack is projected to decline by about 25–40%. Toward the end of the century, the loss of snowpack could reach 30–70% (4,300–11,100 million m^3 of storage) under the low emissions scenario, and a stunning 70–90% (11,000–13,500 million m^3) under the high emissions scenario (Hayhoe *et al.*, 2004).

It remains to be seen whether the new versions of the other major GCMs will show similar changes when downscaled to California. However, it is clear that climate change has the potential to cause some major disruptions to the California water system, starting within the next two or three decades and continuing over the rest of the century, by which time projections show a doubling of population. Climate change is likely to exacerbate the mismatch in the timing and location of precipitation and to sharpen the competition between off-stream and in-stream water uses. The predicted reduction in the snowpack and the earlier timing of snowmelt will greatly complicate the task of managing California's reservoirs, and make for a more difficult tradeoff between filling reservoirs to capture runoff for warm-season uses versus leaving empty space for flood control in the event of a possible late winter storm. Any future adjustment of the current reservoir operations regime in response to this tradeoff also has implications for meeting ecosystem objectives in the system.

In light of the emerging consensus that climate change will have an impact on California hydrology and the management of the state's hydraulic infrastructure, it seems prudent that water management decision-making processes underway need to factor these changes into supporting analyses. At present this is typically not the case. There are several reasons for the apparent reluctance to consider the potential implications of climate change on water resource systems. But the most basic reason relates to the legal frameworks used for project planning, typically the National Environmental Protection Act or the California Environmental Quality Act. Climate change is not considered in most water resources planning efforts because there is a general perception that significant changes in hydrology will not occur within the typical 20–30 year planning horizon of most NEPA and CEQA studies.

While this conclusion may be legitimate within the legal framework used for project planning, it belies the fact that many of the decisions being made have implications that extend beyond this time horizon. A new reservoir, for example, is typically assigned a useful life of 100 years. Investments made to restore damaged ecosystems seek to assure the viability of key, at-risk species in perpetuity, not just for a few decades. Another reason given for discounting climate change in water resource planning and decision making, is the uncertainly inherent is future climate predictions. While there is recognition that new infrastructure and ecosystem restoration investments must perform over more than 20 to 30 years, the feeling is that future scenarios are difficult to delineate within the limits of the "reasonable and foreseeable" standard used to define future scenarios in NEPA and CEQA studies. As stated previously, the preponderance of analysis seems to be converging on the conclusion that climate change is foreseeable.

Even if change is coming, integrating climate change assessment into water resource decision making processes is hampered by a lack of suitable analytical frameworks for rigorously evaluating the impact of a range of future climate scenarios. In California, most analysis conducted in support of water resource planning responds to the question of how the systems might perform differently should (i) a project be implemented and (ii) the past 70 plus years of hydrology repeats itself in the future. Decision-makers are very used to evaluating

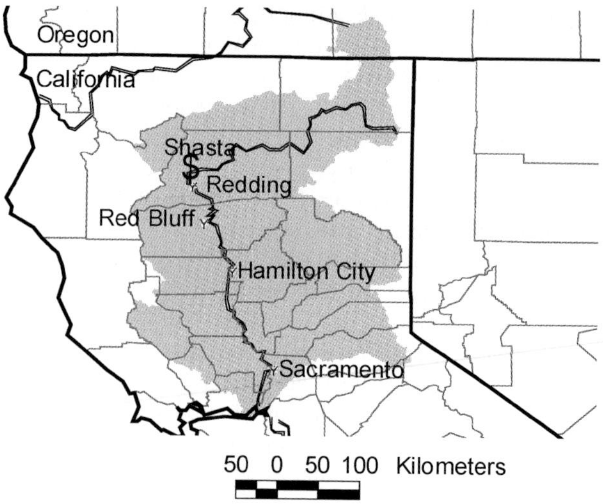

Fig. 1 The Sacramento Valley of California

future projects based on how well they would perform during the 1928–1934, 1976–1978, and 1987–1992 droughts. This reliance on historical hydrology has led to the development of a number of analytical packages that are tightly bound to the historic hydrology and which fail to consider different climate and hydrologic futures.

With the support of the U.S. Environmental Protection Agency, the authors developed an integrated hydrology/water allocation framework for the Sacramento Valley (Figure 1) which apart from being calibrated against historic conditions is unbound from the historic hydrology. While this framework responds to the third challenge, and the emerging consensus about the likelihood of future climate change reduces the uncertainty associated with future climate scenarios, the first reason given for discounting climate change, namely the relatively short planning horizon of most planning studies, remains. After presenting a summary of the integrated hydrology/water allocation framework, this paper identifies specific decision-making processes that are well suited for climate change analysis in the face of this claim and describes the role that the new analytical framework could play in arriving at water management decisions that could be resilient in the face of climate change.

Here it is interesting to note that while the Federal government has not fully embraced the concept that human induced climate change can create major problems in the water sector, the framework described here is gaining traction in California as senior level policy makers are focusing increasing attention of climate change. For example, on June 1, 2005 California Governor Arnold Schwarzenegger signed an executive order stating "that the Secretary of the California Protection Agency shall report to the Governor and the State Legislature by January 2006 and biannually thereafter on the impacts to California of global warming, including impacts to water supply." The current integrated hydrology/water allocation framework will be used as part of this analysis. The executive order also called on California to: by 2010, reduce GHG emissions to 2000 levels; by 2020, reduce GHG emissions to 1990 levels; and by 2050, reduce GHG emissions to 80 percent below 1990 levels.

In spite of this progress, however, actually factoring climate change into water resource planning and decision-making is not universally done. This paper attempts to explain some of the factors that limit this consideration and to provide a method whereby decision-making processes in which such considerations could be successfully implemented can be readily identified.

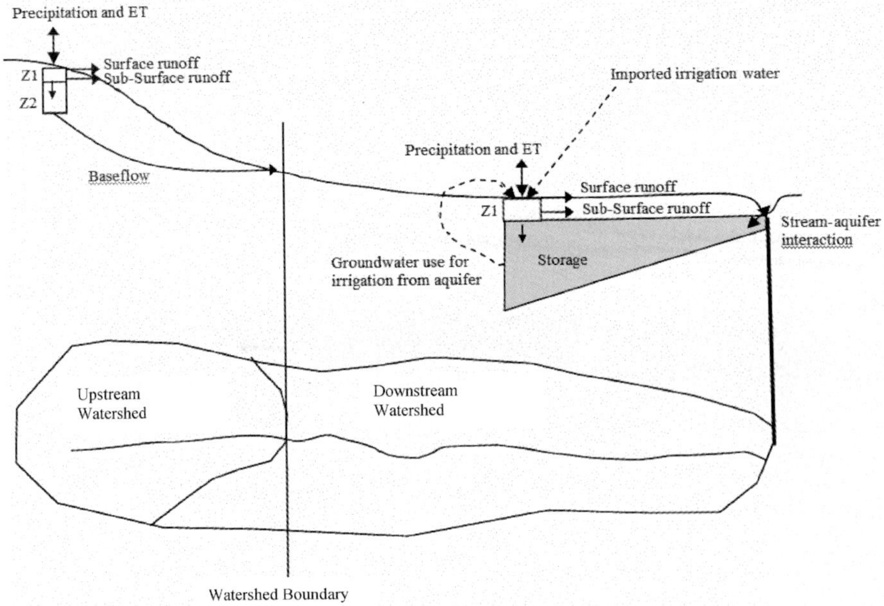

Fig. 2 WEAP conceptual model of upstream and downstream physical watershed processes

The integrated hydrology/water allocation framework

At the most basic level, the integrated hydrology/water allocation framework (Yates *et al.*, 2005a,b), which has been constructed on the Water Evaluation and Planning (WEAP) platform (Raskin, 1992), recognizes that water supply is defined by the amount of precipitation that falls on a watershed. Further, this basic supply is progressively depleted through natural watershed processes, where the watershed itself is the first significant point of depletion through evapotranspiration (Mahmood and Hubbard, 2002). The residual supply, after the satisfaction of evaporative demands throughout the watershed, is the water available to the water management system. Thus, as in the physical realm there is a seamless link in the WEAP framework between climate, land use/land cover conditions, and the management of the water system. This approach also allows for joint management of blue and green water, as described by Falkenmark and Rockström (2004).

Specifically the natural watershed process component of WEAP accounts for two different hydrologic realities (Figure 2). The first is the concept that precipitation in upstream watersheds, with complex topography, steep slopes, and abrupt hills and valleys, contributes to gaining streams with a relatively short time lag (Burness *et al.*, 2004; Eckhardt and Ulbrich, 2003; Winter *et al.*, 1998; and Winter, 2001). Conversely, downstream watersheds with flatter terrain tend to overlie alluvial aquifers linked to river systems to which they can contribute flow and from which they can receive seepage, depending on hydrologic conditions. These groundwater systems also provide storage that can be used to satisfy demands. The WEAP framework also allows for use of surface water supplies imported into a watershed in order to satisfy demand, supplies which are managed by the installed hydraulic infrastructure.

An application of this integrated framework was developed for the Sacramento Valley, which includes three of the primary surface water storage facilities in the California water

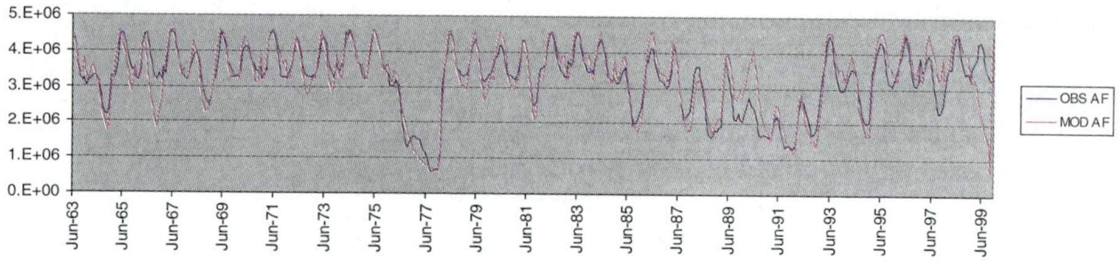

Fig. 3 Observed and modeled end-of-month storage in Lake Shasta

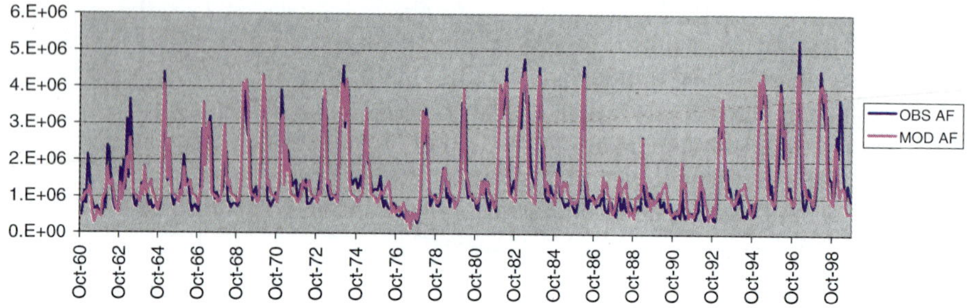

Fig. 4 Observed and modeled monthly Sacramento river flows at freeport

system, Lake Shasta, Lake Oroville and Folsom Lake. The operation of these facilities is based on the assumption that a large portion of the available water supply in the spring months is stored in the higher elevation snow pack. Figures 3 and 4, show the degree to which the calibrated version of the integrated framework was capable of capturing the hydrology and management of the system.

This framework can be applied under future climate change scenarios to investigate how the hydrology could impact associated ecosystem services. Here it is important to point out that there is a great deal of uncertainty associated with future climate scenarios. As such, it would be inappropriate to provide decision makers with a single set of future alternatives. While the Integrated Hydrology/Water Allocation Framework is not automated to generate uncertainty estimates by running an ensemble of future climate scenarios, it can be run in coordination with a tool that can. This is, in fact, an area of ongoing research in California.

An inventory of California water management decision-making processes

Through a series of interviews with informed individuals in the stakeholder community, the authors developed a list of environmental decision making processes currently underway in the California water system that are potentially sensitive to climate change. A sample of ten of those decisions processes are selected to illustrate the range of issues faced by stakeholders and decision makers in the state, as summarized in Table 1. This sample includes decisions ranging from the local to the state, with decision timeframes that are ongoing, on set intervals, or one-time. The impacts of these decisions, while at times viewed in relatively near-term by the decision makers themselves (such as land-use decisions that only consider a 20-year time horizon), are in fact potentially hundreds of years in duration, and longer when extinction of species are considered.

Table 1 3S overview of water management decision processes

	Decision process	Decision summary	Decision timeframe	Impact timeframe
1.	Sacramento river flood control project	State of California is developing a policy related to the financial exposure to flood damage risk.	1–2 years	10–100 + years
2.	Local land use decision-making	Local land use developments must assure that sufficient water supply exists over a 20-year planning horizon.	Ongoing	0–100 + years
3.	Statewide energy planning	California Legislature requires an Energy Policy Report to recommend state policies for current and pressing energy issues.	2 years	2–100 + years
4.	Statewide water planning	The California Water Plan, prepared once every five years, serves as the foundation for local water management decisions.	1–3 years	10–100 + years
5.	Integrated storage investigations	This program is designed to identify promising surface storage opportunities and to quantify both the costs and benefits of new storage projects.	3–5 years	10–100 + years
6.	Ecosystem restoration investments	The Ecosystem Restoration Program (ERP) is tasked to improve habitat and ecological function in the Bay-Delta system and the recovery and support of important at-risk species.	3–5 years	5–100 + years
7.	Yolo bypass shallow water habitat restoration	A consortium will determine if the Yolo Bypass can be operated for improved environmental services without compromising its agricultural water supply function.	1–2 years	5–100 + years
8.	California legislative hearings	The Select Committee on California Water Needs will hold hearings to discuss how federal, state, and local water agencies are planning for climate change.	1–2 years	5–100 + years
9.	Small dam removal (e.g. Battle Creek)	Pacific Gas and Electric, Department of Fish and Game, and community groups will decide whether full or partial dam removal is the most economically and ecologically beneficial approach.	1–2 years	5–100 + years
10.	Hydropower re-licensing (e.g. Yuba, American, and Bear)	Pacific Gas and Electric and community groups will define the most economically and ecologically beneficial strategy for operation of reservoirs.	1–2 years	5–100 + years

In spite of the wide range of decisions, a general observation from these interviews is that few decisions currently being made in the California water arena consider the potential implications of climate change in any formalized manner (with the exception of the legislative hearings which are specifically around climate change considerations). Indeed, in a number of cases the potential impact of climate change is viewed to be too uncertain and too far off to be of much importance relative to the myriad of other factors that influence water management decision-making. In addition, there appears to be a significant disconnect between the water management and the climate change research communities in California. Interviews with climate change researchers typically failed to generate many insights on specific decision-making processes where climate change information was needed, likely because the members of the climate change research community do not actively participate in the decision-making process.

Given the potential impact of climate change, how can better climate change information be introduced into these decision processes? Three criteria were developed to determine what could be the major factors for lack of consideration of climate change in these processes:

- *Sensitivity*: The success or failure of the project could be strongly influenced by climate change;
- *Significance*: The associated potential impacts of climate change are substantial enough to merit a climate change assessment; and
- *Stakeholder support*: Some segment of the stakeholder community has expressed a concern about the potential impact of climate change on the project, thus increasing the chance for climate change to be introduced into the decision-making process.

Together these standards comprise a 3S standard that the authors found useful in assessing the importance of a climate change assessment to a decision making process. Each decision-making process is reviewed against these three criteria, and given a high, medium or low rating, as shown in Table 2. To get a sense of how these criteria are applied, each standard is described for one of the processes.

Starting with the *sensitivity* standard, the Sacramento River Flood Control Project (SR-FCP) is in the process of defining a remedy to the financial exposure to flood damage risk that was assigned to the State of California by a recent court decision related to a significant Central Valley flood event in 1986. In this flood, the Linda levee on the Yuba River failed, releasing a 1.2 m high wall into nearby communities. The levee that failed was approximately 80 years old and was constructed by a local entity using mining debris that was piled up without any compaction. It was also aligned on top of porous remnant channels of the Yuba River without constructing a foundation. In the 1920's the Federal and State governments created the SRFCP, which included numerous existing levees, including the Linda levee, often with limited modifications.

After the failure, a group of flood victims sought damages from the State of California, and following 18 years of litigation the court assigned liability for damage caused by the levee failure to the state, even though the levee was constructed by another entity. While the legal justifications for the decision are somewhat arcane, the fact is that the State of California has been exposed to future liability for the failure of many miles of poorly constructed levees that it acquired during the creation of the SRFCP. Policy-makers are now developing a response to this new financial exposure. This decision-making process is given a 'high' sensitivity standard because the flood risk is real and significant based on the magnitude of current flood events, and could increase due to climate change. This process will lead to a plan to indemnify the state against the damage caused by future levee failures could look significantly different

Table 2 3S ranking of decision processes

	Decision process	Sensitivity	Significance	Stakeholder support	Intervention potential
1.	Sacramento river flood control project	High	High	Low	Medium
2.	Local land use decision-making	Medium	Medium	Low	Medium
3.	Statewide energy planning	High	High	Low	Medium
4.	Statewide water planning	High	High	High	High
5.	Integrated storage investigations	High	High	Medium	High
6.	Ecosystem Restoration investments	High	High	Low	Medium
7.	Yolo bypass shallow water habitat restoration	High	High	Low	Medium
8.	California Legislative hearings	High	High	Low	Medium
9.	Small dam removal	High	High	Low	Medium
10.	Water rights permitting	High	High	Low	Medium
11.	Local flood protection Initiative	High	High	Low	Medium
12.	Hydropower re-licensing	High	High	Low	Medium

if climate change were to be factored into the discussion, as damages could be both higher and more frequent.

One of the more challenging determinations in applying the *significance* standard is around the class of local decision making processes leading to the approval of new residential, commercial, industrial and mixed-use real estate developments. Historically land-use decisions in California have been made by cities and counties with the assumption that a local water supplier would expand its service area to include the new development. This changed in 1992 when Contra Costa County approved the residential development project and identified the East Bay Municipal Utility District as the water provider. East Bay MUD objected claiming that is had insufficient supplies to meet projected demand within its existing service area. In response, the California Legislature enacted laws in 1995 (SB 901) and 2002 (SB 610 and SB 221) that sought to build an assurance of sufficient water supply into land-use decision-making in California.

The combined implication of these laws is that cities and counties must include a Water Supply Assessment (WSA) in the documents considered in the approval of real estate development projects. The basic premise is to assure that sufficient water supply exists over a 20-year planning horizon, even in the case of "multiple dry years." While the details of what constitutes an acceptable WSA are being worked out, frequently through litigation, they are being prepared for projects currently under consideration. In reality climate change is probably not a significant factor in the approval of these projects as other planning considerations such as transportation and educational infrastructure dominate the discussion. Further no single project can likely be assigned the responsibility for a potential failure of the statewide water delivery system under a dramatically different climate and hydrologic future. However, the accumulation of these local land use decisions could be affected by climate change. However, this is a discussion better suited to higher level water planning dialogues, such as the statewide water planning described below. Therefore this decision process was ranked medium in terms of significance.

The majority of decision making-processes are ranked low according to the *stakeholder support* standard. Looking at the Statewide Energy Planning process, the Integrated Energy Policy Report is called for on a biennial basis by the California Legislature. The report seeks to: identify historic and current energy trends; forecast and analyze potential future energy developments; and recommend new policies for current and pressing energy issues facing the state. The most recent version was published in 2003 and work is underway to prepare the 2005 edition. One mandated component of the Integrated Energy Policy Report is the Electricity and Natural Gas Assessment Report, which among other objectives seeks to assess trends in electricity and natural gas supply, demand, and wholesale and retail prices for electricity and natural gas and assess the adequacy of electricity and natural gas supplies to meet forecasted demand growth. This study helps to inform generation and demand decisions that could be made within the next two years by analyzing their possible intended and unintended consequences through the coming decade.

While there is a recognition that climate change may have a long-term impact on both the overall demand for electricity and the supply generated by installed hydroelectric capacity, this process is geared towards relatively short-term adjustments in the California energy sector. The stakeholders involved with the preparation of the report have many complex considerations to balance in planning these short-term adjustments which limits potential enthusiasm for climate change assessment. The stakeholders involved with this planning dialogue do not necessarily see the value in adding additional complications to the process.

Each of the ten decision processes had the 3S standard applied. Three examples of decision making processes that ranked highest according to the 3S standard, and are therefore good potential candidates for intervention, are the process of updating the Statewide Water Planning, which is already considering climate change, the Integrated Storage Investigation that has some stakeholder interest in incorporating climate change into the decision process, and Ecosystem Restoration Investments in the Central Valley, which has no stakeholder directly interested in looking into climate change. Each of these processes is presented in further detail, including how the new climate change analytical framework could play a role in improved decision-making, in the following sections.

Statewide water planning

The California Department of Water Resources is mandated by the Legislature to update the California Water Plan (Bulletin 160) once every five years. This document serves as the statewide foundation upon which a myriad of local water management decisions are made (Significance). The last update was published in 1998 and the next edition was scheduled for release in 2003. To date, it has not been released due to the fact that the approach taken in developing the document has undergone major reform since Bulletin 160–98 was published. Historically, the approach to develop projections of future demand and to compare these to the yield provided by currently installed water infrastructure under average and dry conditions (Sensitivity). The analysis typically lead to an assessment of how much additional supply development was required to meet anticipated demand. Further, Bulletin 160 was historically developed by DWR staff with only limited input from the public.

This has changed with Bulletin 160–2003, which has adopted a new portfolio approach to water planning which has its origins in financial planning. Much like an investor would analyze the potential value of a financial portfolio by making different assumptions about the performance of individual assets, the new Bulletin 160 will consider the future, and by nature uncertain, role that a range of factors will play in determining "future" balances between water supply and demand in California. DWR has been guided in this transformation by an advisory panel comprised of over 70 stakeholders. For the first time, one of the potential factors that may influence these futures, climate change, has been recognized and considered (Stakeholder support).

According to information released by the Department of Water Resource, in addressing global climate change in the current update of Bulletin 160, rather than focus on causes of global climate change, the update will look at the potential impacts of climate change on water resources in California and potential strategies for adapting to these changes. The word commonly used to describe this approach is "qualitative." The department suggests, however, that future updates of the Water Plan will contain more intensive evaluations of climate change as more data become available, modeling techniques are improved, and management strategies implemented. The intention is to develop a more quantitative assessment of climate change in future editions of Bulletin 160.

The qualitative assessment of climate change in the current version of the Bulletin 160 will be contained primarily in a two papers included in a chapter on climate change in the document referred to as the Reference Guide. The first of these papers is a survey of the literature documenting the current understanding of global climate change and its potential impact on California. The second paper is a compilation of data for California that attempts to describe the extent to which the climate shifts may already be underway and to lay out important markers that can be used to monitor future changes in climate and hydrology. The decision was made, however, not to include climate scenarios in the analysis of various future portfolios included in this edition of the document, but instead, to spend some time evaluating various analytical platforms for potentially including this analysis in the next version of the document.

The DWR staff responsible for selecting an analytical platform for future analysis is considering the integrated hydrology/allocation climate change assessment tool developed by the authors and understands its unique integration of watershed response and water management. With a next phase of support from the U.S. Environmental Protection Agency (EPA) the authors will inform the decision regarding the ultimate selection of a model by developing a study on water management tradeoffs associated with future climate change. This will be accomplished by running the tool under a variety of climate scenarios and priority/preference landscapes and the development of a matrix of tradeoffs. The results of these analyses will then be provided to DWR decision makers. Ongoing meetings between the authors and DWR staff are guiding the formulation of the climate change scenarios that will be investigated using the integrated framework.

Integrated storage investigations

The CALFED Bay-Delta Program is an initiative of several federal and state agencies designed to develop and implement a plan to better balance the off-stream and in-stream uses of water in California. As part of the CALFED Record of Decision published in 1999, a commitment was made to launch the Integrated Storage Investigation (ISI). This program is designed to identify promising surface storage opportunities and to quantify what stand to be both the substantial costs and benefits of new storage projects (Significance). Storage programs are part of the CALFED water management strategy that combines storage with program actions such as conservation, water transfers, and habitat restoration. Together these complementary actions will contribute to meeting CALFED's water supply reliability, water quality, and ecosystem restoration objectives. The analytical test of performance typically applied in this assessment is how CALFED actions will perform during the dry periods that characterize California hydrology (Sensitivity).

Since its inception, the ISI has successively narrowed the field of candidate surface storage projects to a current list of five projects. These include:

- Raise Shasta Dam on the Sacramento River
- Construct an off-stream reservoir in the Sacramento Valley
- Construct an in-Delta storage facility by converting a Delta island to a reservoir
- Construct an off-stream reservoir in the San Joaquin Valley
- Raise or Replace Friant Dam on the San Joaquin River

Studies are currently being conducted to assess the viability of each of these projects with goal of developing draft environmental documentation by the end of 2006.

The potential off-stream storage facility in the Sacramento Valley is located in the sparsely populated valley in the Coast Range Mountains. The name of the single community in the valley, Sites, is used to describe the Sites Reservoir project. This facility, which could have a capacity of up to 2,500 million m^3, will be operated by diverting water from the Sacramento River. Water will be returned to the system from storage either by delivering it to water users on the west side of the Sacramento River in exchange for their normal Sacramento River diversions or through the construction of new conveyance works from the propose reservoir back to the Sacramento River. One issue of concern is the impact that the potential diversions and returns will have on the flow regime in the Sacramento River and on the important in-stream benefits supported by this flow regime. A group of stakeholders has spent over two years designing a required flow regime that could be used to guide the operation of Sites Reservoir. This group recognizes that climate change could significantly impact the components of this flow regime (Stakeholder support), but to date has not had the ability to bring climate change directly into the process.

The benefits associated with a major water storage project such as Sites Reservoir will depend on the characteristics of future climatic and hydrologic regimes. Using the next phase of support from the U.S. EPA the authors will use the integrated hydrology/allocation climate change assessment framework to simulate the operation of Sites Reservoir under a variety of climate future climate scenarios. The results of this analysis will be provided to interested stakeholders in the context of the larger decision-making process. As the authors are actively involved with the applications of analytical tools less suited for climate change assessment to the analysis of the Sites Reservoir storage option, this will allow for useful benchmarking of the integrated framework against other models in current usage in California. In modeling rich environments such as California this technical benchmarking is a critical step in generating support for climate change analysis.

Ecosystem restoration investments

The Ecosystem Restoration Program (ERP) of the California Bay-Delta Authority (formally CALFED) has been created to meet several important objectives. These can be summarized as improving habitat and ecological function in the Bay-Delta system and the recovery and support of important at-risk species. Since its inception seven years ago the ERP has invested tens of millions of dollars (Significance) in a variety of ecosystem restoration projects designed to help assure regulatory compliance for other aspects of the CALFED program, such as water supply and flood control. These projects fall into six broad categories related to the ERP goals that include at-risk species, ecological processes, harverstable species, habitat restoration, non-native invasive species, and environmental water and sediment quality. One critical issue in assessing the utility of these investments is whether over the decades-long time scale anticipated for the realization of a return on ecosystem restoration investment, climate change will have an impact on the design and ultimate success of a particular investment (Sensitivity).

The Yuba River offers an excellent example of this issue. The Yuba drains a watershed of approximately 3,500 square kilometers from the crest of the Sierra Nevada to the confluence of the Feather River near Marysville and Yuba City in the northern Central Valley. The north fork of the Yuba River flows into Bullards Bar Reservoir above the confluence of the north and middle forks. Further downstream, the middle and south forks of the Yuba River flow into Englebright Lake, which provides water-based recreational benefits; 55 million m^3 of stored water-right capacity; and hydroelectric generation to meet the annual energy needs for 50,000 homes. The height of Englebright Dam effectively blocks fish migration although biological data suggests that the Yuba River above Englebright historically had habitat that supported anadromous fish species. The Upper Yuba River may present an opportunity to improve habitat for native anadromous fish species whose populations are in decline, while developing a comprehensive plan that will restore ecological health, improve water management and provide positive benefits to the public.

In 1998, the ERP recommended a studies program to determine if returning steelhead trout and spring-run salmon to the Yuba River was feasible by changing Englebright Dam. Through active public involvement and collaborative efforts, stakeholders agreed on key issues and concerns to be addressed in the studies, including upstream and downstream habitat, water quality, sediment, flood risk management, water supply and hydropower, and economics. Study plans were developed for each issue and consultants were engaged to implement the plans. The implication is that if the studies reveal that the restoration of anadromous fish to the Upper Yuba is feasible, then additional funds will be invested to make the necessary structural and operational changes in the system.

As the Yuba River is a classic snowmelt driven Sierra Nevada watershed, there is a strong possibility that climate change will have an influence on the hydrologic conditions in the basin, and that these conditions may have a bearing on the viability of any proposed anadromous fish recovery strategy. Under the second phase of support from the U.S. EPA the authors will refine the current formulation of the Yuba River watershed in the hydrology/allocation climate change assessment tool and add climate change considerations to the ERP analysis. This information will be provided to stakeholders in the hope of introducing climate change into the process. Stakeholders interested in climate change will collaborate with the authors in developing the refined model representation of the Yuba River system and in the definition of future climate and watershed management scenarios.

Conclusions

Future climate change has the potential to substantially alter the hydrologic regime within which water management in California takes place. This nesting of water management within a hydrologic regime motivated the development of the integrated hydrology/water allocation climate change assessment framework embedded in WEAP. This tool has been unbound from past hydrology and is driven solely by the climate signal that will evolve over the course of the coming century. It is uniquely suited to introducing climate change assessment into water management decision-making processes and an understanding of tradeoffs.

Not all water management decision-making processes, however, are necessarily amenable to the introduction of a climate change impact assessment at this time, as awareness of the importance of climate change generally remains low. The authors have developed and applied the 3S standard which weights the Sensitivity of the project under consideration to climate and hydrologic variability, the Significance of the project in terms of the contemplated investment, and the degree of Stakeholder support for a climate change impact assessment. While numerous decision-making processes fail to rank high on all of the 3S thresholds, three have been identified where the application of the integrated hydrology/water allocation framework is warranted: statewide water planning; the integrated storage investigation; and ecosystem restoration investments. These three processes have expressed varying degrees of interest in including climate change analyses into their processes, with the Statewide Water Planning having the most, and the Ecosystem Restoration Investments having the least level of interest. Future support from the U.S. EPA will be used to conduct climate change impact analyses in support of these planning dialogues. By working to introduce climate change analyses into these processes, we can learn about barriers to inclusion of climate change research more broadly.

Introducing climate change into decision-making processes represents both a challenge and an opportunity. The challenge is to convince decision-makers for water policy that it is in their interest to consider climate change in their decisions, although they are still not entirely convinced is needed. The opportunity is to begin to move climate change research from the academic to the public policy arena – one that is taken on directly in the approach presented here. It is heartening, that at the highest level of state government, there in an increasing interest in better understanding the potential impact of climate change. What is needed, however, is more than just high level assessments. Instead, each individual water management decision should consider the potential impact of climate change. The use of the integrated framework as part of the collaborations described in this paper is a first step in implementing this recommendation.

References

Brekke LD, Miller NL, Bashford KE, et al. (2004) Climate change impacts uncertainty for water resources in the San Joaquin River Basin, California. J Am Water Resour Assoc 40(1):149–164

Burness S, Chermak J, Brookshire D (2004) Water management in a mountain front recharge aquifer. Water Resour Res 40:2160

Dettinger MD, Cayan DR (1995) Large-scale atmospheric forcing of recent trends toward early Snowmelt runoff in California. J Clim 8(3):606–623

Dettinger MD, Cayan DR, Meyer M, et al. (2004) Simulated hydrologic responses to climate variations and change in the Merced, Carson, and American River basins, Sierra Nevada, California, 1900–2099. Clim Change 62(1–3):283–317

Eckhardt K, Ulbrich U (2003) Potential impacts of climate change on groundwater recharge and streamflow in a central European low mountain range. J Hydrol 284(1–4):244–252

Falkenmark M, Rockström J (2004) Balancing water for humans and nature: the new approach in ecohydrology. Earthscan Press

Gleick PH (1987) The development of a water-balance model for climate impact assessment: modeling the Sacramento basin. Water Resour Res 23(6):1049–1061

Gleick PH, Chalecki EL (1999) The impacts of climatic change for water resources of the Colorado and Sacramento-San Joaquin River Basins. J Am Water Resour Assoc 35(6):1429–1441

Hayhoe K, Cayan D, Field CB, Frumhoff PC, Maurer EP, Miller NL, Moser SC, Schneider SH, Nicholas Cahill K, Cleland EE, Dale L, Drapek R, Hanemann RM, Kalkstein LS, Lenihan J, Lunch CK, Neilson RP, Sheridan SC, Verville JH (2004) Emissions pathways, climate change, and impacts on California. Proc Nat Acad Sci 101(34):12422–12427

Lettenmaier DP, Sheer DP (1991) Climatic sensitivity of California water resources. J Water Resour Plann Manage January/February 117(1):108–125

Lund JR, et al. (2003) Climate Warming & California's Water Future. A report for the California Energy Commission. Center for Environmental and Water Resource Engineering, University of California, Davis. Sacramento, California. Available in Internet at http://cee.engr.ucdavis.edu/faculty/lund/CALVIN/

Mahmood R, Hubbard K (2002) Anthropogenic land-use change in the North American tall grass-short grass transition and modification of near-surface hydrologic cycle. Clim Res 21(1):83–90

Miller NL, Bashford KE, Strem E (2003) Potential impacts of climate change on California hydrology. J Am Water Resour Assoc 39(4):771–784

Raskin P, Hansen E, Zhu Z (1992) Simulation of water supply and demand in the Aral Sea Region. Water Int 17:55–67

Stewart IT, Cayan DR, Dettinger MD (2004) Changes in snowmelt runoff timing in western North America under a 'business as usual' climate change scenario. Clim Change 62(1–3):217–232

VanRheenen NT, Wood AW, Palmer RN, et al. (2004) Potential implications of PCM climate change scenarios for Sacramento-San Joaquin River Basin hydrology and water resources. Clim Change 62(1–3):257–281

Winter T (2001) Ground water and surface water: the linkage tightens, but challenges remain. Hydrol Process 15(18):3605–3606

Winter T, Harvey J, Franke OL, Alley W (1998) Groundwater and Surface Water, A single resource. U.S. Geological Survey Circular 1139, Denver, Colorado

Yates D, Purkey D, Galbraith H, Huber-Lee A, Sieber J (2005b) WEAP21 a demand, priority, and preference driven water planning model: part 2, Aiding freshwater ecosystem service evaluation. Water International. (in press) (http://weap21.org/)

Yates D, Sieber J, Purkey D, Huber-Lee A (2005a) WEAP21 a demand, priority, and preference driven water planning model: Part 1, Model Characteristics. Water International. (in press) (http://weap21.org/)

Water Resour Manage (2007) 21:331–349
DOI 10.1007/s11269-006-9056-9

ORIGINAL ARTICLE

Involving stakeholders in integrated river basin planning in England and Wales

Paula Orr · John Colvin · David King

Received: 20 April 2006

Abstract The European Water Framework Directive provides a new impetus to manage river catchments in more integrated, joined-up ways. This article looks at the role of stakeholders in integrated catchment management. Taking the work of the Environment Agency as a case study, the article begins by looking at recent successes at managing water related issues and the role of stakeholders in this. It then looks at ways in which water environments continue to be vulnerable, particularly to diffuse pollution, some development practices and climatic changes. It argues for the need for more integrated management responses, characterised by collaborative and inter-disciplinary learning to manage the interdependencies, complexities and uncertainties of catchments as integrated systems. This will require both the strengthening and streamlining of current approaches to stakeholder engagement, as well as the development of new approaches. The article concludes by outlining recent work by the Environment Agency to shape these new arrangements for stakeholder engagement, and by reflecting on the lessons learned from this.

Keywords Integrated catchment management · Public participation · Social learning · Stakeholder involvement · River basin management plans · Water Framework Directive

1. Introduction

In Europe, there is a very long tradition of stakeholder involvement in water management. For example, in the Netherlands the founding of the Polder Boards in the 10th Century recognised the importance of the involvement of local landowner organisations; while in Spain, water irrigation associations as a form of management by users can be traced back to Roman and Arab times (Patel and Stel, 2004).

In the UK, over the past 30 years there has been a gradual intensification of stakeholder and particularly public involvement in water related issues, prompted in part by public fears

P. Orr · J. Colvin · D. King
Environment Agency, Rio House, Waterside Drive, Aztec West, Almondsbury, Bristol, BS32 4UD
e-mails: paula.orr@environment-agency.gov.uk; john.colvin@environment-agency.gov.uk; david.king@environment-agency.gov.uk

and concerns and in part by wider changes in patterns of governance. With the introduction of more integrated approaches to water management as a result of the Water Framework Directive, the role of stakeholders in water management is undergoing further changes (Rees *et al.*, 2005).

In England and Wales, the Environment Agency has been designated as the sole 'competent authority' for implementing the Water Framework Directive (WFD). This means that the Environment Agency is solely accountable to the EC for reporting on progress towards implementation (Defra, 2003). However, the Environment Agency is in no position to implement the WFD on its own and will rely on the collaboration of a wide range of other partners if it is to ensure successful implementation.

This paper addresses some of the challenges involved in shaping national policy on integrated water management and some of the lessons learned along the way. In setting out the case for more integrated catchment management, we look at what integration means to us, and how the ways in which stakeholders work together lies at the heart of a more integrated approach.

The findings reported in this paper are the result of an 'action research' process (Reason and Bradbury, 2001) primarily involving the first two authors (Orr and Colvin) as members of the Environment Agency's river basin planning project team. As a member of the Environment Agency's team of directors, the third author (King) has had a less 'hands on' but nonetheless key role in the learning process. These findings therefore reflect a process of learning and self-evaluation ('formative evaluation') by the Environment Agency rather than an external evaluation through more objective scientific study or by stakeholders ('summative evaluation').

The paper starts by outlining the Environment Agency's current arrangements for stakeholder engagement in water management. We consider why these arrangements might need further development in order effectively to address the challenges raised by the Water Framework Directive and other drivers for improved water management. We then outline the process we have been through over the past 18 months, of working with stakeholders to shape a new set of proposals from the Environment Agency for stakeholder engagement in integrated river basin management. We end the paper with some reflections on the lessons learned from this process.

2. Current arrangements for stakeholder engagement

2.1. Recent achievements in water management in England & Wales

To put the current role of stakeholder engagement into context, we need to look briefly at current arrangements for water management in England and Wales, including the achievements these have brought to society and the environment.

Tackling point source pollution is one major area of achievement. As a result of our regulatory activity over the past decade, the chemical and biological quality of water in our rivers and lakes has improved markedly, particularly as a result of improved effluent quality from sewage works and industrial discharges (Environment Agency, 2002a). The microbiological quality of bathing waters has similarly improved (Defra, 2002). In the 21st century we can now find salmon in the Tyne, Tees and industrial South Wales rivers where they were absent for much of the 20th century, due to major reductions in pollution. Otters are now returning to rivers where they have been absent for decades.

The amount of water abstracted for the public water supply is slowly decreasing, thanks to a reduction in leakage from our supply systems, although this masks major regional differences[1]. Farmers have increased the capacity of winter storage reservoirs for irrigation tenfold between 1982 and 2001.

Regulating for improvements in water quality and quantity is a major area of responsibility for the Environment Agency. In addition to this, we manage or contribute to a raft of other water or water related management plans. For example we:

- manage flood risk through the development of Catchment Flood Management Plans (CFMPs) and Shoreline Management Plans (SMPs);
- protect fish populations through Fisheries Action Plans (FAPs) and Salmon Action Plans (SAPs);
- contribute to the enhancement of biodiversity through Biodiversity Action Plans (BAPs);
- contribute to the management of nitrates from farming in Nitrate Vulnerable Zones.

A recent success story is the way in which our urban rivers have become a focus for urban living, rather than something to "back onto". As our rivers become cleaner (and safer), "facing onto water" increasingly provides a focus for urban regeneration schemes. Rivers that are cleaner and safer, and able to support wildlife, offer an aesthetic focus for regeneration and urban living, bringing quality of life as well as development and other economic benefits (Environment Agency, 2002b).

Healthy rivers also provide valuable opportunities for recreation. It is estimated that 3% of our adult population enjoy fishing, canoeing, boating and windsurfing. Our inland waters also provide recreational benefit for many more, such as walkers and birdwatchers. Grouped together recreational use of water and watersides is significant. The 2002/3 GB Leisure Day Visitor Survey (Countryside Agency) reported that over 250 million recreational visits were made to water with an associated spend of £2.5 billion (Environment Agency, 2005a).

2.2. Who are our 'stakeholders'?

Much of this success derives from the ways in which we work with our stakeholders. We already do much of our work collaboratively and have reaped the benefits. We draw on the wealth of knowledge coming from different sources, we can co-ordinate our plans and actions with those of others and we are more likely to recognise potential conflicts early on when there is still time to find solutions.

This work brings us into contact with a very broad range of organisations and individuals, who have many different types of 'stake', views or interests in the water environment. Some organisations, like the water companies or the Ports Authority, share responsibilities for managing the water environment. Others have stakes as users of the water environment, for example for water abstraction or for recreation, or because they are affected by the state of the water environment, for example aesthetically.

In planning our work with stakeholders we distinguish between four different categories: stakeholder regulators; professional stakeholder organisations; local stakeholder organisations; and members of the public (Table 1). These categories are not watertight – there are overlaps between them – but they help us plan for different types of interactions with what is otherwise a very broad and mixed grouping of organisations and individuals.

[1] In the south east, the increasing number of household is leading to greater demand for water.

Table 1 Categories of stakeholder for river basin planning and management

Type of stakeholder	Definition	Explanation
Co-deliverers	Agencies and institutions	All organisations with statutory powers to implement the basic measures needed to deliver River Basin Management Plans.
Professional stakeholder organisations	Professional organisations	Public and private sector organisations, professional voluntary organisations and NGOs. This category can include academics, industry, insurance, business, and conservation organisations with paid professional staff. All those professional organisations or individuals acting in a professional capacity who use the water environment and whose activities impact on it
Local stakeholder organisations	Local groups: non-professional organised entities, operating at the regional/local level	Communities centred on place e.g. tenants, residents and amenity associations, site action groups; communities centred on interest, e.g. farmers, fishermen, bird-watchers, with no paid staff.
Members of the public	Individuals representing themselves, not groups or business groups	Individual residents, users, workers in area, business owners, landowners, farmers, visitors from outside area.

2.3. Current arrangements for stakeholder engagement

Over recent years we have adopted a more consistent approach to working with others, based on:

- early engagement;
- providing clear information;
- being transparent about what we are doing and how others can contribute;
- listening to others and understanding their needs and interests;
- providing opportunities for involvement that make efficient use of time and resources and allow stakeholders to have an effective input.

In each Agency region and in Wales our work is supported by the following Statutory Committees: Flood Defence Committees, Environmental Protection Advisory Committees and Fisheries, Ecology and Recreation Advisory Committees. These Committees are made up of representatives of a range of interests, including business, local and regional authorities, academics and environmental non-governmental organisations (NGOs). Water is a concern for all Committees. Our Statutory Committees have three main roles:

- they provide advice from different perspectives to help us work more effectively;
- with their wide networks of contacts, they help us initiate and develop new partnerships and influence the agendas of others;

– they monitor and advise us on how well we are delivering, challenging us to continuously improve our operations.

Advice and support from our Committees has helped us to develop our thinking on river basin planning, and we will continue to draw on their expertise and experience in the future.

In addition to engagement with our Statutory Committees, we have many other relationships and mechanisms for involving stakeholders. The overall decrease in trust in public authorities means that we need to work harder to get people interested in and supportive of our activities, especially at the local level. Our Building Trust with Communities programme is providing training and support to enable staff around the country to develop stronger relationships with the communities they serve (Environment Agency, 2004a). While no two places or contexts are the same, sharing experience is helping to spread common principles and good practice.

On some subjects such as water abstraction and flood risk management we find it useful to bring stakeholders together to discuss problems and possible solutions. Since 2001 we have used local Stakeholder Groups to help us develop our Catchment Abstraction Management Strategies (CAMS), and a similar approach is being taken in drawing up Catchment Flood Management Plans (CFMPs). On the Humber, close collaboration between the Agency and around 30 members of the Stakeholder Forum meant that the Humber Estuary Shoreline Management Plan was approved with little conflict (Environment Agency, 2001).

We cannot afford just to wait for other organisations to come to us with their needs. In many places we are also actively involved in bringing environmental issues into wider debates and decisions, for example in relation to development and regeneration, particularly through Local Strategic Partnerships (LSPs) in England and Community Strategic Partnerships (CSPs) in Wales (Porter *et al.*, 2005). The Agency currently contributes to some 75% of LSPs/CSPs and in many places we are part of strong local networks (Environment Agency, 2004b).

Other networks and fora give us the opportunity to talk and listen to and work with groups of diverse stakeholders with common interests. Associations like the Coastal Fora or the Rivers Trusts that exist in many parts of England and Wales are important institutions that help us to find solutions to complex problems. In tackling flooding, the National Flood Forum provides an important point of contact for communities around the country who are at risk from flooding. It also provides a forum in which the Agency can talk with these communities and look at ways in which together we might improve solutions to flood warning and flood risk management.

The Agency links in to and supports these networks wherever possible. Table 2 gives examples of how the Agency is currently engaging with stakeholders around water based issues in both Agency-led and stakeholder-led fora.

3. Moving towards integrated catchment planning

3.1. Current challenges

Despite recent achievements, substantial challenges remain. Diffuse water pollution is one of the major challenges we face. Diffuse water pollution arises from many small point sources, that in themselves may be minor, but collectively can have a severe environmental impact. Diffuse pollution often follows rainfall and affects both surface water and groundwater. Diffuse pollution results from widespread activities in rural and urban areas and may be transported via the atmosphere as well as directly onto land or into water. Diffuse pollutants

Table 2 Examples of existing agency consultation and engagement

Work with Agency-led stakeholder groups
Catchment Abstraction Management Strategies (CAMS) – Consultation with Stakeholder Groups
A stakeholder group of around 8 to 10 key stakeholders is set up in the early stages for each CAMS as the main mechanism for active engagement. For example, for the Don and Rother CAMS the group had seven members: one from British Waterways, one representative each for farming interests, fisheries, local industry, a local authority ecologist, a university representative and the local water company (Environment Agency, 2003).
The main role of the Stakeholder Group is to provide information. The group has opportunities for discussion, but the final decision rests with the Agency.
Other roles of CAMS Stakeholder Groups include:
– To assist in the sustainability appraisal and express views on what the water status should be;
– To provide cost benefit information and any additional information on the implications of the different options;
– To discuss the responses received on the formal consultation on the draft CAMS document.
The Humber Estuary Shoreline Management Plan – working with partners in a small liaison panel
The Environment Agency encouraged the setting up of a steering group of 25 members in 1997. The group was drawn from a wide range of bodies, including Local Authorities, statutory and voluntary nature conservation and heritage organisations, Internal Drainage Boards and Associated British Ports. The Agency also created a liaison panel consisting of six non-Agency members, including a local farmer, an English Nature representative, local authority and industry interests, along with four Agency staff, which was used for detailed work and to debate issues at an early stage. The larger steering group provided their views when issues had been more fully addressed.
The approach was successful in ensuring that potential problems were highlighted early on. In some cases external organisations drafted key sections of the Plan to reflect their concerns. The consultation approach allowed progress in a situation where there was considerable potential for conflict. Active engagement on this project was combined with formal consultation on documents. The Agency published its Shoreline Management Plan ('Planning for the Rising Tides') in 2001 (Environment Agency, 2001).
Work with local stakeholder-led initiatives
West Country Rivers Trust – Cornwall Rivers Project
The West Country Rivers Trust, in conjunction with the Agency, Royal Holloway College, BDB Associates and South West Water, with backing from a range of voluntary organisations, is developing a sustainable way of protecting Cornwall's rivers through "empowering local communities". This is being done by direct contact with stakeholders in 10 geographical areas, through the use of a web site and by working with stakeholders on issues of importance in their area. The Project brings groups together, using free field visits to discuss on-site land use, to develop site-specific management plans and to consider options to improve land use.
River Valley Initiatives (RVIs)
These were established by the Mersey Basin Campaign, in conjunction with local businesses, community organisations, local government and residents, to raise awareness of local waterways. The Darwen RVI consists of 10 partners, including United Utilities, Lancashire County Council, the Agency, British Waterways and local councils. The RVI encourages active involvement in issues related to the river, assists groups in the clean up of stretches of the riverbank and promotes measures for water quality improvement. The RVIs arrange school visits in some of the most deprived wards of the North West, reaching children from sectors which have not previously been engaged. When the children go home they pass on information to parents and guardians.

include nitrates, phosphates, and pesticides, the main sources for which include agriculture and run-off from urban areas.

We cannot easily control diffuse pollution by traditional permitting methods and the Environment Agency is developing an overall approach to deal with the problems. Our

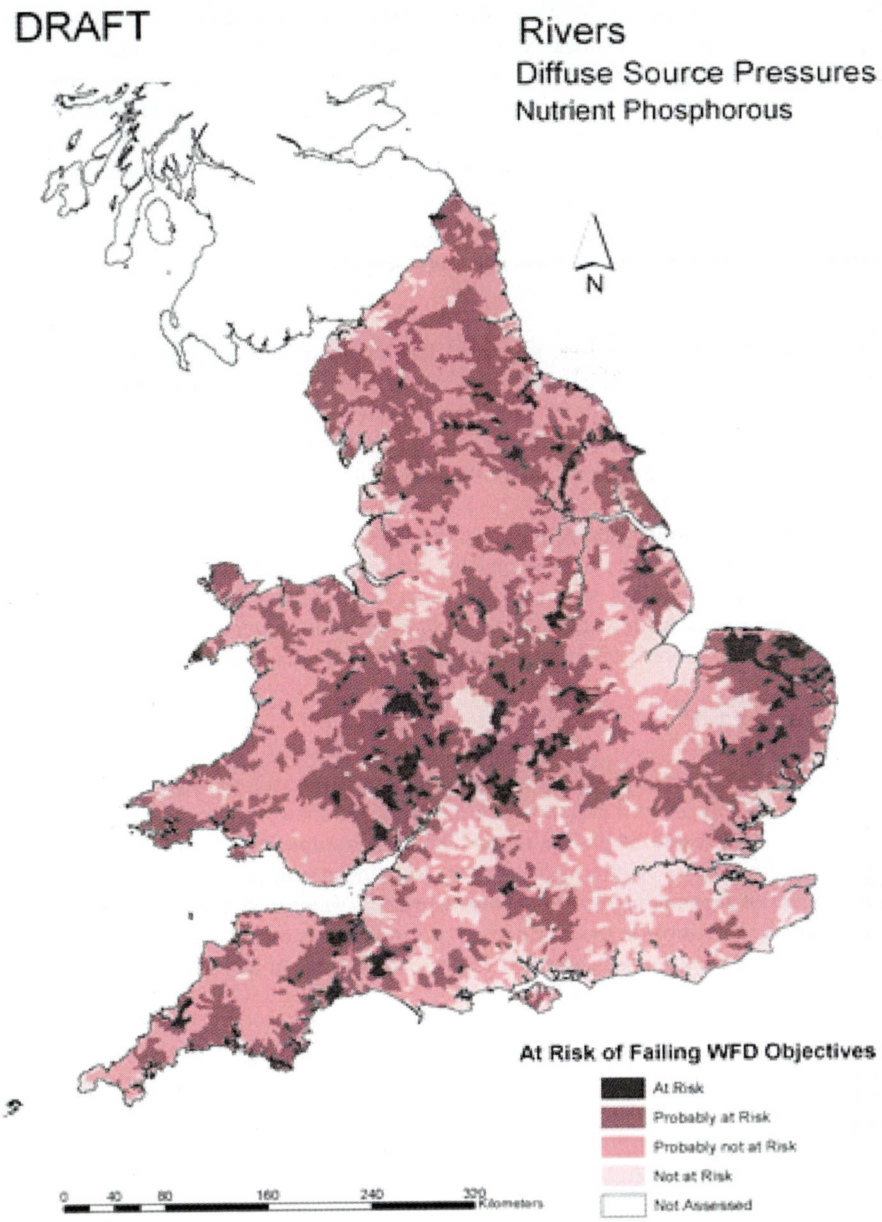

Fig. 1 Catchments in England and Wales at risk of not achieving WFD objectives in 2015 due to diffuse phosphorus

recent WFD "characterisation" maps show that many of our water bodies are currently at risk of failing to achieve the objectives of the WFD by 2015 (Figure 1). Much of this risk can be ascribed to diffuse water pollution (Environment Agency, 2004c). We are currently assessing where action to control diffuse water pollution is needed through the river basin planning process and further risk assessment "characterisation" within this process.

In many parts of England and Wales, there will be further challenges as a result of climate change and land use development. Predictions of how the UK's climate may change as a result of greenhouse gas emissions have important implications for how we manage our water resources. Recent scenarios suggest that by the 2080s winters will become warmer and wetter, with more intense periods of rainfall; summers may become warmer and dryer, especially in the south east; and extreme weather events such as periods of heavy rain and coastal storm surges will become more frequent (Hulme *et al.*, 2002).

In terms of land use development, the strong growth in expected household numbers poses a further challenge for water supplies and water quality, especially in the south and east of England. Not only is this the driest part of the country, but it is where growth is expected to be greatest.

There is still much to be done to restore fish stocks and biodiversity to our rivers and other water bodies. Despite the increases on some previously polluted rivers, salmon stocks remain severely depleted, with 70% of rivers failing to meet their conservation limits in 2002 (Environment Agency, 2004d). The UK biodiversity strategy shows that many aquatic species and habitats have suffered major declines; a number of these are now the subject of biodiversity action plans.

While changes to water quality and quantity have both contributed to these declines, the dredging and straightening of rivers and the drainage of wetlands has also damaged the wildlife they supported. It is now recognised that flood management needs to be integrated with other river uses, with restoration projects to restore the natural form and function of rivers, while still providing appropriate standards of flood management.

3.2. What is needed?

Many people have been arguing for a more integrated approach to catchment management and integration appears as a key theme in the Water Framework Directive. We have begun to move towards greater integration in the UK – for example, our AMP process has started to bring together water quality and quantity. Nonetheless, at present we and others use a large number of separate plans for different aspects of water management (Table 3). Recent work on the River Ribble in the North West of England has estimated that the Environment Agency itself is working with over 20 water-related plans at any one time (Environment Agency, 2004e).

Table 3 Examples of agency and others' catchment related plans that run in parallel rather than in a joined up way

	Plan	Organisation responsible for producing the plan	Focus of the plan
AMP	Asset Management Plan	Environment agency	Discharges to water
CAMS	Catchment Abstraction Management Strategy	Environment agency	Water abstraction
CFMP	Catchment Flood Management Plan	Environment agency	Flooding
NVZ	Nitrate Vulnerable Zone	Defra	Nitrates
SAP, FAP	Salmon Action Plan; Fisheries action plan	Environment agency	Fisheries issues
BAP	Biodiversity Action Plan	Environment agency/others	Biodiversity enhancement

There is clearly also a need to link water and land use planning more carefully. The Water Framework Directive has assumed major importance in the water and engineering sectors but received relatively little attention in the land use planning sector (White and Howe, 2003).

Integration, then, is needed in a number of areas:

- *Integration of water plans* – as shown in Table 3, current management approaches remain fragmented;
- *Much better linkages to other planning processes* – especially with land use planning; but also to sustainable communities plans;
- *Whole catchment approach* – strategic plans tend to retain a local focus on sites/schemes rather than looking at bigger system;
- *Improved assessment of costs and benefits* – externality costs of agriculture are currently ignored and multiple values poorly linked.

Perhaps one of the reasons why we have failed over the years to reach the holy grail of integrated catchment management is that it requires us to develop new management approaches. This is because the more issues that are considered within a single planning process – i.e. the more sources, pathways, impacts and potential responses that are included – the greater the number of inter-dependencies that must be understood, and the more complex the picture that is required.

The benefit of such a picture is that it provides a far better representation of what's really going on. The difficulty is that as the picture gets more complex, not only is it potentially more difficult to create and to understand, but it may also throw up greater uncertainties, raising new controversies.

This isn't just theory – this is what we have also been finding out for ourselves over the last couple of years. In managing our Water Framework Directive Programme we have found that we have needed to develop new ways of managing the inter-dependencies and complexities raised by a wealth of different initiatives and research projects (Collins *et al.*, 2005).

We are learning that while we should be able to manage some of this complexity through the development of new 'whole catchment modelling' techniques, more sophisticated modelling on its own is insufficient. We also need to take a 'social learning' approach, where we learn together how to make sense of complex problems, and adapt our ways of managing these in the light of what we have learned (Box 1) This in turn requires new ways of working with our stakeholders – both internal and external.

In fact, the Common Implementation Strategy (CIS) guidance for public participation under the WFD already points us in this direction:

> "The implementation of the WFD should be done together although the text of the Directive is not explicitly requiring an active participatory approach. The future will also require a more intersectoral approach and a broader view on water management, crossing established boundaries and embankments" (EU, 2003. p. 56)

We have come to recognise that for integrated catchment management to work, we need to work with stakeholders:

- to build better, joint understanding of river management problems;
- and to build better, joint understanding of potential river management solutions;
- so that we can ensure *effective co-delivery* of jointly agreed solutions.

While not all measures for delivering ICM will involve these types of collaborative agreements, the measures that are agreed for each catchment – which in most cases will involve a

Box 1 Social learning in river basin management

The value of a social learning approach to integrated catchment management has received considerable research attention in recent years (Pretty and Chambers, 1994; Pahl-Wostl, 2002; Mostert, 2003; Craps, 2003; Collins, 2004; Jiggins and Roling, 2004). These authors emphasise that in the process of resource management, social involvement (e.g. the generation of social capital, the development of new social practices) is as important as content management (e.g. the use of models to predict the effect of measures to achieve good ecological status for a river). The outcomes of the management process are not only technical qualities such as an improved state of the environment but also relational qualities such as an improved capability of actors in a basin to solve conflicts and come to co-operative agreements. Social learning is seen as an iterative and ongoing process that comprises several loops and enhances the flexibility of the socio-ecological system and its ability to respond to change.

The EC SLIM research project has further highlighted the value of social learning towards the co-creation of knowledge into the causes of, and the means required to transform, situations of uncertainty, complexity, interdependence and controversy. Drawing on several case studies, the SLIM researchers have highlighted the role of stakeholders and stakeholding, the role of enabling institutions and of facilitation skills, and the role of conducive policies in fostering social learning for integrated water management (SLIM, 2004). This project particularly highlights the value of social learning as a policy option.

The Environment Agency has been working with the SLIM team since late 2003 in order to begin to embed practices of social learning within the Agency's culture (Collins *et al.*, 2005)

Box 2 Water for life and livelihoods: the Environment Agency's vision for River Basin Planning in England & Wales

We will ensure sustainable water use and a vibrant water environment for wildlife and people, integrating the management of land and water by:

- creating a more integrated, long-term approach to river basin planning and management;
- working closely with our partners and providing increased opportunity for stakeholder engagement;
- aiming to achieve environmental, social and economic benefits concurrently;
- directing limited resources to where they will bring about the greatest benefit (Environment Agency, 2005b).

mix of regulatory, economic and social instruments – will stand a better chance of succeeding if they have been jointly developed and agreed.

4. Shaping future arrangements for stakeholder engagement

4.1. Setting a strategic framework

While the Agency has a long history of consulting and engaging with stakeholders, the move to integrated catchment management under the Water Framework Directive requires a more ambitious approach. We will need not only to ensure that people have access to the information used in river basin planning and management, but also to encourage stakeholders to get actively involved. This is expressed in our vision for River Basin Planning in England & Wales (Box 2, second bullet).

4.2. How stakeholders have been involved in developing our ideas

To turn this vision into a set of practical proposals, we have worked closely with our stakeholders to explore different options and possibilities. We felt it was important to involve our stakeholders in this process from the outset, not only in order to 'walk our talk', but also as a direct means of learning to work in new ways. In developing our proposals, we drew both on practical experience within the Agency and on the results of a number of collaborative projects around England and Wales:

- We have supported research and trials into innovative methods for engaging with stakeholders in the SMURF, PURE (for example, Box 2), Water4All, Stockbridge and CYCLEAU projects[2];
- We have gained experience from the EU pilot project in the Ribble River Basin (NW England) focusing on public participation and river basin management (Environment Agency, 2004e);
- We have learnt from work being carried out by other organisations and institutions such as the Rivers Trusts and the Mersey Basin Campaign's River Valley Initiatives.

During 2003–2004 we also held three rounds of workshops to test our thinking with stakeholders working at different scales and in different parts of England and Wales (Rees *et al.*, 2005). These workshops covered:

- Review of the advantages and disadvantages of initial models for stakeholder involvement (two workshops with local and regional stakeholders in the Ribble River Basin and one workshop with national stakeholders participating in Defra's WFD Stakeholder Group for England).
- Testing specific approaches to involving others in river basin planning (four workshops: two with local and regional stakeholders in the Ribble and Tamar River Basins, one with regional stakeholders in London and the South East of England, and one with national, regional and local stakeholders in Wales).
- Preliminary consultation on proposed arrangements for River Basin Planning and Stakeholder Engagement (two workshops with national stakeholders in England and Wales and three workshops with stakeholders from the English regions).

Common themes emerging from these discussions referred to the need for different approaches to engagement at different scales:

- At national level, the existing Stakeholder Group, convened by the Department of the Environment, Food and Rural Affairs, should have a continuing role;
- At River Basin District level, the Environment Agency should take a consistent approach across the country to working with stakeholders;
- At Catchment and water body level, there are existing networks that provide an effective mechanism for involving many stakeholders in river basin management. The Environment

[2] SMURF (Sustainable Management of Urban Rivers and Floodplains): EU Life partnership project led by the Environment Agency investigating how the principles of river basin planning, encapsulated in the Water Framework Directive, can be applied to highly modified and degraded urban river catchments (www.smurf-project/info).

PURE North East: EU Interreg project looking at how different kinds of partnership contribute to sharing knowledge and improved environmental stewardship (www.teamrevival.org; Doyle, 2005).

Water4All: Sustainable management of groundwater through spatial planning. The project is based on the River Slea in Lincolnshire (Lovett *et al.*, 2004).

Stockbridge Pathfinder: Environment Agency project to explore local work with communities and other stakeholders to improve responses to dealing with both the aftermath of flooding and longer-term flood prevention (Wilkinson and Colvin, 2005).

CYCLEAU: EU Interreg project seeking innovative approaches to planning and managing natural water resources in the coastal zone. The focus is (i) on combining resource planning and management in the coastal zone with whole area catchment planning and management and (ii) on involving local communities in the management of the catchment. The focus in the UK is on 8 catchments in the South West (www.cycleau.org).

Table 4 Administrative scales for the water framework directive

Term	Definition
National	National administrative area
River basin districts	EU reporting unit – River basin management plans cover one River basin district
Catchments	This may be a whole river basin* or a smaller catchment in the case of long rivers such as the Thames
Waterbody	Individual management unit

*A unit comprising all rivers that exit into a single estuary

Agency should where possible build on these dynamic but sometimes fragmented processes, to increase understanding and develop capacities to manage water environments;

- The Environment Agency should take a risk-based approach to involving local communities and individuals, matching our action to the degree of concern in different locations about particular problems.

4.3. Future arrangements for stakeholder engagement

As a result of this process of research and learning, we have developed the following proposals for a framework for stakeholder engagement in River Basin Planning (Environment Agency, 2005c):

4.3.1. Working at a number of different scales

The new arrangements will need to operate at the scales established in the Water Framework Directive (Table 4 describes the different scales; Figure 2 shows the boundaries of the nine River Basin Districts in England and Wales and the two districts that cross the border between England and Scotland). These scales provide reference points and will not necessarily be the only focus for participation. We want to encourage people to discuss and get involved in work on the water environment in ways that make sense to them.

A key element of the framework will be its capacity to co-ordinate planning at a number of different scales (Figure 3). The scales described here are linked to one another. We want to ensure that stakeholder involvement at each scale informs the others, so that participants get a better understanding of each others' perspectives and are able to contribute effectively to decision-making. A key role for the Agency will be to facilitate links between the different planning levels.

4.3.2. Arrangements at the national scale

At this scale, organisations and institutions working across England and Wales will contribute to shaping national policy and instruments (e.g. economic instruments, regulatory measures). This will strengthen commitment to common policy principles; we hope it will also encourage organisations to promote support for these principles among their own members. A wide range of national stakeholder interests will need to be represented. The existing WFD Stakeholder Group set up by Defra allows for closer engagement at this level and we would like to see a similar body set up in Wales and co-ordinated by the Welsh Assembly Government.

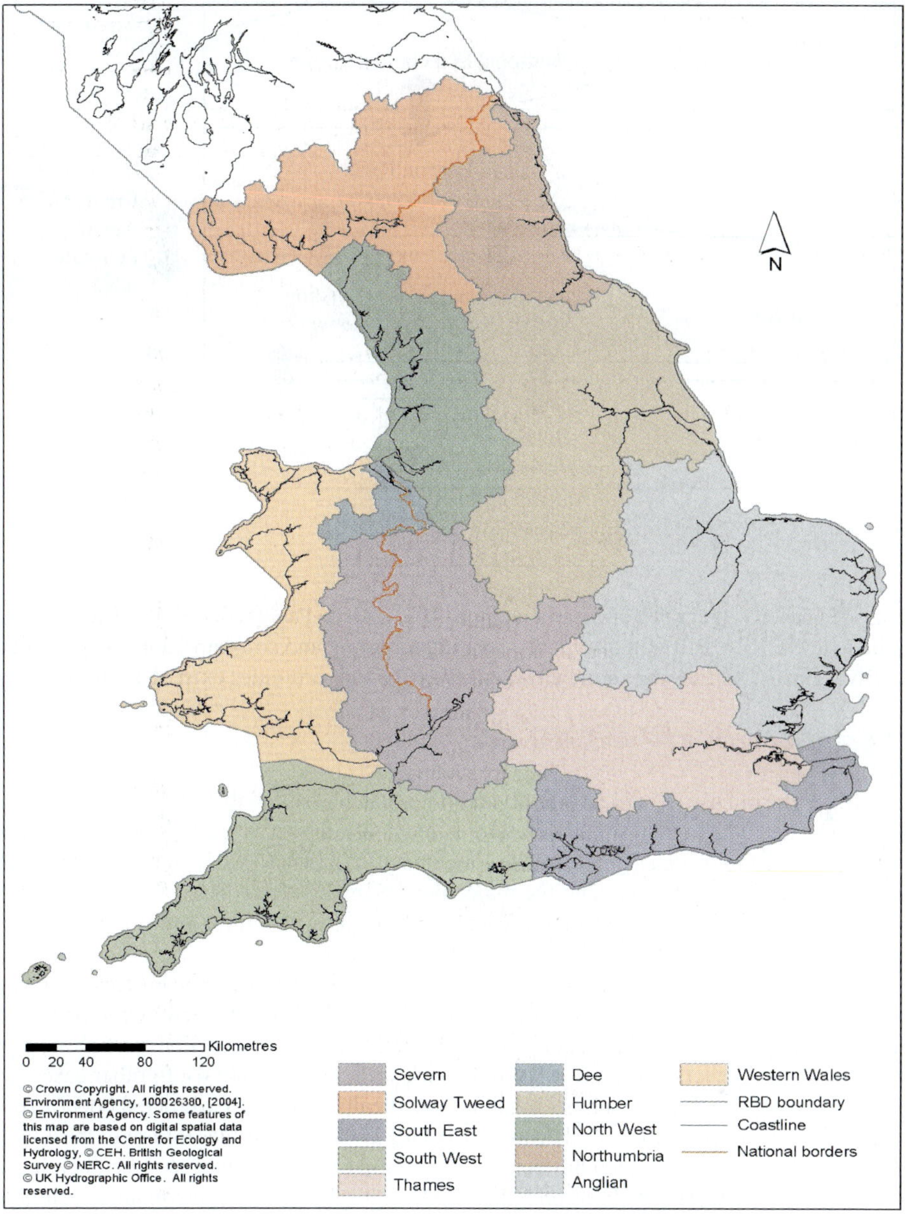

Fig. 2 River Basin Districts in England and Wales and cross-border districts between England and Scotland

4.3.3. Arrangements at the River Basin District scale

At this scale, we propose to work in partnership with a small number of agencies and institutions with statutory powers to implement the basic measures needed to deliver RBMPs, as well as other key stakeholders as appropriate. We are often already working with these bodies, and we will build on these relations to form a cohesive Liaison Panel that ensures

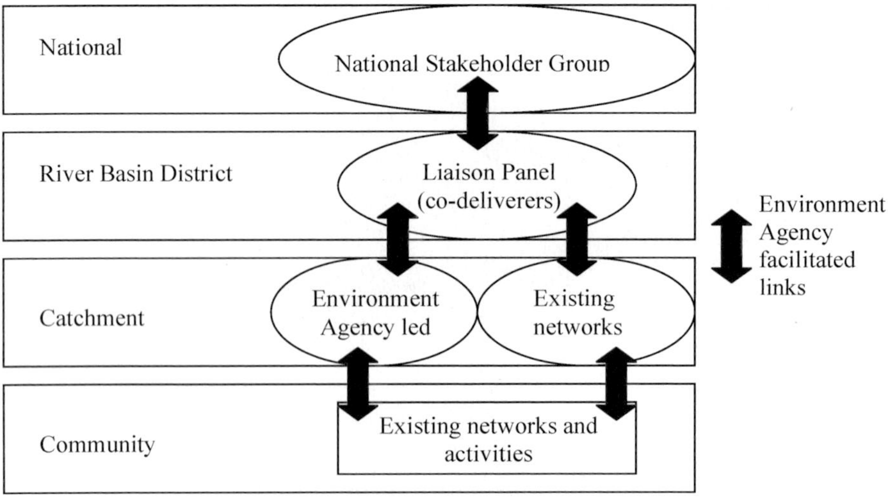

Fig. 3 Linking different levels of involvement

plans are properly integrated and implemented. This Liaison Panel will help us to develop and implement the RBMP. It will also play a role in promoting and co-ordinating engagement at the River Basin District scale and below, ensuring that opportunities exist at all these scales.

4.3.4. Arrangements at the catchment scale

Each River Basin District will be sub-divided into 'catchments' – the total number of these is still to be decided – with joint stakeholder agreements known as 'catchment frameworks' will be produced for each catchment. The Catchment Framework will be the basic "building block" for river basin planning and the majority of planning effort (80–90%) is likely to be expended here. Most of our existing plans are developed at this level and it is the scale at which we effectively engage a wider range of stakeholders.

At this scale the primary emphasis will be on the use of existing arrangements to engage with a range of groups who have an interest in or are likely to be affected by the plans. We will be flexible in adapting arrangements to different circumstances. For example, where Coastal Fora and Rivers Trusts exist we will work with them (Figure 4). In other cases, we may bring groups and organisations together specifically to help us with elements of river basin planning. One example of this way of working is the Stakeholder Forum in the Ribble Pilot River Basin Project, which has been successful in bringing groups together, helping to develop a shared understanding of problems, and identifying creative and realistic solutions (Environment Agency, 2004e).

4.3.5. Arrangements at the community scale

At this scale, we will target discussion with individuals and local networks to where the need or risk across the catchment is greatest and within the resources available. In these cases, we want local communities to discuss problems and their potential solutions. We are developing experience of involving communities in looking at local river management problems, through research projects in different parts of England and Wales (see Box 3).

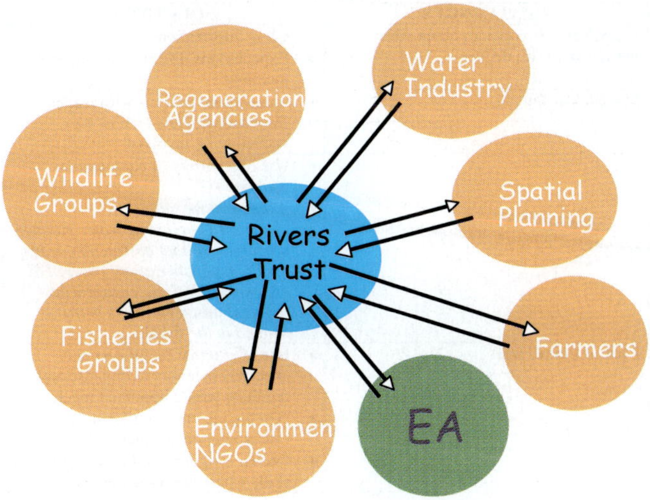

Fig. 4 Figure illustrating an arrangement for working with stakeholders within a catchment through networks and fora facilitated by a local Rivers Trust

Box 3 PURE North East

The Environment Agency is working with partners from communities in four catchments in the North East of England: Seaton Burn, Ouseburn, the River Team and the Skinningrove beck. The Seaton Valley Forum brings local people together with scientists from the University of Newcastle to understand water pollution problems caused by farm wastes and sewer outflows and to develop a programme for river restoration. In the Lower Ouseburn Valley, Newcastle City Council is leading an initiative to involve community groups in the preparation of a management and heritage education plan for the catchment. The Agency is working on a programme to share the lessons from the experience of the four partnerships and improve the effectiveness of each.

Members of a community have local knowledge based on experience developed over time and continually evolving. Where problems related to heritage and environmental stewardship arise, this understanding is of equal or greater importance than the expertise of outside 'specialists'. The PURE project provides an opportunity for an exchange of information. For example, the work in Seaton Burn is broadening understanding of the history, landscape and river system, leading to a programme of restoration and the development of new proposals for environmental actions (Doyle, 2005).

We expect engagement to be issue-led and proportionate to the scale of the problems. It is envisaged that the information generated, including proposals for appropriate measures, will be fed upwards into the Catchment Framework or RBMP as appropriate.

5. Some reflections and lessons learned

There is a wealth of lessons we can draw from our work on these issues over the past couple of years. Here we highlight two sets of lessons – those that we learnt from our work with the Ribble Pilot River Basin, and those that have been gleaned from the work of our national Water Framework Directive team.

Work with the Ribble Plot has provided an opportunity to reflect both within the Agency and with other partners on approaches to engagement. In terms of Agency experience, the project has convinced staff that "public participation is vital in planning and engaging the

CHALLENGES	SOLUTIONS
Unrealistic expectations about what participation can deliver	• Clear information • Expectations discussed at stakeholder meetings
Stakeholder fatigue	• Stakeholder analysis with feedback from stakeholders on how they wanted to be involved
	• Use of a variety of participation methods • *A longer-term solution to this challenge is to join up water planning processes, so that abstraction, water quality and fisheries are covered by one plan and in one engagement process: this is currently being trialled in the Ribble*
Lack of trained staff	• Use of external facilitators and experts, especially to organise and facilitate large / important meetings • "On the job" training for staff in stakeholder engagement • Recognising the need for engagement skills when recruiting new staff

Fig. 5 Ribble Pilot River Basin – Challenges and solutions for public participation

public at an early stage of the process and is important in gaining their confidence and trust to complete the project" (Environment Agency, 2004e).

The Ribble Pilot faced some significant challenges in involving stakeholders. Figure 5 sets out the major difficulties and how these were resolved:

Stakeholder analysis was particularly important in identifying all the different interests who needed to be involved. The method used was to create a data base of all known interested parties and to ask them what aspects of river basin planning interested them, how they would like to be involved in planning and what other groups should be involved. This meant that the project team contacted an ever-wider group of stakeholders and avoided the problem of only working with familiar faces; they were also able to use the engagement methods preferred by their stakeholders. One example is the publication of a paper newsletter as well as the electronic version: although the Agency originally saw electronic media as more efficient and less resource intensive, many stakeholders said that they would prefer to receive a paper copy, so this is now being produced.

For stakeholders, the main problems are ones of time and resources. They need to be involved early on so that they can influence the development of the process, but for many, particularly for small voluntary organisations or private companies, fielding someone to attend meetings over a period of time can be a heavy resource commitment. One of the advantages of a tiered system of engagement, with information and views feeding both up and down, is that it allows people to be involved at scales relevant to their own interests. Nevertheless, tiering is unlikely to completely solve the resource gap, and other ways will need to be found of supporting the participation of organisations that don't have the necessary resources.

Experience of involving stakeholders both in the UK and other parts of the world was discussed in an international conference in October 2004,[3] which agreed the following principles of active involvement in River Basin Management (Box 4).

[3] Ripple Effect Conference, Bolton, North West England, 12–14 October 2004.

Box 4 Principles of Active Involvement in River Basin Management agreed at the Ripple Effect Conference, Bolton, October 2004

- Good involvement takes time, start early!
- Develop and share a sense of ownership for the river basin
- Work to build and maintain trust with your partners
- Undertake "mapping" of stakeholders to find out more about them and their interests
- Learning from mistakes is as important as sharing successes
- Listening is as important as talking
- Be passionate about your cause, passion persuades
- Work with each other and build a common vision for your basin, to put the management plan into context
- Nobody can do it alone. True partnership leads to shared responsibility and decision making for shared actions
- Where cultures and traditions vary, agree key messages and adapt to their needs.

Most of the principles derived from the experience of involving the public in the Ribble Pilot are not unique to the Ribble and reinforce learning from experiences of public involvement in other places and also outside the fields of water and environmental management. What makes the lessons so rich and powerful is that they involved experiential learning on the part of all involved, which in turn contributes to a development of individual and institutional capacities for the future.

Because these lessons are not derived from "objective" research (involving, for example, the formulation and testing of hypotheses under specified conditions) they do not transfer in the same way as the facts generated by a hard science model. This raises the challenge of finding new ways of sharing learning about changing and complex relationships and institutional structures.

Recognising the limitations of many management systems which seek to reduce uncertainty and control complexity by breaking projects and programmes down into their component parts, the Environment Agency tested the application of a social learning approach on the development of a River Basin Planning strategy with support from the Open University. In the first stage, activity focused on defining the scope and parameters of the RBP strategy. The Agency's RBP team found the use of systems thinking combined with a learning approach to be a useful one for clarifying the issues the RBP strategy must address (Collins *et al.*, 2005).

The same approach was taken to organising workshops to discuss the developing strategy with external stakeholders. The structure of the workshops acknowledged that stakeholders bring different perspectives to the table, and tried to develop shared understandings. While feedback from stakeholders on the workshops was generally positive, it was felt that the Agency had not reflected this stakeholder input adequately in its RBP strategy. This underlines the need for ongoing social learning and the development of capacities for new ways of working.

The work of the national WFD team has provided us with interesting insights about the changes we will need to make in the way we develop integrated catchment management. We will have to invest resources in building better working relations with external partners who we will need to help us deliver many of the Directive's objectives. Because our resources are limited, we will need to select these partners through careful prioritisation.

Supporting these efforts to build external partnerships will also involve more thorough investment in partnership working within our own organisation. This is not merely a question of resource allocation – we need to plan for cultural change. Current practices that may require change include project management and catchment modelling.

Admitting that we don't know all the answers and that integrated catchment management presents us with a new and complex set of questions, has been an important first step in shaping change and learning to think and work differently.

6. The next phase of learning and research

We plan to take these lessons forward in a number of ways. Firstly, we recognise the need to design our future catchment science programme in a way that embeds the development of an integrated catchment modelling framework within a social learning process (Environment Agency, 2005d). This is likely to involve us in a substantial shift in the way that we undertake the science, with a much stronger focus on interdisciplinary working (Environment Agency, 2004f):

> '. . .a social learning approach will frame the whole integrated catchment science programme, supporting interdisciplinary research both within the different work packages, and in the way these are inter-related. . .' (Environment Agency, 2005d, p. 7)

Secondly, significant work will be needed to develop a more effective interface between the Environment Agency's new catchment science programme and our river basin management process, at both policy and operational levels. At present, institutional barriers between our science, policy and operational teams tend to hinder this interface. However, a start has recently been made at addressing these barriers (Environment Agency, 2005e).

Finally, the Environment Agency will need to consider ways in which it might adapt current approaches to programme and project management in the light of the lessons discussed in this paper. For example, Winter and Checkland (2003) argue that conventional project management theory is essentially rooted in 'hard' systems thinking, which is only one way of looking at this area of practice. They contrast this with a 'soft' systems image of project management, which emphasises 'process' over 'product'. They go on to argue not that one approach is better than the other, but rather, that complex and uncertain situations call for an appropriate mix of the two.

These ideas represent new cultural challenges for the Environment Agency. It is too early to say to what extent our institutional and cultural practices can or will adapt. But in recognising what is needed, a start has been made.

References

Collins K (2004) The Tweed Forum and Tweed catchment management plan: a SLIM-UK case study. SLIM (Social Learning for the Integrated Management and Sustainable Use of Water at Catchment Scale) Case Study Monograph 10 (accessed at http://slim.open.ac.uk)

Collins K, Ison R, Blackmore C (2005) River Basin Planning project: social learning. science report to the Water Framework Directive programme. Environment Agency, Bristol

Craps M (ed) (2003) Social learning in river basin management. WP2 Report of the HarmoniCOP project, Leuven, www.harmonicop.info

Defra (Department for the Environment, Food and Rural Affairs) (2002) Directing the flow: priorities for future water policy. HMSO, London

Defra (Department for the Environment, Food and Rural Affairs) (2003) The Water Environment (Water Framework Directive) (England and Wales) Regulations. HMSO, London

Doyle A (2005) Developing effective partnerships: case studies from the PURE North East Interreg project. Environment Agency, Leeds
Environment Agency (2001) Planning for the rising tides. Environment Agency, Leeds
Environment Agency (2002a) Rivers and estuaries: a decade of improvement. Environment Agency, Bristol
Environment Agency (2002b) Our urban future. Environment Agency, Bristol
Environment Agency (2003) Don and Rother Catchment Abstraction Management Strategy. Environment Agency, Leeds
Environment Agency (2004a) Building trust with communities: a toolkit for staff. Environment Agency, Bristol
Environment Agency (2004b) Environment Agency working with local strategic/community strategic partnerships. Environment Agency, Bristol
Environment Agency (2004c) River Basin characterisation: draft pressure and impact assessment outputs. CD ROM, Environment Agency, Bristol
Environment Agency (2004d) The nation's fisheries. Environment Agency, Bristol
Environment Agency (2004e) Ribble Pilot River Basin "Public Participation and River Basin Planning – early experiences". Environment Agency, Warrington
Environment Agency (2004f) Solving environmental problems using science: Science Strategy 2004 onwards. Environment Agency, Bristol
Environment Agency (2005a) Conservation, Access and Recreation Report 2004/05. Environment Agency, Bristol
Environment Agency (2005b) Water for life and livelihoods: a strategy for River Basin Planning – a consultation. Environment Agency, Bristol
Environment Agency (2005c) Framework for stakeholder engagement. Environment Agency, Bristol
Environment Agency (2005d) Integrated catchment science: a research strategy. Environment Agency, Bristol
Environment Agency (2005e) Notes of the integrated catchment science workshop. 25–26 April 2005, Environment Agency, Bristol
European Union (2003) Common implementation strategy for the Water Framework directive (2000/60/EC) Guidance Document No 8, Public Participation in Relation to the Water Framework Directive. European Commission, Brussels
Hulme M, Jenkins GJ, Lu X, Turnpenny JR, Mitchell TD, Jones RG, Lowe J, Murphy JM, Hassell D, Boorman P, McDonald R, Hill S (2002) Climate change scenarios for the United Kingdom: The UKCIP 2002 Scientific Report. Tyndall Centre for Climate Change Research, University of East Anglia, Norwich
Jiggins J, Roling N (2004) Key informant studies II: water conservation project in North Brabant and Limburg (2nd generation project). SLIM (Social Learning for the Integrated Management and Sustainable Use of Water at Catchment Scale) Case Study Monograph 10 (accessed at http://slim.open.ac.uk)
Lovett A, Dockerty T, Hiscock K, Johnson P, Sandhu C, Sunnenberg G (2004) Understanding movements of pollutants through a catchment area (River Slea). Report on sub-project 2: Stakeholders and scenarios. University of East Anglia, Norwich
Mostert E (2003) The challenge of public participation. Water policy 5:179–197
Pahl-Wostl C (2002) Towards sustainability in the water sector: The importance of human actors and processes of social learning. Aquatic sciences 64:394–411
Patel M, Stel JH (2004) Public Participation in River Basin Management in Europe: a national approach and background study synthesising experiences of 9 European countries, www.harmonicop.info
Porter S, Warburton D, Wilkinson D (2005) Developing effective engagement with local and community strategic partnerships. Science Report E2-057/SR6, Environment Agency, Bristol
Pretty JN, Chambers R (1994) Towards a learning paradigm: new professionalism and institutions for sustainable agriculture. In: Scoones I, Thompson J (eds) Beyond Farmers First
Reason P, Bradbury H (2001) Handbook of Action Research: Participative inquiry and practice. Sage, London
Rees Y *et al.* (2005) Developing a national strategy for public participation in river basin planning for the WFD. Environment Agency, Bristol
The SLIM Project (2004) Social Learning as a Policy Approach for Sustainable Use of Water: A field-tested framework for observing, reflecting and enabling. SLIM (Social Learning for the Integrated Management and Sustainable Use of Water at Catchment Scale) Framework (accessed at http://slim.open.ac.uk)
White I, Howe J (2003) Planning and the European Water Framework Directive Journal of Environmental Planning and Management 46:621–631
Wilkinson D, Warburton D, Porter S, Colvin J (2005) Joining Up: Stockbridge Pathfinder. Science Report E2-057/SR4, Environment Agency, Bristol
Winter M, Checkland P (2003) Soft systems: a fresh perspective for project management. Civil Engineering 56:187–192

Water Resour Manage (2007) 21:351–373
DOI 10.1007/s11269-006-9057-8

ORIGINAL ARTICLE

Integrated assessment of water resources: Australian experiences

B. F. W. Croke · J. L. Ticehurst · R. A. Letcher · J. P. Norton · L.T. H. Newham · A. J. Jakeman

Received: 20 April 2006

Abstract It is widely accepted that water resource management demands an integrated assessment of resource use options, including local and regional impacts on the environment and stakeholders. Multiple issues, stakeholders and scales of system behaviour must be considered, as well as the key disciplines within and between the human and natural sciences. Modelling is a critical tool in integrated assessment. It enables effects of policy interventions, climate forcing and demographics to be predicted (although with some uncertainty), and provides a means of expanding understanding of river basin behaviour. It also acts as a vehicle for social learning among various interest groups. This paper discusses the various frameworks and methods being used for integrated modelling, and their suitability and unfulfilled potential for these purposes. The frameworks include coupled component models, systems dynamics models, metamodels, risk-assessment approaches, Bayesian decision networks, agent-based methods, expert systems and other heuristic knowledge-based techniques. Specific software platforms are not considered but the lessons from software development and implementation are clearly spelt out. The paper presents three Australian case studies in integrated assessment. They vary in their range of catchment/watershed sizes, hydroclimatology, issues of concern and stakeholders engaged. Two of them utilise a coupled component modelling framework and the third a Bayesian decision network approach. The paper illustrates the value, problems and lessons of integrated assessment and modelling. In particular it proposes some

J. L. Ticehurst · L. T. H. Newham
Integrated Catchment Assessment and Management (iCAM) Centre, The Australian National University, Canberra

B. F. W. Croke (✉) · R. A. Letcher · J. P. Norton
Integrated Catchment Assessment and Management (iCAM) Centre, and Department of Mathematics, The Australian National University, Canberra
e-mail: barry.croke@anu.edu.au

A. J. Jakeman
Integrated Catchment Assessment and Management (iCAM) Centre, and Centre for Resource and Environmental Studies, The Australian National University, Canberra

ways to address the challenges of assessing options to obtain more sustainable basin-wide outcomes.

Keywords Integration frameworks · System dynamics · Metamodels · Risk assessment · Bayesian decision networks · Agent-based · Expert systems · Knowledge-based · Integrated modelling · Stakeholders · River basin management

1. Introduction

Integrated management of a catchment, watershed or river basin requires integration (i.e. joint consideration) of: (i) different objectives and their related outcomes, e.g. economic efficiency, social equity and ecological integrity; (ii) all water resources (surface water, groundwater, estuaries); (iii) multiple water- and land-related issues; (iv) different types of water use (agricultural, ecological, domestic, industrial, recreational); (v) all water users, everyone affected by water use and interest groups generally. For such complex integration, modelling must play a key facilitating role, whatever its form: qualitative or quantitative, traditional or novel, or typically a mix of these forms. It must meet the challenge of representing, but not over-representing, the complexity and spatial and temporal variation within catchments and river basins. The modelling must cover a range of sciences including. hydrology, ecology, agriculture, forestry, economics, demography, psychology, sociology and politics, and a range of categories of people affected. The evolving discipline of Integrated Assessment (IA) aims to deal with such situations. According to The Integrated Assessment Society (www.tias-web.info), IA is defined as "Integration of knowledge from different disciplines with the goal to contribute to understanding and solving complex societal problems, that arise from the interaction between humans and the environment, and to contribute in this way to establishing the foundation for sustainable development. Modelling and participatory processes should include stakeholder groups and the public at large." IA has been summarised by Jakeman and Letcher (2003) as:

- being problem-focussed, using an iterative, adaptive approach that links research to policy;
- possessing an interactive, transparent framework that enhances communication;
- being enriched by stakeholder involvement and dedicated to adoption;
- connecting complexities between the natural and human environment, recognising spatial dependencies, feedbacks and impediments; and
- attempting to recognise missing essential knowledge.

IA depends crucially on public participation, and on iteration of the development steps as more is learnt about the focus system. IA may be conceptualised as in Figure 1.

Integrated Scenario Modelling (ISM) is a core activity of many IA exercises. It involves constructing a model to approximate the system under study. The model allows simulation of the impact of changes in input drivers (scenarios), e.g. climate and human activities, on outputs (indicators) representing the states of the system. In this context, a state may be a sustainability target or an indicator that summarises an output, usually a simple function such as an average, a maximum or minimum level or a trend. The ISM component is at the centre of IA but must be extended in three ways, so that policy and management are linked to the controllable human activity inputs, stakeholders participate in all relevant parts of the process, and scientists and IA practitioners manage that process and its communication aspects. ISM outputs also play an essential part in multicriteria analysis, formal or informal. As summarised by Jakeman and Letcher (2003) ISM provides:

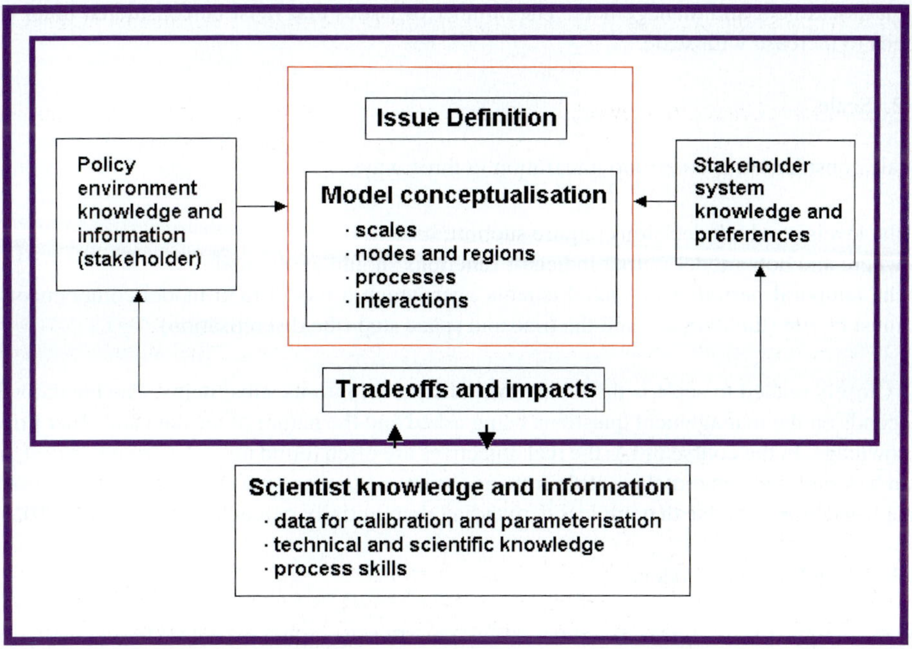

Fig. 1 Diagrammatic representation of integrated assessment

- a way of investigating and explaining tradeoffs;
- a readily accessible collection of models, methods and visualisation tools;
- a focus for integration across researchers and stakeholders;
- a training and education function;
- an exploratory aid capable of adoption and further development by stakeholders;
- a permanent summary of the project methods; and
- a means of making the management analysis transparent.

2. A hierarchy of modelling problems

The aspects of integration were characterised above from a management perspective. While still recognising that the model is for management and related purposes, this section focuses more on integration from a modelling perspective. In river basins as in other environmental systems, a hierarchy of problems require IA. The hierarchy is defined by several qualifiers: issues, scales, disciplines, and interest groups. One could begin with any of these, but the issues really determine the nature of the other qualifiers. Problems lower in the hierarchy have fewer issues of concern, scales to be considered, disciplines to invoke and interest groups to involve.

2.1. Issues

The issues that arise in river basins tend to be interrelated, especially when a problem has offsite or downstream effects, and often they are conflicting, so that one needs to integrate their assessment and management. The number of issues that must be considered in an IA tends to increase with scale.

2.2. Scale

Scale considerations enter into integration in three ways:

- the levels at which decisions require support;
- where and how model-output indicator functions should apply; and
- the temporal periods and spatial extents over which a model (and model componnents) must be run (the domain) and the time and space step (the discretisation).

Closely related to scale is the breakdown of a model into its subsystems. The breakdown depends on the management questions being asked and the nature of the data and other prior knowledge. In the course of IA, the real objectives are often found not to be broadly based, or can be simplified and remain useful for management, so that one may be able to ask questions in a modelling exercise that are less demanding than initially perceived (Jakeman, 1989).

2.3. Disciplines and models

At the simplest level, integration often refers merely to coupling models. More holistic use of the term 'integration' refers to a model or assessment that includes a mix of physical, social, economic, ecological or community perspectives. A mix of disciplinary approaches can lead to new insights into a management issue that could not be reached from a single viewpoint and, importantly, can lead to identification of critical gaps in disciplinary research and knowledge.

2.4. Interest groups

Environmental management and improving the sustainability of a system always involve trading impacts between different interest groups. Hence virtually all environmental management involves some level of conflict resolution and will require input from, and final adoption of the management solution by, a range of stakeholder groups. IA thus requires the integration of perspectives and input from a broad range of interest groups, some affected by the initial management issue and others affected by possible solutions.

3. Reasons for and objectives of integrative modelling

Broad reasons for integrative modelling include:

- improving system understanding;
- discovery, acquisition or elicitation of knowledge;
- social learning among interest groups;
- prediction or simulation;
- forecasting;
- management and decision-making; and
- discovering limitations, inconsistencies and gaps in data.

These motives are not independent but should be considered by the IA modeller with regard to the appropriate framework, model type, available knowledge and data, and the desired outcomes.

The model developer (see Jakeman *et al.*, 2006) should be clearly aware of the:

- questions and issues that the model is being developed to address;
- interest groups, including the clients or end-users for the model;
- outputs required;
- forcing variables or drivers;
- accuracy expected of the model;
- temporal and spatial scales;
- time frame for the completed model, e.g. for a decision;
- effort and resources available for modelling.

4. State of integrative methods

There are many different integration approaches, reviewed by Letcher and Jakeman (in press) and Letcher *et al.* (in press), but they share some aspects. Bayesian Networks, Agent-Based Models and Expert Systems, described below, might be considered as artificial intelligence or knowledge-based techniques. However their individual use as integrating frameworks warrants specific mention.

4.1. System dynamics

System dynamics (e.g., Deaton and Winebrake, 1999) is a modelling approach that investigates and manages complex feedback systems (e.g. aquatic food webs). Many authors consider it to be a philosophy of model development rather than an integrated approach. Nodes in the conceptual framework generally represent state variables, while the links or arrows between nodes represent functions transforming one state variable to the next. The conceptual frameworks for system dynamics models often contain feedback loops. These loops may be very complex, or may only represent perceived 'plausible' connections. Thus, system dynamics models are most commonly used to improve systems understanding and to compare simulation responses, rather than for decision-making and policy development. However, in theory the latter could be achieved.

4.2. Bayesian networks

Bayesian networks consist of a series of nodes and links that conceptualise a system, but feedback loops cannot be included in this approach. They are fundamentally a decision-making tool.

The nodes in the system are variables. The links are defined by conditional probability distributions (Borsuk *et al.*, 2004), thus providing a measure of the certainty in the causal relationship between each pair of nodes. This integrated approach differs from others that use deterministic, rather than probabilistic, methods to determine the relationship among variables (Borsuk *et al.*, 2004). The implicit ability to account for uncertainty means that Bayesian networks are able to make use of 'soft' sources of data, such as expert opinion, where observation data are not available (Pearl, 1988).

Bayesian Decision Networks also include decision variables, which allow for management options to be implemented, and utility variables, which represent the benefit or cost of a particular decision (see for example Sadoddin *et al.*, 2003).

4.3. Metamodels

Metamodels are essentially a simplified overall description of the processes within complex systems or more complex models. Data-mining techniques and regression are often used to develop metamodels.

Bouzaher *et al.* (1993) suggest that metamodels approximate, and aid in the interpretation of, simulation models. The mere size of the output from complex models can make them difficult to view and interpret. Metamodels can provide look-up tables or simpler functions to represent the information found by the more detailed models.

In integrated modelling, metamodels can be used to replace a complex model, or complex components of a model, completely. In the latter case, the metamodels can be coupled into an integrated system.

4.4. Risk assessment approaches

Risk assessment basically provides "a principled way of organising what we know about the world, particularly about its weak spots and creaky joints" (Jasanoff, 1993). Risk is defined as the probability of an outcome multiplied by the severity of its consequence, leading to the potential quantification of risk in a number of ways. This notion can also be extended to positive as well as adverse impacts and the characterisation of uncertainty.

Kammen and Hassenzahl (1999) present much of the central theory and methods including order-of-magnitude estimation, cause-effect calculations, exposure assessment, fault-tree analysis, and managing and estimating uncertainty.

4.5. Coupled component models

Coupling component models involves combining models from different disciplines to arrive at an integrated outcome. Conceptually each node in the framework represents a model of a particular issue. The links between models pass the generated data. The links maybe manually linked externally to the original models, or may be more tightly linked where the component models share inputs and outputs (e.g., Merritt *et al.*, 2004; Letcher *et al.*, 2004). Coupled component models are generally able to incorporate feedback loops.

Coupled component models can account for non-trivial temporal and spatial discretisation. This is particularly relevant in catchment and river basin management, where it is important to be able to isolate the impacts of upstream nodes and prior stream conditions. Two examples of this approach are presented in a later section.

4.6. Agent-based models

An agent- or actor-based model is essentially a type of coupled component model. It focuses on the interactions between agents (individuals) in a system (e.g., Brown *et al.*, 2004), where agents adapt to changes to their environment. A system in which two or more agents exist at the same time, share resources and communicate with each other, is called a multi-agent system.

Agent-based models are efficient at identifying large-scale outcomes resulting from often simple, local interactions between individuals. For this reason, and because they tend to be hypothetical, agent-based models are usually applied in social and ecological science.

4.7. Expert systems

An expert system is a type of qualitative model where prior knowledge is encoded into a knowledge base and then logic is used to infer conclusions (Davis, 1995). The knowledge base determines the success of the system (Forsyth, 1984). Given a problem, the expert system simulates the problem-solving task(s) (Kidd, 1987). The conceptual diagram for an expert system refers to questions about the nature of the system directed at the user. The response to these questions then dictates the route down which the procedure looks for a solution.

5. Other artificial intelligence methods

Innovation in methods that deal better with uncertain and/or qualitative knowledge is essential for progress in making sustainability assessments. Artificial intelligence techniques have a long history of promise but do not seem to have delivered yet in this context. They are potentially particularly useful for knowledge acquisition; see, for example, the short discussion on data-mining techniques below. According to Sell (1985), there are four primary sources of knowledge: literature, human specialists, existing models and examples. Schmoldt (1998) initiates attempts to organise the different methods of knowledge acquisition by specifying a conceptual approach known as linguistic-based knowledge analysis to develop lexicons, syntax and semantics for a domain. While many natural-resource domains present unique sets of problems in acquiring existing expertise, he argues that there are enough commonalities to be exploited by sharing knowledge acquisition experiences.

Likewise, the emerging field of environmental informatics has much promise. It combines artificial intelligence, GIS, software frameworks for modelling and linking models, and user interfaces, but effective capture of knowledge has lagged far behind software implementation. As with artificial intelligence, environmental informatics also needs urgently to be complemented by user-driven IA providing priorities, structure and efficiency. Without such application-oriented discipline, environmental informatics runs the risk of becoming capable of doing a wide range of things, all badly.

A taxonomy of knowledge-acquisition categories (and particular techniques within each category), as provided by Schmoldt (1998), includes: unstructured and structured interviews (using free association, analytic hierarchy processes, psychological scaling, sorting and knowledge diagramming); questionnaires; automated tools; problem solving (familiar cases, limited information cases, tough cases); machine learning (induction from examples, neural networks, genetic algorithms); and protocol extraction (goal decomposition, forward scenario simulation, verbal protocol, retrospective protocol).

Data mining is a discipline that covers some of these techniques and merits brief individual mention for illustrative purposes. It also has enormous potential for knowledge extraction in both socioeconomic and biophysical applications.

5.1. Data mining

Data mining can be defined in many different ways. Following Spate *et al.* (2003) it can be the discovery of interesting, comprehensible and previously unknown rules, trends or characteristics from data. While much of the knowledge sought with data-mining techniques could, in principle, be extracted by detailed and rigorous visual or statistical investigation, that may involve impracticably large time and computation costs. This is especially so when prior knowledge is not good enough to point to where to look in large data sets. The quite large number of techniques and tools available includes clustering, classification, association rule extraction, dominant mode analysis and time series similarity measures.

6. Opportunities and challenges

6.1. Incorporating key considerations in integrated modelling

Some of the modelling considerations listed below should be addressed more routinely in the management of natural resources (Jakeman and Letcher, 2003). Few examples in the literature demonstrate systematic attention to them.

- Climate variability and episodes
- Modelling process complexity
- Beyond business-as-usual scenarios
- Modelling long leads and lags
- Narrowing modelling objectives
- Assessment of model sensitivity and uncertainty
- Error accumulation
- System representation

6.2. Quality assurance for credible models

Quality assurance refers to the standards and protocols for model and data reporting and distribution (see stars.net.au). It is contended that this type of regulation is required to enhance the credibility and ultimate utility of models. Uniform reporting on the limitations and assumptions in models and associated input data would empower model users and decision makers to use models appropriately.

It is suggested that model credibility can be enhanced by:

- a serious two-way modeller-stakeholder dialogue;
- appropriately rigorous model evaluation tests;
- sensitivity and uncertainty assessments; and
- peer reviews of models at their various stages of development.

Refsgaard *et al.* (2005) classify guidelines "according to how much focus is put on the dialogue between the modeller and the water manager as: (Type 1) Internal technical guidelines developed and used internally by the modeller's organisation; (Type 2) Public technical guidelines developed in a public consensus-building process; and (Type 3) Public interactive guidelines developed as public guidelines to promote and regulate the interaction between the modeller and the water manager throughout the modelling process."

6.3. Recognising broad objectives

It should be remembered that the broad objective of modelling is to increase understanding of the directions and magnitudes of change under different management options. The ability to differentiate the relative impacts from change compared to the 'do nothing' scenario is all that really can be supported in the current modelling environment. Treating simulation outputs as accurate predictions is unrealistic. To assist in policy implementation and decision making, the ability to ascertain at least a qualitative measure (high, medium, or low) of certainty in the outputs is required. Ideally, predictions would be produced with a quantitative confidence level, but in most situations this is impracticable at present. Currently, methods for quantifying uncertainties have severe limitations. Similar comments go for sensitivities of model outputs with respect to variation in parameters of the model itself or its drivers and boundary conditions. Norton *et al.* (2003) and Jakeman and Letcher (2003) discuss new research required to address this deficiency.

6.4. Participatory modelling

Public participation can be defined as direct involvement by the public in decision-making (Mostert, 2006). There are several reasons for organizing public participation including the possibility of:

- more informed and creative decision making;
- greater public acceptance and ownership of the decisions;
- more open and integrated government;
- enhancing democracy; and
- social learning, the ultimate objective, to manage issues.

According to Mostert (2006), there are three principles of social learning:

- reflection on goals – participants are encouraged to reflect on why they want to attain the goals they are seeking, as there may be better ways to achieve what they want;
- reciprocity – participants must realise that they are interdependent, and should consider others in their actions and information exchange; and
- respect for diversity – participants should acknowledge that other participants may have different cultures, backgrounds and interests.

He also notes that public participants must be open and honest about their opinions, issues and concerns, as defensive behaviour will hamper the coordinated approach. The process for including public participation in decision making must be well organised and sensitive to cultural differences in order to maximise the potential benefits. This requires leadership and facilitation to ensure that all participants are given opportunity to speak freely. To further the benefits, public opinion should also be considered when designing relevant scientific research and in the presentation and dissemination of results.

IA can be a useful mechanism for public participation. For example, public consultation during the development of a decision-making model means that different perceptions can be acknowledged and accounted for within the model structure. The final product is then more likely to provide output useful to the public. Conflict that arises during discussion in model development may identify areas of disagreement or knowledge gaps, thus indicating that it should be a priority in the project to collect data and improve understanding of that issue. Such conflict resolution is usually received positively by most stakeholders, as they feel that their concerns were listened to and addressed by the process. Hare *et al.* (2003) present a recent comparison of different participatory processes.

The output that IA produces to assist in decision-making should be dynamic, so that the tools evolve and remain usable as the knowledge, systems understanding, goals, and conflicts of the participants develop, or as the landscape responds and changes. In this way, tools should account for short-term learning and goals, in the light of long-term visions of sustainability.

6.5. Adaptive management

Adaptive management (Holling, 1978) and active adaptive management (e.g., Allan and Curtis, 2003) are principles to improve environmental management through learning. They can be used to develop management-revision principles, experiment design, outcome indicators and monitoring practices. Designed assessment and evaluation of current and past experiments can identify past successes and failures, and the knowledge gaps.

Adaptive management implies feedback from monitored outcomes to revised actions, as in control engineering. Some essential issues confronting adaptive management can be identified by examining what factors are crucial in employing designed feedback in control engineering:

- determining what behaviour dominates the system's response under control, and specifying simple requirements for that response;
- observability (ability to see, through monitored outputs, the behaviour which is to be controlled) and controllability (ability to influence, through available inputs, the behaviour which is to be controlled);
- comparison between measured and desired output to determine error and generate corrective control action;
- "actuator constraints", i.e. limitations on the action that can be applied to the system; and
- consideration of robustness of control-system performance over a range of uncertainty in the model and the system's environment.

Such ideas are fundamental in design of feedback systems but it is surprising how little discussion their relevance to environmental modelling and management has had.

6.6. Targeting disciplinary gaps

To utilise fully the benefits offered by IA, there needs to be greater integration of all disciplinary resources. The following case studies represent just a tiny fraction of potential examples. The social sciences, for instance, can offer insight and information into the decision-making and adoption processes previously ignored in many scenario-based models. In particular, social survey data, linking information about decision-making and adoption to the biophysical and socioeconomic characteristics of farmers, industries or households, is crucial to developing more sophisticated ISM and other policy analyses. Public participation in model development and testing is one way to account for the social considerations in a particular system (e.g., Haslam *et al.*, 2003).

Artificial intelligence techniques offer an interesting and useful alternative to theory-based models of biophysical and socioeconomic processes. Theory-based models developed to maximise profit, for example, are difficult to validate. However, where 'soft data' (i.e. interview and survey results) are used to govern decisions, the model performance and outcomes can be easily tested through further public participation and validation. Comparison of the output from the two approaches (theoretical and artificial intelligence) could indicate whether the greater complexity in the theoretical approach is warranted, and how robust current techniques are.

7. Australian case studies

In this section we present three examples of IAM projects in Australia to illustrate the potential positive impacts of the IA process and the modelling therein. The key components of the IA for each are summarised in Table 1.

7.1. Water allocation project (Namoi and Gwydir valleys, NSW)

The sustainability of groundwater stores and streamflow in some of Australia's agricultural regions is under pressure from an increase in the number of irrigators and their irrigation requirements. In this project a decision-support system (Letcher *et al.*, 2004; Letcher and Jakeman, 2003) was developed. This DSS is accessible to industry representatives and state government agency staff, for considering the impact of water allocation policies on water users and the environments of the Namoi (40,000 km^2) and Gwydir (30,000 km^2) River basins in NSW.

To achieve these outcomes the model includes the ability to differentiate spatially and temporally between:

Table 1 Comparison of integrated assessment approaches for three Australian case studies

	Namoi/gwydir	Ben chifley	Coastal lakes
Spatial scale (km^2)	40 000	985	23 to 1660
Dominant land use activity	• Rotational based irrigated and dryland options	• Sheep and beef cattle grazing, forestry	• Urban settlement, • Nature reserves/national park, recreational water sports, various agricultures depending on the lake
Key water management concern	• Water allocation as it relates to agricultural production and environmental flows	• Land management impact on water quality for recreation and town water supply	• Land and water management impacts on coastal lake health for social, economic and environmental use
Modelling approach	• Coupled	• Coupled	• BDN
Management options/ policies that can be explored with the model	• Off-allocation • Groundwater allocation reductions • Sleeper license activation, • Volumetric conversion • Daily flow extraction rules • Environmental flows	• Riparian revegetation or clearing, • Establishment of buffer strips with stock exclusion • Engineering river or gully stabilisation • Land use change • Change in emission from point sources	• Lake entrance openings • Location of urban development • Urban development controls (stormwater and sewerage management, flood mitigation) • Agricultural management practices (fertiliser application, buffer strip controls) • Riparian management

- changes in regulated, unregulated, groundwater and supplementary allocations or entitlements;
- changes in "commence" and "cease to pump" thresholds (in river flow) for unregulated and supplementary water, as well as changes in daily extraction limits, including the option of multiple pumping regimes throughout the year;
- changes in carryover rules for regulated water, unregulated water and groundwater, including the option of no carryover;
- changes in the cost of water for different systems; and
- the influence of climate on the impact of these changes.

These issues were identified in consultation with stakeholder groups (see Letcher *et al.*, 2003).

The model accuracy is sufficient for a decision-based tool, but the complexity is kept as low as possible so it can be used in an integrative framework. The decision support system enables the evaluation of tradeoffs between socioeconomic indicators and environmental flows. More specifically, it represents impacts and effects from:

- allocation of groundwater;
- change in the number of irrigation licences;
- environmental flow of the streams; and
- the capacity of farmers to adjust their practices.

Stream health (flow) indicators, agricultural production and how these vary across the basins are used to gauge basin health.

Spatial variation is addressed by dividing the basin into regions of similar groundwater policy, surface water policy and production type. These regions for the Namoi river basin are shown in Figure 2a. The integration of socioeconomic and environmental factors is achieved using a coupled model approach. Hydrological, policy, economic and extraction submodels are integrated to address the given concern. The daily streamflow into a region under particular climatic conditions is simulated and fed into the policy model. This gives the total volume of water available for irrigation in each month, which is used in the economic model to determine farmers' decisions on water management, irrigation practices and crop planting (areas and types). The total water extracted from the stream is then calculated and the remaining water flow is available for input into the downstream region.

The results presented in Figure 2 show the strength of this decision support tool in integrating economic and environmental factors spatially to assist in decision-making for the Namoi river basin. The tool was run given various percentages of sleeper (currently unused) licence activation throughout the basin. Figure 2b shows that the regions in the top of the basin (A, B, D and G) receive a continuous increase in their profit (to up to 3.0%) with increased water allocation, but in the lower basin (regions M, N and O), when over 40% of sleeper licences have been activated some regions begin to decrease in profit. Thus the expansion of irrigation in the upper basin comes at a cost to the lower basin users. Figure 2c shows the impact on streamflow under the same scenarios presented in Figure 2b. There is up to 15% reduction in non-zero median flows in the upper basin (regions B, D, E, F, G and H), the greatest reductions being created when the percentage allocation of sleeper licences exceeds 55%. There is also up to 5% reduction in non-zero median flows in some regions in the lower basin (J, K and L) when licence activation exceeds 60%. Thus for a comparatively small increase in profit in the upper basin, the profits in some regions of the lower basin will decrease, and there will be a strong reduction in non-zero median streamflow in the upper

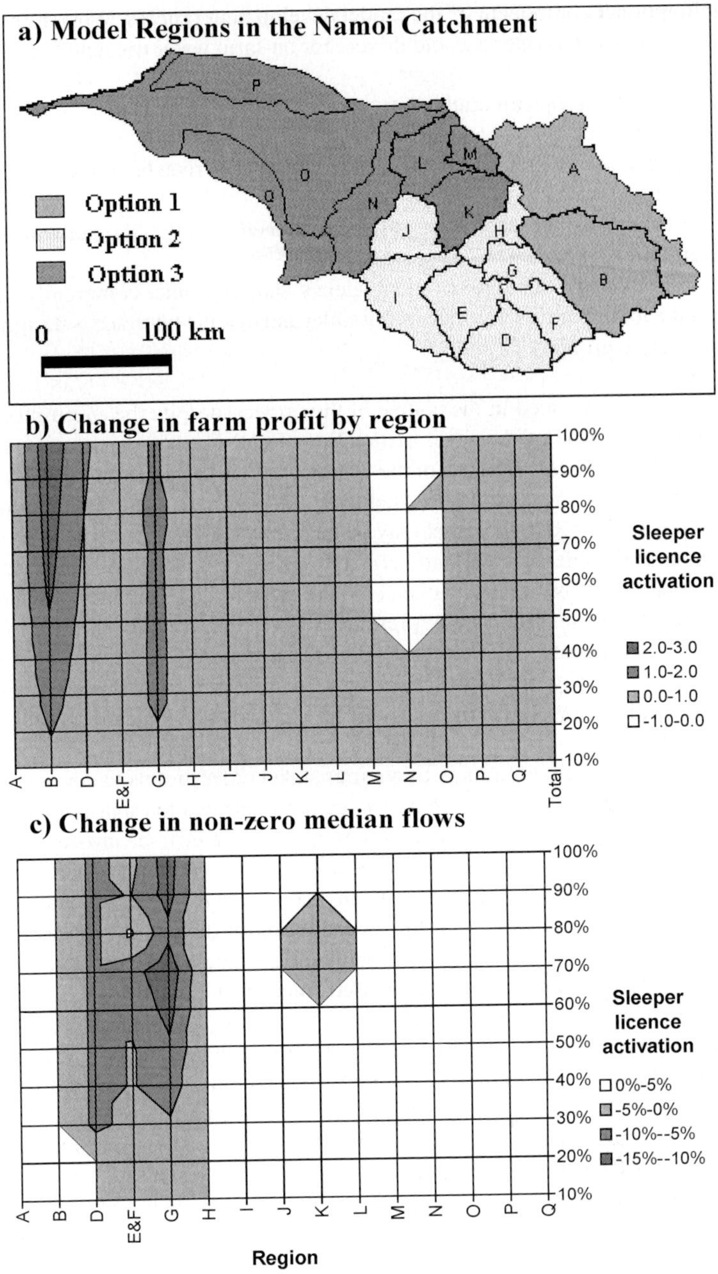

Fig. 2 Example of results from the Namoi decision support tool

basin. These results effectively show the implications of policy changes for the economic and environmental quality of the basin.

This water allocation project has been a collaborative undertaking with input from NSW Agriculture, the NSW Department of Infrastructure, Planning and Natural Resources, the Australian Cotton Cooperative Research Centre and irrigator groups. Stakeholder participation was an important component of the model development process. Workshops were held to identify the controls on water use and drivers for on-farm water use, and to refine and test the model.

Outcomes from the project include:

- delivery of decision-support tools for the Namoi and Gwydir river basins, accessible through user-friendly interfaces;
- training workshops for industry, state government agency staff and other community representatives; and
- feedback from industry, state government agency staff and other community members to ensure that the models are as accurate as possible, and that the interface is designed to meet the needs of these groups.

Many obstacles were faced in the course of the project, despite the successful outcomes presented above. Because of the scale of the Namoi and Gwydir catchments, there is an inherent diversity of interest in the use of the waterways, with a large number of stakeholders involved. Consequently, the consultation had to be targeted, but is still believed to be representative. However, the resulting model may be biased by the level and type of consultation. There is also considerable political and economic pressure within the Namoi and Gwydir catchments, leading to substantial criticism of water reforms and estimates of impact as well as conflicts over water use. This type of pressure is likely to affect any participatory process undertaken in these catchments. There are also large differences in the views of individual farmers and government agency representatives, so community participation relying on representatives from the groups had to involve mechanisms for revealing and incorporating these different opinions.

Initially there was an unwillingness by various stakeholder groups to commit early on to supporting the project. All groups essentially required clarification that one or more other groups had agreed to support the project before they would agree to be involved. This tended to slow the initial model development, through lack of active community participation from some groups in the early stages. However, it can also be viewed as an advantage because it meant that no particular stakeholder group 'took control' of the whole project, so there was less room for bias in the early, perhaps more vulnerable, stages of the project.

Once stakeholder participation had increased, there was evidence of a trend of over-commitment by individuals, a situation likely to occur in many catchments throughout Australia. The problem is that people who have an interest in participating in such projects, and are active and respected within the catchment, often commit extensive time and resources to the project. This of course is desirable to a point. But if one key person commits to too many projects, then they can become over-committed and spend too much time in meetings and workshops. This can be at the expense of keeping in touch with the catchment and the community they represent, and potentially may damage their own financial and personal interests.

7.2. Pollutant management project (Ben Chifley Dam Basin, NSW)

A concern of many water managers is elevated nutrient levels in water supply reservoirs. Increases in nutrient levels may increase the likelihood of algal blooms. The objective of this project was to develop a modelling system to assist in the management of diffuse pollution (nutrients and sediment) inputs into reservoirs. The project was focused in the Ben Chifley Dam catchment, a 1,000 km^2 basin in the central tablelands of NSW (Newham *et al.*, 2004a).

In order to address this problem, the IA model had to account for the key controlling factors on dam water quality (climate and associated hydrologic factors, topography, land use and management of the riparian zone) as well as climate variability and the socioeconomic costs of remediation and adoption. It also had to be sensitive to spatial and temporal patterns in order to target remediation works.

Again a coupled model approach was selected, which included hydrologic, sediment, nutrient and economic sub-models. The resultant IA model is called CatchMODS. In CatchMODS, the basin is divided into stream reach and sub-basin elements. Various riparian management and land use changes can be applied to each. Figure 3a shows the sub-basins and stream reaches used for the Ben Chifley Dam basin, and the larger management sub-basins. Given the defined land use and riparian conditions, the hydrologic model is run for the highest reach of the basin. The mean annual flow, mean annual base flow, median overbank discharge, and bankfull stream flow are calculated and used as input into the sediment and nutrient sub-models. The resultant sediment and nutrient outflows are routed into the next downstream reach.

Outputs from the modelling system include mean annual estimates of the fluxes of total suspended sediment, total nitrogen, total phosphorous and stream flow, together with the fixed and ongoing costs and change in agricultural gross margin associated with the implementation of each management scenario. Figure 3b shows the spatial variation in suspended sediment export for the basin, given the current land use and management conditions. Outputs such as this allow model users to identify areas where remediation to control inputs should be targeted. Figure 4 shows the estimated costs and potential sediment reduction of channel remediation options for the Lower Campbells River sub-catchment. Such outputs enable model users to explore the tradeoffs between various remediation options for a single sub-catchment and also to compare the effectiveness of remediation between catchments.

Community consultation and participatory activities were incorporated into the process of model development. Input was sought from a range of interest groups, including State and Local government, individual landholders and community-based natural resource management organisations. The development of environmental models provided a very useful focus for communication activities between scientists, basin managers and the broader basin community. In this way, the impact of policy and management strategies could be effectively evaluated, and stakeholders and researchers engaged in a two-way interaction to facilitate the adoption of practices to improve basin sustainability (Newham *et al.*, 2004b). A weakness of the participation activities in the case study was that the focus of the delivery of the IA model was a single organisation, the Ben Chifley Catchment Steering Committee. This resulted in rapid progress in development of the model but presented difficulties in the broader adoption of the results of the research following a significant restructure of natural resource management agencies in the catchment.

7.3. Sustainability assessment of coastal lakes (pilot study for eight lakes in NSW)

Coastal lakes are ecosystems of significant value, generating ecological, social and economic benefits enjoyed by a large proportion of the NSW community. They are under increasing pressure from encroaching urban development and intensification of agricultural practices, which threaten the very qualities that make the lakes so inviting. This project aimed to develop a tool to assist in planning and management of the basins of eight coastal lakes in NSW, in order to promote the sustainability of their function (Newham *et al.*, 2004c). A separate tool is developed for each lake, following the process outlined below.

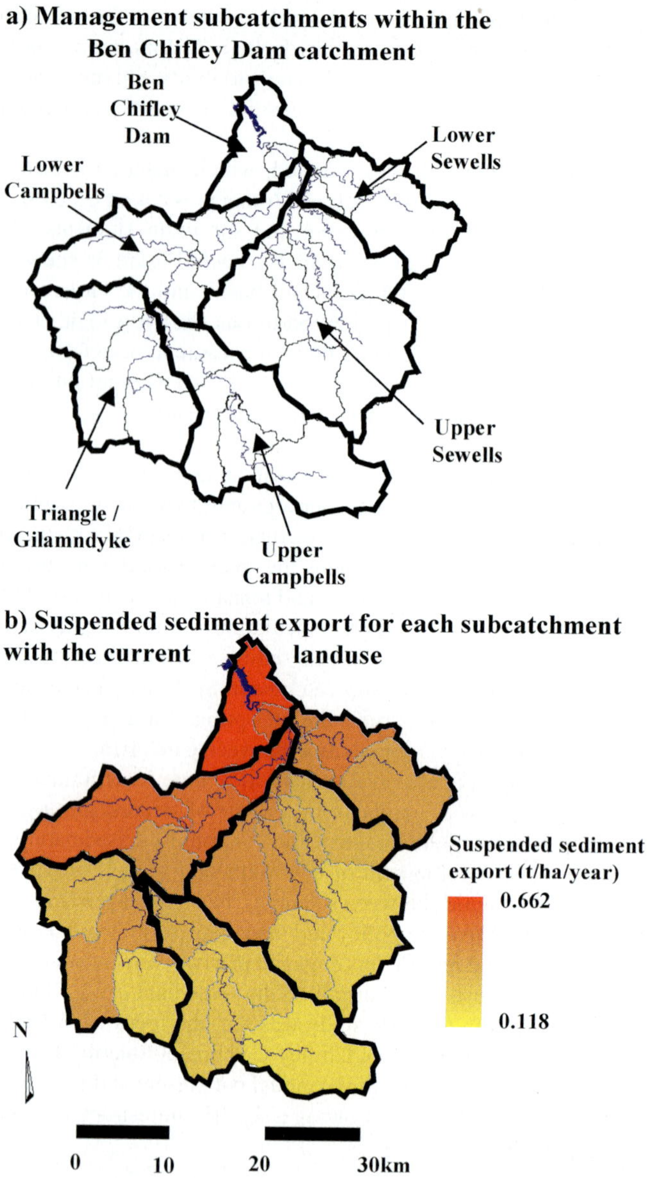

Fig. 3 Example of results from the CatchMODS tool for the Ben Chifley Dam basin

To develop a useful tool, the key constraints, issues and drivers impacting on each lake's health had to be identified by reading management plans and literature and through workshops with stakeholders. This identified many social, environmental and economic issues, which operate at various spatial and temporal scales. To account for this complexity and the time constraints on the project, a Bayesian decision network modelling approach was chosen.

Local stakeholders, including government agencies, shire council and farmers for each lake, were invited to assist in the development of potential management scenarios, such as

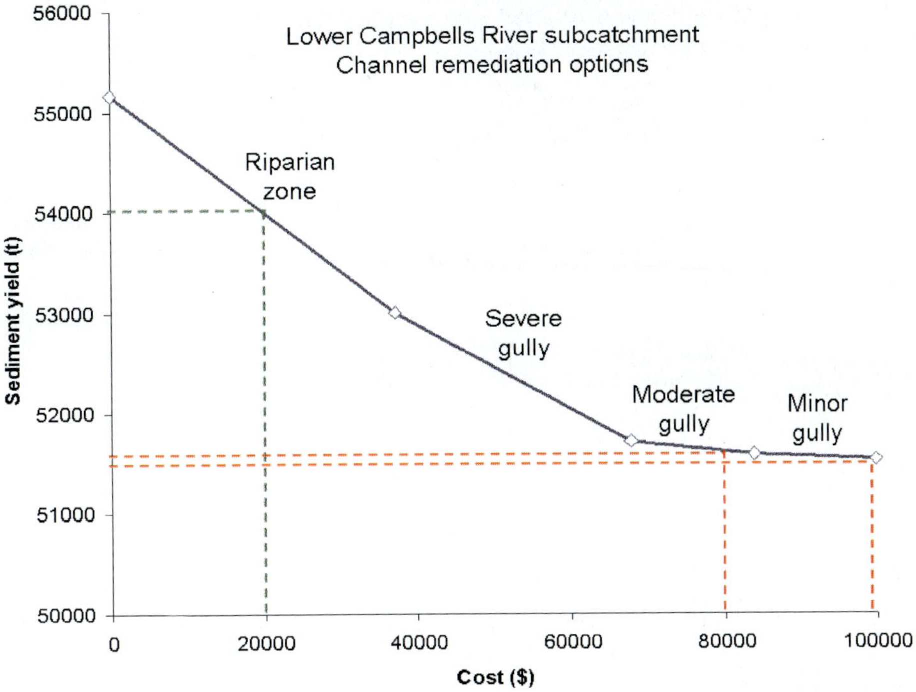

Fig. 4 An example output of CatchMODS from the Ben Chifley Dam subcatchment, of the costs and potential sediment reduction of channel remediation options

urban development, stormwater management, riparian zone management, and agricultural practices, which impact on lake health. They were also asked to identify the important economic, social and environmental costs/benefits that they deem important, (e.g. threatened species, recreation, tourism, oyster production) which may be impacted upon by the lake health. Given stakeholder knowledge and sought expert opinion, the key environmental indicators between the identified management scenarios and costs/benefits were linked to form the Bayesian decision network framework. The framework for one of the lakes, Cudgen Lake, is shown in Figure 5. This approach clearly illustrates the interrelatedness of the issues and values within the catchment. The BDN approach allows each variable within the model to be represented at an independent scale, so for each the scale can be selected that is most representative of that process, value or level required for decision support. The probability densities for each link were filled using model simulation, analysis of existing data and expert knowledge.

The key output from this modelling approach is an easy-to-use model interface that can assist council planners and other decision makers in making trade-offs between social, economic and environmental factors that affect their lake's health. The framework developed in this original pilot study can be easily altered. Thus if more detailed data become available to compile probability densities, or additional issues are identified that should be considered in the management of the lakes, the model can be updated. This dynamic approach allows users to iterate easily and further develop the tool so it can always meet their current needs.

The key challenge for this project was the time constraint on project completion, approximately one year. This limited the time that could be spent in building a relationship

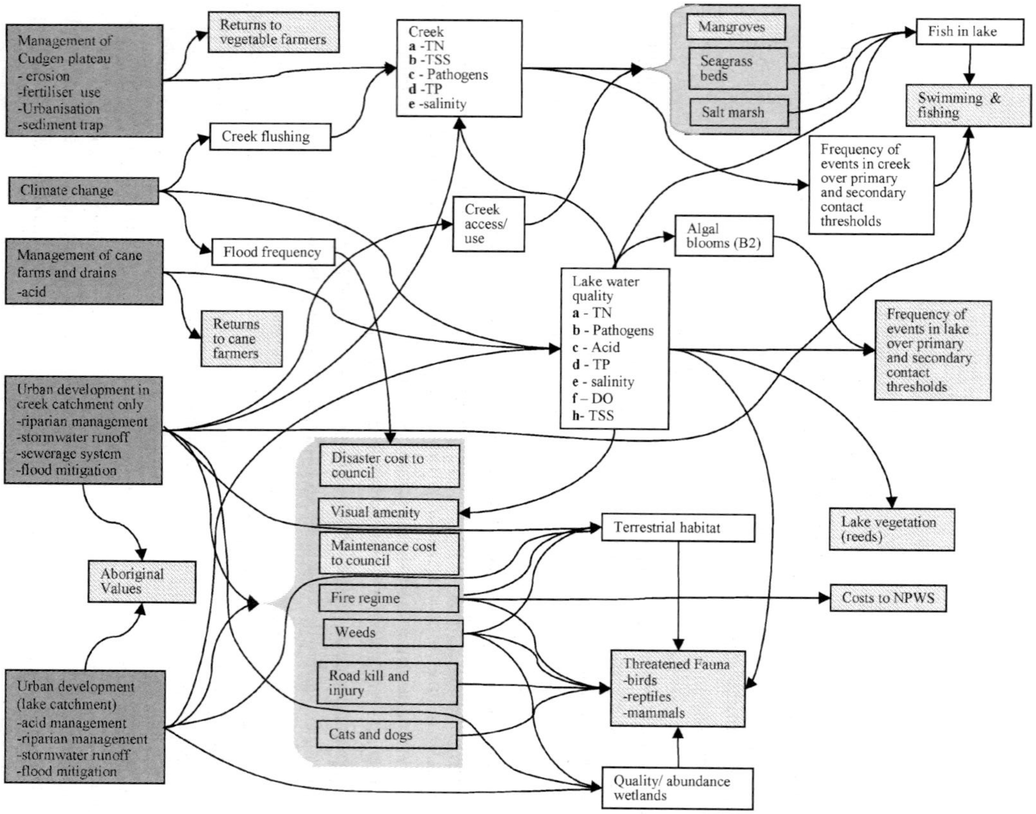

Fig. 5 Conceptual framework for the Cudgen Lake Bayesian decision network

and confidence with the stakeholders. Nonetheless, almost all community stakeholders were enthusiastic about the project and provided as much input as they could in the allocated time.

Another key constraint for the BDN approach is the inability or unwillingness of experts in the various disciplines to provide model input at a general level appropriate for the model. Many experts appeared to concentrate so much on the complexity of the system that they were unable to agree on and make generalised statements about the impact upon that system of given input changes. For example someone with detailed knowledge of the dynamics of fish populations by species might be unable to group fish populations or identify key indicator species in order to answer general questions on the change in fish populations given changes in water quality, fish catch and urban development. This inability restricts integrative models as presented here to existing, narrowly disciplinary research on the social, ecological and economic values, biasing the model and drastically reducing the potential of the integrative approach.

8. The general problems and impacts

IA or ISM projects may assist in directly achieving sustainability, but it is important to recognise that the most useful outcome may be in the learning experience of researchers, IA practitioners and stakeholder groups. Providing different stakeholders groups with opportunity to state their concerns and thoughts on management in an open unbiased forum can lead to improved understanding of the integrated nature of sustainability by participants in the project, an outcome worth achieving. However, it is important to note that the process is open to manipulation by the stakeholders consulted. People who are keen to participate may do so to 'control' the output and results. Their views will be heard and considered, while those who do not 'scream the loudest' may not be adequately represented. Some of the benefits of stakeholder participation in the development of decision management tools are evident from the case studies discussed above. However, successful participatory research requires substantial commitment of time and goodwill from the individuals and organisational groups involved. Even if these are attained in the establishment stages of a project, stakeholder participation can suffer if the key people driving the project for the community, or driving groups within the community, leave the catchment or change jobs.

Issues of commitment come from both sides. For example, people can be reluctant to provide input to a project because the information that they provide is then taken from their control and there is a fear that it might be used to their detriment. One possible means to avoid this, as was done in the Namoi/Gywdir project, is to introduce stakeholders early in the process and involve them in model development, the assumptions used and the input data. Thus they have a large input into model development and even if the model results are not favourable for them, they are less likely to criticise the model. It is important not to show preliminary results, as this can undermine stakeholder confidence in the researchers. There is a danger that select people may become too committed to the community consultation process and potentially become removed from the community that they are meant to represent. An intermediate level of commitment is desirable but difficult to achieve.

Valuable outputs from the process (e.g. increased knowledge of the environmental systems; enlightened or creative management suggestions; increased acceptance and uptake of models to support management decision making; improved access to and interpretation of existing data; and enhanced cross-agency collaboration) are often not readily measurable outcomes. This makes it difficult to attract funding, as funding bodies tend to target projects that deliver visible results.

Adoption can be used to gauge the success of a tool or model. However, what is the definition of adoption? Adoption of integrative tools is multifaceted, dynamic and evolves over time. Typically adoption is considered to be the use of a tool or software following completion. However, adoption is also achieved if stakeholders embrace and employ the concepts identified in a tool through increased understanding of the interactions within a system. It is often difficult to gauge the extent of this type of adoption.

It is suggested that for successful participatory activities in support of integrated basin modelling:

- consultation activities are included as a feature of the entire project cycle;
- multiple mechanisms are developed for groups and individuals to be involved;
- input is invited from a broad range of organisations and individuals, and
- sufficiently long time scales are made available to develop trust.

If a long time scale is not possible, it is recommended that at least one key contact is made in the study basin to help in accelerating the development of relationships between the local stakeholders and the research party.

The Namoi Basin research over the last four years involved IA of water allocation options. The strong stakeholder focus in all stages of model development and testing has led to strong industry and government support for future work in the Namoi and Gwydir Basin. Pursuing model development with a clear and open process has resulted in stakeholders making and continuing to make valuable contributions and improvements to the model development and function. This has led to a sense of empowerment and ownership of the model by the stakeholders. With their increased understanding of the system that they live in or work with, it is hoped that they will use the model as a tool to assist decision-making into the future and not misuse the results for personal gain.

There were difficulties in the Namoi and Gywdir study because of government restructuring, resulting in some participants changing employment roles or leaving the catchment. This changed the key catchment 'drivers', setting back the participatory process.

The Ben Chifley Dam study developed a software tool based on science as accepted by stakeholders. The strong stakeholder participation and the simple nature of the model tool has generated strong interest from the regulators to make the software available for basin managers right across the state of New South Wales. This project too ran into difficulty in adoption because initial consultation was focused on one particular catchment group. This group was decommissioned because of a government restructure, making it difficult to push the adoption and continuation of the project and approach within the catchment.

IA exercises promote engagement by stakeholders and systems thinking, which can only assist people in appreciating one another's perspectives. This is particularly evident in the coastal lake project. The Bayesian decision network approach enables stakeholders and decision makers to see, with a glance at the framework, the variables which their management decisions may impact upon. After running the model they are able to identify the likely impacts of that management decision. This increase in awareness of the impacts on other lake users is likely to promote greater synthesis in lake management within the community. Again there are difficulties due to the restructuring of the lead agency. Another major obstacle is having experts commit to presenting their expert opinion on a particular topic in a simple manner. Often, experts with disciplinary focus pay too much attention to the complexities inherent in the system, and are unwilling to make the generalisations necessary for integrated studies.

Consequently, it is difficult to find people who have a good knowledge of how the system works but who are still happy to make an expert judgement on the general changes expected.

In these three projects the ISM tools and exercises have also provided information and insights that would not have otherwise been possible. For example, in the Namoi, the initial stakeholder focus was on allocation of water in unregulated parts of the basin, but through ISM their concern additionally focused on other access rules for irrigation water. It also clarified that the timing of impacts was important, not merely the average impact. The cumulative impact after a series of dry years manifested itself as a major concern. The estimation of the impacts and the possible tradeoffs would not have been possible without models that in this case are predominantly quantitative.

All of the IA projects in which we have participated have reinforced the conviction that software development must be undertaken with a clear picture of the target audience, specific issues and uses. In none of the cases discussed here do the tools developed aim to provide recommendations to the land managers or to provide an "optimal" answer. That would inevitably require the model developer to determine what the optimal response is, tending to alienate stakeholders and thus hinder adoption. Instead these tools present potential outcomes or trade-offs that should be considered in managing the system. Thus while a sophisticated, object-oriented software platform may be both useful and desirable in some circumstances, in other cases a spreadsheet-based model may be more useful for these purposes. Different software products aimed at different audiences can also be useful outcomes of a project. On the other hand, software development should not be the primary objective. The software is a tool to enhance communication and interaction between different disciplinary teams. It should be a focus of the project primarily only in so far as it encourages communication of ideas and enhances understanding of the integrated nature of the problem.

9. Conclusions

Some broad conclusions can be drawn from the situations described above:

9.1. Targeted modelling

This research has clearly demonstrated that modelling, as used in IA, must be targeted to the issues, stakeholders and users in order to optimise its chance of success. This is most easily achieved through stakeholder participation throughout the whole project. A tool developed in isolation from stakeholders may not represent their concerns or yield useful outputs, reducing its value. Also, a lack of involvement of stakeholders in the IA process is likely to engender animosity and risk the tool reaching a complexity beyond the users' capabilities and beyond what is needed to achieve the desired outcomes.

9.2. Politics, conflicts and participatory processes

The stakeholders involved in the participatory process will inevitably bias the resultant tool. If one group voices its opinion more loudly, or is able to present its opinion better and so is more easily understood, it is likely that the tool will be biased towards its interests. At best this bias will result in an inequitable outcome. If that group has insufficient political and community backing, it may lead to management advice which cannot be adopted or will exacerbate conflict.

9.3. Adoption of participatory tools

9.3.1. Software

The software developed for effective decision support tools must present the results so they can be easily viewed and interpreted by the users. The tool must be targeted to the inputs available, outputs necessary and the goals identified by the future users. Contrary to common thinking, the more integration represented in the model, the simpler the model needs to be to allow for testing. Finally the model should not be developed to become a final, fixed product. The tool should be capable of being reused and easily developed in the future to allow for social, economic and physical changes.

9.3.2. Doing integrated assessment and modelling

Several integration frameworks already available have demonstrated potential to provide strong support for decision making. For example, at one extreme Bayesian networks (Section 7.3) allow interest groups to conceptualise their system and populate it with a wide variety of knowledge types and sources. At another extreme, coupled component models (Section 7.1) are most useful where users needs to overlay a hydrological network on their system to facilitate decisions at defined spatial locations, and where one has resources to add complexity to the system over time as confidence is gained. It is now timely to accelerate our application of such frameworks and synthesise the scientific and practical lessons from them.

Acknowledgements We are grateful for the funding of the three IA projects discussed in the paper by the Cotton Research and Development Corporation, the NSW Environmental Trust and the NSW Department of (Infrastructure Planning and) Natural Resources. We are also indebted to our colleagues in the various agencies and industry groups and in the wider community for their generous collaboration and input to the projects.

References

Allan C, Curtis A (2003) Learning to implement adaptive management. Nat Res Man 6:23–28

Borsuk ME, Stow CA, Reckhow KH (2004) A Bayesian network of eutrophication models for synthesis, prediction and uncertainty analysis. Ecol Mod 173:219–239

Bouzaher A, Lakshminarayan PG, Cabe R, Carriquiry A, Gassman PW, Shogren JF (1993) Metamodels and nonpoint pollution policy in agriculture. Water Res 29:1579–1587

Brown DG, Page SE, Riolo R, Rand W (2004) Agent-based and analytical modelling to evaluate the effectiveness of greenbelts. Environ Mod Software 19:1097–1109

Davis JR (1995) Expert systems and environmental modeling. In: Jakeman AJ, Beck MB, McAleer MJ (eds) Modelling change in environmental systems. Wiley, Chichester, pp 505–517

Deaton ML, Winebrake JJ (1999) Dynamic Modelling of Environmental Systems. Springer, New York

Forsyth R (1984) The expert systems phenomenon. In: Forsyth R (ed) Expert systems: principles and case studies. Chapman and Hall, London, pp 3–8

Hare M, Letcher R, Jakeman AJ (2003) Participatory modelling in natural resource management: a comparison of four case studies. Integrated Assessment 4:62–72

Haslam SA, Eggins RA, Reynolds KJ (2003) The ASPIRe model: actualising social and personal identity resources to enhance organisational outcomes. J Occup Organ Psychol 76:83–113

Holling CS (ed) (1978) Adaptive Environmental Management and Assessment. Wiley, Chichester

Jakeman AJ (1989) Application of identification and system parameter estimation to environmental problems: some recent examples. In: Barker HA, Young PC (eds) 7th IFAC symposium on identification and system parameter estimation. York, 3–7 July 1985, 1, Oxford:pergamon Press, pp 445–450

Jakeman AJ, Letcher RA (2003) Integrated assessment and modelling: features, principles and examples for catchment management. Environ, Mod Softw 18:491–501

Jakeman AJ, Letcher RA, Norton JP (2006) Ten iterative steps in development and evaluation of environmental models. Environ Mod Softw 21:602–614

Jasanoff S (1993) Bridging the two cultures of risk analysis. Risk Anal 13:123–129

Kammen DM, Hassenzahl DM (1999) Should we risk it? Exploring environmental, health and technological problem solving. Princeton University Press, Princeton, N.J., 404 pp

Kidd AL (1987) Knowledge acquisition an introductory framework. In: Kidd AL (ed) Knowledge acquisition for expert systems: a practical handbook. Plenum Press, New York, 194 pp

Letcher RA, Jakeman AJ (2003) Application of an adaptive method for integrated assessment of water allocation issues in the Namoi River Catchment, Australia. Integr Assess 4:73–89

Letcher RA, Croke BFC, Jakeman AJ, Merritt WS (2006) An integrated modelling toolbar for water resources assessment and management in highland catchments: model discription. Agricultural Systems 89:106–131

Letcher RA, Jakeman AJ Croke BFW (2004) Model development for integrated assessment of water allocation options. Water Resour Res 40:W05502, 15 pp

Letcher RA, Jakeman AJ (in press) Integrative frameworks for natural resource management. Envron Mod Softw

Merritt WS, Croke BFW, Jakeman AJ, Letcher RA, Perez P (2004) A biophysical toolbox for assessment and management of land and water resources in rural catchments in Northern Thailand. Ecol Mod 171:279–300

Mostert E (2006), Participation for sustainable water management. In: Giupponi C, Jakeman A, Karssenberg D, Hare M (eds) Sustainable management of water resources: an integrated approach. Elgar, pp. 153–176

Newham LTH, Letcher RA, Jakeman AJ, Kobayashi T (2004) A framework for integrated hydrologic, sediment and nutrient export modelling for catchment-scale management. Environ Mod Softw 19:1029–1038

Newham LTH, Letcher RA, Jakeman AJ (2004b) Integrated assessment and modelling for catchment management: An Australian case study in participation. In: Henze M (ed) Proceedings of international symposium: uncertainty and precaution in environmental management. Copenhagen

Newham LTH, Ticehurst JL, Rissik D, Letcher RA, Jakeman AJ, Nelson P (2004c) Assessing the sustainability of NSW coastal lakes using a Bayesian decision network approach. In: Proceedings NSW coastal conference: seachange? The Delicate Balancing Act. Lake Macquarie 9th–12th November, pp 63–69

Norton JP, Nathan R, Podger G, Vertessy R (2003) Sensitivity assessment needs of complex simulation models for integrated catchment management. In: Post DA (ed) Proceedings modelling and simulation society of Australia and New Zealand. MODSIM 2003, 4:1667–1672

Pearl J (1988) Probablistic reasoning in intelligent systems: networks of plausible inference. Morgan Kaufmann Publishers, Inc. San Francisco, ISBN 1-55860-479-0

Refsgaard JC, Henriksen HJ, Harrar WG, Scholten H, Kassahun A (2005) Quality assurance in model based water management – review of existing practice and outline of new approaches. Environ Mod Softw (in press)

Sadoddin A, Letcher RA, Jakeman AJ (2003) A Bayesian decision network approach for salinity management in the Little River Catchment. NSW, In: Post DA (ed) Proceedings modelling and simulation society of Australia and New Zealand. MODSIM 2003, 3:953–958

Schmoldt DL (1998) Knowledge acquisition using linguistic-based knowledge analysis. AI Appl 12(1–3):1–20

Sell P (1985) Expert systems: a practical introduction. John Wiley and Sons, Inc, New York

Spate JM, Croke BFC, Jakeman AJ (2003) Data mining in hydrology. In: Post DA (ed) Proceedings modelling and simulation society of Australia and New Zealand. MODSIM 2003 4:422–427